Perceptual Audio Evaluation–Theory, Method and Application

Perceptual Audio Evaluation – Theory, Method and Application

Søren Bech *Bang & Olufsen a/s, Denmark*
Nick Zacharov *Nokia Corporation, Finland*

John Wiley & Sons, Ltd

Telephone (+44) 1243 779777

Email (for orders and customer service enquiries): cs-books@wiley.co.uk
Visit our Home Page on www.wiley.com

Other Wiley Editorial Offices

John Wiley & Sons Inc., 111 River Street, Hoboken, NJ 07030, USA

Jossey-Bass, 989 Market Street, San Francisco, CA 94103-1741, USA

Wiley-VCH Verlag GmbH, Boschstr. 12, D-69469 Weinheim, Germany

John Wiley & Sons Australia Ltd, 42 McDougall Street, Milton, Queensland 4064, Australia

John Wiley & Sons (Asia) Pte Ltd, 2 Clementi Loop #02-01, Jin Xing Distripark, Singapore 129809

John Wiley & Sons Canada Ltd, 22 Worcester Road, Etobicoke, Ontario, Canada M9W 1L1

Wiley also publishes its books in a variety of electronic formats. Some content that appears in print may not be available in electronic books.

Library of Congress Cataloging-in-Publication Data

Bech, Søren.
Perceptual audio evaluation : theory, method and application / Søren Bech, Nick Zacharov.
p. cm.
Includes bibliographical references and index.
ISBN-13: 978-0-470-86923-9 (cloth : alk. paper)
ISBN-10: 0-470-86923-2 (cloth : alk. paper)
1. Sound – Recording and reproducing. 2. High-fidelity sound systems – Testing. 3. Sound – Measurement. I. Zacharov, Nick. II. Title.
TK7881.4.B34 2006
621.389'33 – dc22 2006006475

British Library Cataloguing in Publication Data

A catalogue record for this book is available from the British Library

ISBN-13: 978-0-470-86923-9
ISBN-10: 0-470-86923-2

Typeset in 11/13.6pt Palatino by Laserwords Private Limited, Chennai, India
Printed and bound in Great Britain by Antony Rowe Ltd, Chippenham, Wiltshire
This book is printed on acid-free paper responsibly manufactured from sustainable forestry in which at least two trees are planted for each one used for paper production.

To Emilie and Manu

Contents

Preface

THE importance of sound quality and its perception is increasing in daily life–for better or worse. We communicate using landline or mobile phones, we enjoy music reproduced at home, in the car, walking along the street. Yet we get annoyed by the noise of traffic, other people talking on their mobile phones or others using music players on the bus, and so on. In fact, sound is becoming such an important factor in people's daily lives that some envisage that 'silent places' will become the luxury of tomorrow.

The increasing importance of sound quality has spurred a growing awareness of the meaning of 'good' or 'bad' sound quality, how it is quantified and how it can be controlled. As a consequence many organisations, both private and public, have started research programmes or research teams in an effort to understand how sound and its perceived qualities affect our daily lives.

Research in what constitutes 'good' or 'bad' sound quality has been conducted for decades in the academic environment. While in the field of product sound quality, this topic is commonplace, the quantification of perceived quality of reproduced sound is a relatively new field in the audio industry. As a consequence a large number of people, usually engineers, have suddenly been faced with the task of quantifying the perceived audio quality of a new device, for example, a loudspeaker or mobile phone. To further complicate the task, not only should the perceived quality be quantified, but it is often desired that the results be objective as they will form the basis for serious company decisions. A good example is the decision by major broadcasters to use low bit-rate coding technology in their transmission systems. The installation of such systems represented a major financial decision and one of the major factors, apart from financial aspects, was the perceived sound quality of the selected systems.

Objective quantification of perceived sound quality is, as the reader probably knows or will learn, an exercise that involves many scientific disciplines from audio recording and production, electronic engineering, signal processing, room acoustics, electroacoustics, experimental psychology through to statistics. Very few engineers or professionals from other disciplines master all of the aforementioned fields and as a result many newcomers struggle with the theoretical and practical aspects of performing perceptual audio evaluations.

Almost everyone listens to sound most of the time, so there is often an opinion that the evaluation of audio quality must be a trivial matter. This frequently leads to a serious underestimation of the magnitude of the task associated with formal evaluations of audio quality, which can lead to compromised evaluations and consequently the poor quality of results. Such a lack of good scientific practise is further emphasised when results are reported in journals or at international conferences and leads to a spread of scientific darkness instead of light.

The authors have worked in both academia and industry and have experienced through their own research the time and effort required to find the relevant information in the various scientific fields to perform perceptual evaluation of audio. They have experienced, as advisors, the struggle students experience when having to learn and gather information from scientific fields with different scientific paradigms and traditions. While often experiments are well performed and reported, the authors have also had the depressing experience at conferences to witness poorly planned, conducted or analysed experiments, which by marginal effort could have been turned into valid and valuable information. Finally, they have had the challenging task of convincing managers of the benefits of conducting properly designed and executed experiments instead of relying on the opinions of other managers or CEO's.

The idea thus emerged that to remedy this situation a book was needed containing the basic information from all the scientific fields included in the process of perceptual audio evaluation.

As a result this book is written for the professionals and the engineers in the industry, academics, students at all levels and others who, for the first time, are faced with the task of conducting a perceptual evaluation of audio quality. It is also written for the advisors or managers, who are to demand or authorise such evaluations, to provide them with insight into the complexity and effort involved with such tasks. Finally, it is written for those experienced in conducting perceptual audio quality evaluations, who need a handy reference where all the different aspects of the entire process are assembled.

The basic idea of the book is that the reader, after having consulted the book, should be capable of planning and executing the majority of auditory evaluations encountered in research and industrial settings, and further to analyse the results and report the conclusions at a level, which is required for a scientific publication. For those readers who move on to more advanced methods or types of experiments, a comprehensive list of references has also been included.

The information included and the experience gathered, and hopefully conveyed, within the book is based on the input from many people in the field. We could have chosen to mention them all by name here in the preface, but the list would have been very long and we feel that it is more important to let them speak through their publications that have all been referenced and discussed in the following chapters. It is also important to emphasise that any errors that you may find in the book are entirely the responsibility of the authors and not their colleagues. This book represents the views and opinions of the authors and, therefore, does not necessarily represent the views of, or is an endorsement of the book's contents by, their employers.

Organisation of the book

Chapter 1: Introduction A short introduction to perceptual audio evaluation is given as well as the motivation for and the boundaries of this text. A brief introduction to the role of standardisation and predictive models is provided here.

Part 1: Experimental considerations The planning of a perceptual evaluation includes a large number of issues. They can roughly be divided into two major groups: the experimental and technical considerations.

The experimental considerations include all the non-hardware or more psychological related issues whereas the hardware or engineering related issues are discussed in the 'Technical considerations' section.

Chapter 2: Definition of research question and hypothesis This chapter includes an overview of topics related to the definition of the research question, the initial conditions, and the testable statement. The two basic principles of scientific argumentation, the principles of empiricism and rationalism are discussed in detail and in relation to audio evaluations. A number of other principles are also discussed.

Chapter 3: Fundamental of experimentation This chapter discusses the fundamental issues that need to be considered during the planning process. This includes the gathering of previous knowledge–a literature study, considerations in relation to why or whether a perceptual evaluation is needed, definition of research question, hypothesis, and initial conditions, definition of the statistical design and model, pilot experiments, documentation of the experimental setup, statistical analysis of the results, conclusions and their reporting.

Chapter 4: Quantification of impression This chapter discusses the dependent variables of an auditory experiment. They are divided into two main groups, one related to the definition of the question to the subject and another to the definition of how the subject is going to report the answer. Two basic types of measurements of subject's impression are discussed: perceptual measurements where the subject is asked to scale the sensorial strength for specific auditory attributes and affective measurements where the subject is asked to be in an integrative state of mind and report an overall impression. Various methods for elicitation of the specific attributes to be evaluated are discussed for both direct and indirect elicitation. Various types of affective measurements are also discussed. Finally, various forms of scaling methods such as direct and indirect scaling methods and their associated bias phenomena are discussed. Examples of scales used for audio evaluation are presented.

Chapter 5: Experimental variables This chapter discusses the independent variables of an auditory experiment. These are the experimental variables controlled by the experimenter and they are related to the signal category (natural or synthetic signals), the recording and storing of the signals, the reproduction system (amplifiers, loudspeakers etc.), the listening room, and the subject. The subject variable, being an important but quite a complex variable, is discussed in detail and aspects such as how to categorise and use the different subject types, how to select subjects for permanent listening panels, how to pre- and post-select subjects, how to evaluate their performance in various types of test methodologies and how to train them are discussed.

Chapter 6: Statistics This section is divided into two major parts. The first concerning the statistical planning of the experiments and the second the statistical analysis of the response data. The basic issues in statistical design of experiments such as using balanced Latin squares, controlling the Type I and II errors, and estimating the number of observations needed for a given resolution are discussed and an example is given. The statistical analysis section discusses the classification of the data into the statistical type, and how to test if the data fulfils the standard statistical assumptions, how to test the relationship between an estimated mean value and the true population mean, and how to test the relationship between two or several estimated mean values. Finally, the analysis of variance model is introduced and the various tests associated with such an analysis. The implications and consequences of using fixed or random variables in the model are discussed and the tests of the various assumptions are introduced. Finally, a statistical check list is introduced to guide the experimenter through the statistical analysis process.

Part 2: Technical considerations This part of the book will focus upon practical and sometimes hardware related issues that are required to perform listening test. Topics will include electroacoustic considerations, calibration and test planning, administration and reporting matters.

Chapter 7: Electroacoustic considerations The focus of this chapter will be upon rooms for performing perceptual audio evaluations and associated electroacoustic issues. Listening rooms are discussed in general and the IEC, ITU-R, and EBU standardised rooms in more detail. This includes topics such as floor area, volume, shape, aspects ratios between room dimensions, reverberation time, distribution and level of reflection and back ground noise level. Listening booths and other listening spaces are also considered. The positioning of loudspeakers and listener(s) are also discussed for the various reproduction modes. The topic of a separate loudspeaker for low frequency reproduction is included as is the issue of an accompanying picture. Some details will be provided regarding performance requirements associated with electroacoustic devices for listening tests, for example, loudspeakers.

Chapter 8: Calibration This chapter addresses the matter of calibration in both the physical and perceptual domains. This includes calibration of the absolute reproduction level, within system levels, between system levels, and between stimulus levels. A number of level and loudness metrics are introduced and discussed in some detail. Their application to signal and system calibration is included and recommendations are given regarding which to use for a particular application. The issues of the preferred listening level and standardised reference reproduction levels are discussed. The complicated issue of calibration of a separate low frequency channel (LFE) are discussed as are the measurement and calibration of headphone systems for both free- and diffuse field applications. The issue of a 6 dB level difference between headphone and loudspeaker reproduction, for equal perceived level is also discussed.

Chapter 9: Test planning, administration and reporting This practical chapter discusses some of the essential details associated with the administration and logistics of planning and running perceptual audio evaluations. Topics include experimental planning, logistic considerations, ethical considerations including maximum noise or music exposure levels, administration of subjects and their familiarisation with the experimental setup and task. Also included are examples of verbal and written instructions, and the issue of subjects familiarisation with the stimuli is discussed. A detailed list of software programs and systems for controlling the experiment and collection of the subject responses, is presented. Finally, the matter of reporting is discussed with details of key elements associated with good reporting practises.

Part 3: Applications This part will provide the example applications of listening methods discussed so far.

Chapter 10: Commonly encountered experimental paradigms This chapter will provide a short review of the commonly encountered test paradigms from standardisation bodies. Brief summaries of the ITU-T recommendation P.800, ITU-R recommendation BS.1116-1, and ITU-T recommendation BS.1534-1 are given with details of key experimental factors.

Acknowledgements

THE first author would like to express his sincere thanks to Associate Professor (retired) O. Juhl Pedersen. Through his inspiring and excellent lectures, he initiated my interest in acoustics and psychoacoustics in particular. He also started my research career by initiating my PhD application and he initiated the Archimedes project–one of many interesting projects that we have since collaborated on.

Henrik Staffeldt and Floyd Toole are thanked for supporting my initial work and for always being willing to discuss important matters. Laurie Fincham is thanked for his thoughtful comments and for introducing me to the Audio Engineering Society and all its fine and dedicated people. Jan Voetmann is thanked for the many hours of good discussion and company while travelling to conferences, and Peter Petersen for the good and inspiring collaboration when establishing the funding for the Sound Quality Research Unit at Aalborg University. All my colleagues in the acoustics department at B&O are thanked for always being good colleagues and for the many good hours at B&O.

The second author would like to express his gratitude to Professor Matti Karjalainen and his laboratory for creating an exciting and dynamic environment for creativity, fun and academic excellence in audio, which has been highly contagious. He and all the lab members have certainly been an inspiration.

Gaëtan Lorho is thanked for providing valuable comments to early version of the book and technical input. All my colleagues within the

Nokia Research Center and the Nokia Audio Quality Assurance team are thanked for their efforts over the years and also for assistance in the preparation of this text. Members Nokia Information Services, in particular Juha Riihioja and Susanne Luoto, are thanked for significant support in digging out old and obscure papers/references with great efficiency.

Both authors are indebted to all those persons who assisted in gathering, contributing, and commenting on the various aspects of the text. George Gratzer, author of *Math in LaTeX*, is thanked for providing the LaTeX style sheets, which have been slightly modified by the authors for the preparation of this book. Alistair Smith is thanked for additional assistance with typesetting in LaTeX. Schuyler Quackenbush of Audio Research Labs is thanked for providing the STEP–subjective training and evaluation program, examples of which are presented herein.

We would also like to express our sincere thanks to the members of the gastro-acoustical society (GAS) with whom we have had so many enjoyable hours of discussion: Durand Begault, Karlheinz Brandenburg, Elizabeth Cohen, Louis Fielder, David Griesinger, Tom Holman, Bill Martens, Francis Rumsey, Jan Voetmann, Wieslaw Woszczyk, and many more.

The Nokia Foundation is thanked for providing funding support towards the preparation of this text.

Additionally, we would like to express our sincere thanks to all those hard working students, scientists, engineers and others working in the field for the numerous hours of learning and good experiences that we both have had for many years.

Our wives, Ida and Paula are thanked for putting up with us on the long and winding road leading to this book–and for taking care of everything during the finalisation of the whole project. Our sincere thanks also to Emilie and Manu for tolerating that we were not available at all times as would be expected.

...and to everyone else who provided assistance in this project...

Søren Bech and Nick Zacharov, April 2006

CHAPTER

1

Introduction

THE aim of this book is to study the topic of perceptual evaluation of audio. Audio is very multidimensional in nature as is its perception. In order to study the nature of audio, it is possible to measure the physical characteristics of an audio signal in the acoustic or electrical domains. However, this characterisation of the physical audio signal does not tell us how the human auditory system will interpret and quantify it. In order to do this, a direct measurement of the human perception of the audio signal would be needed, as illustrated in Figure 1.1(a), but this is not yet possible.

An alternative means of assessing how listeners perceive an audio signal would be to ask them to quantify their experience. This is the most common form of perceptual evaluation that often takes the form of a formal listening test, as illustrated in Figure 1.1(b), and forms the core subject matter of this text. The manner in which listeners quantify their experience can occur on one of two levels, as defined

Perceptual Audio Evaluation – Theory, Method and Application Søren Bech and Nick Zacharov

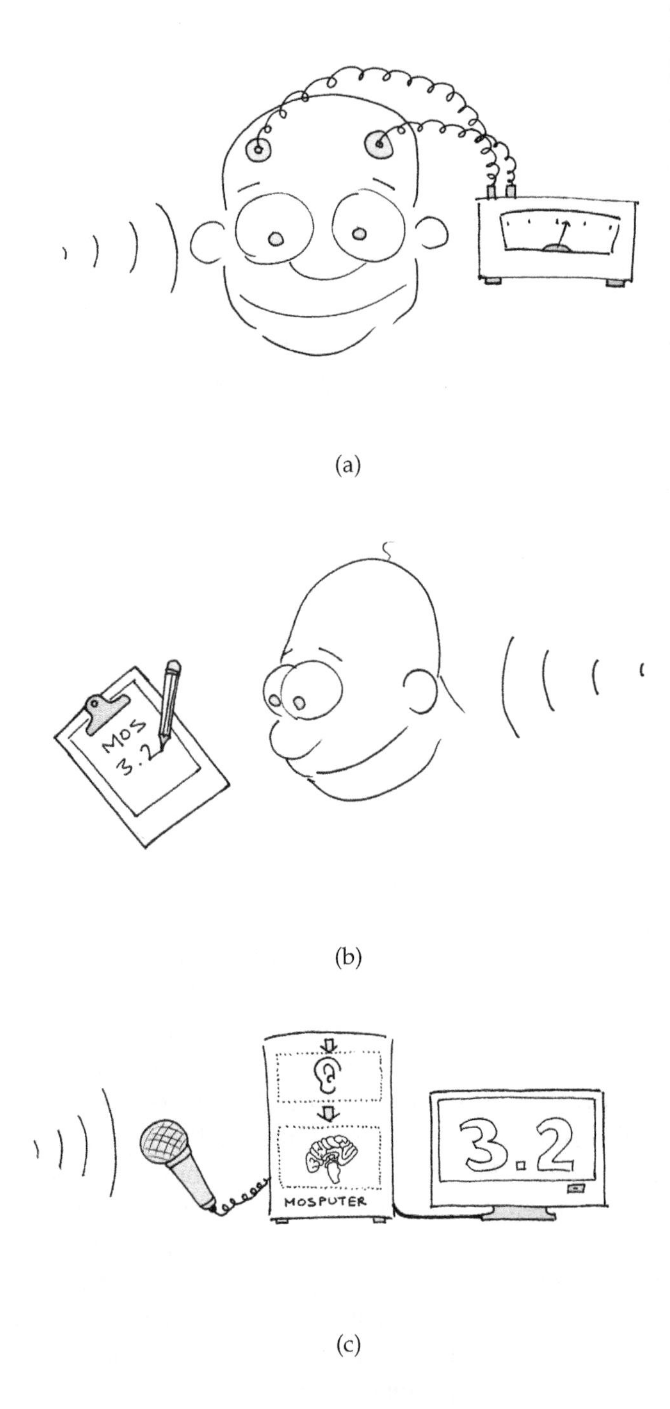

(a)

(b)

(c)

Figure 1.1: Examples of different forms of perceptual evaluation. Reproduced by permission of Heikki Niemelä.

through the filter model to be introduced in detail in Section 4.1. The two measurement levels are defined as follows:

Perceptual measurement An objective quantification of the sensorial strength of individual auditory attributes of the perceived stimulus

Affective measurement An objective quantification of an overall impression of the perceived stimulus.

This text will address both of these measurement levels, though emphasis will be placed on perceptual measurement, as an objective means of evaluating the perception of an audio signal.

An alternative means of estimating the perceptual evaluation is through the use of predictive modelling techniques. This approach employs a perceptual model of the human auditory and cognitive system to predict the human response to an audio signal as illustrated in Figure 1.1(c). This approach is highly effective, as the aim of such predictive models often is to simulate the response of listeners participating in a listening test. Additionally, such models are very efficient compared to listening tests and can provide predictions of the perceptual evaluation in a very short time. As a result of time and cost efficiency, such methods are gaining popularity. However, as these methods are still under development, they cannot yet replace listening tests in many cases. A brief introduction to such perceptual evaluation methods, some of their applications and limitations is provided in Section 1.3.

1.1 *Motivation for listening tests*

Having established that one of the main means to study perceptual audio characteristics is through listening tests, it is now important to understand when such procedures should be applied. As will be illustrated during the course of this book, listening tests are arduous to perform and, as a result, are only undertaken when there is a clear need and benefit.

When considering the evaluation of audio and its qualities, the researcher should consider the different methods that are available. He should thus ask himself:

- Whether a measure of the physical signal will provide sufficient information?

- Does a direct measurement of the perceived audio quality exist?
- Can a suitable predictive model of perceived audio quality be identified for the assessment of the stimuli under consideration?

If it is established that the answer to all of these questions is no, then the researcher will need to consider performing a listening test, in which case he/she may read on.

In some cases, the experimenter will be working in a field that is well established and for which the norm is to perform a perceptual evaluation. Examples of such areas are audio and speech codec performance. In such cases, the researcher would benefit from the effort and expertise employed in the specification leading to a standardised approach for performing the listening test. The role and application of standards and recommendations will be discussed further in Section 1.2.

Having established that a listening test is required for perceptual evaluation, it is pertinent to have an idea of what to expect from the results. This is healthy for both the experimenter and his management to ensure that everyone knows what to expect.

Listening tests can provide some of the following pieces of information:

- Identify whether or not audio stimuli are perceptually identical.
- Establish whether a sample is perceptually equivalent, superior or inferior to another sample with regard to audio quality.
- Define to what degree a sample is superior to another in terms of audio quality.
- Establish which audio system is preferred.
- Establish whether an audio system is acceptable for a given task.
- Rate the performance of audio systems in a detailed manner employing a number of perceptual attributes.
- Define the absolute audio quality of an audio system.

Listening tests are not directly able to:

- Identify and locate the problem parameter of an audio algorithm.
- Identify which system will score highest in a Hi-Fi magazine evaluation.
- Identify what technical aspects of a competitor's audio system make it superior.

- Define how developers should improve their systems to obtain significant audio quality improvement.

However, with the use of advanced data mining techniques, it is possible for listening test data to be employed with other technical information to answer most of these latter issues. Such methods are, however, outside the scope of this book.

To guide the researcher through the process of identifying the correct approach to perceptual evaluation, the block diagram in Figure 3.1 (Chapter 3) is provided. The steps illustrated in this diagram form the basis for the chapters and sections of this book. Additionally, the mindmap provided in Figure 1.2 gives a detailed visualisation of the book structure to assist the reader in navigation.

Lastly, a short note regarding the scope of this book. This book is intended to cover audio and telecommunication applications and associated perceptual evaluation procedures. The focus has been placed on perceptual audio evaluation practices, although other methods are introduced in brief. The emphasis has been placed on listening-only procedures. As a result, topics such as conversational quality, discourse analysis, and so forth, are outside the scope of this text.

1.2 Role of standardisation

In the field of audio, a number of standards or recommendations exist that cover a wide range of topics from measurement devices through to perceptual evaluation methods for telecommunications or audio systems. A large number of standards pertain to perceptual evaluation of audio, either from the listening test perspective or relating to predictive models, which are discussed in the next section.

Typically, standards are developed when there is a large-scale need to address a problem within a field or industry. The need is usually driven by several parties when there is a perceived benefit from establishing a commonly agreed upon approach to addressing the problem.

Standards provide the benefit of an agreed upon approach, developed by experts in the field from both industry and academia. In terms of perceptual evaluation, this means that a methodology has been developed and verified as being applicable to the domain defined for that standard. This is of great benefit, as the experimenter does not have to develop his own evaluation method but can rely upon the

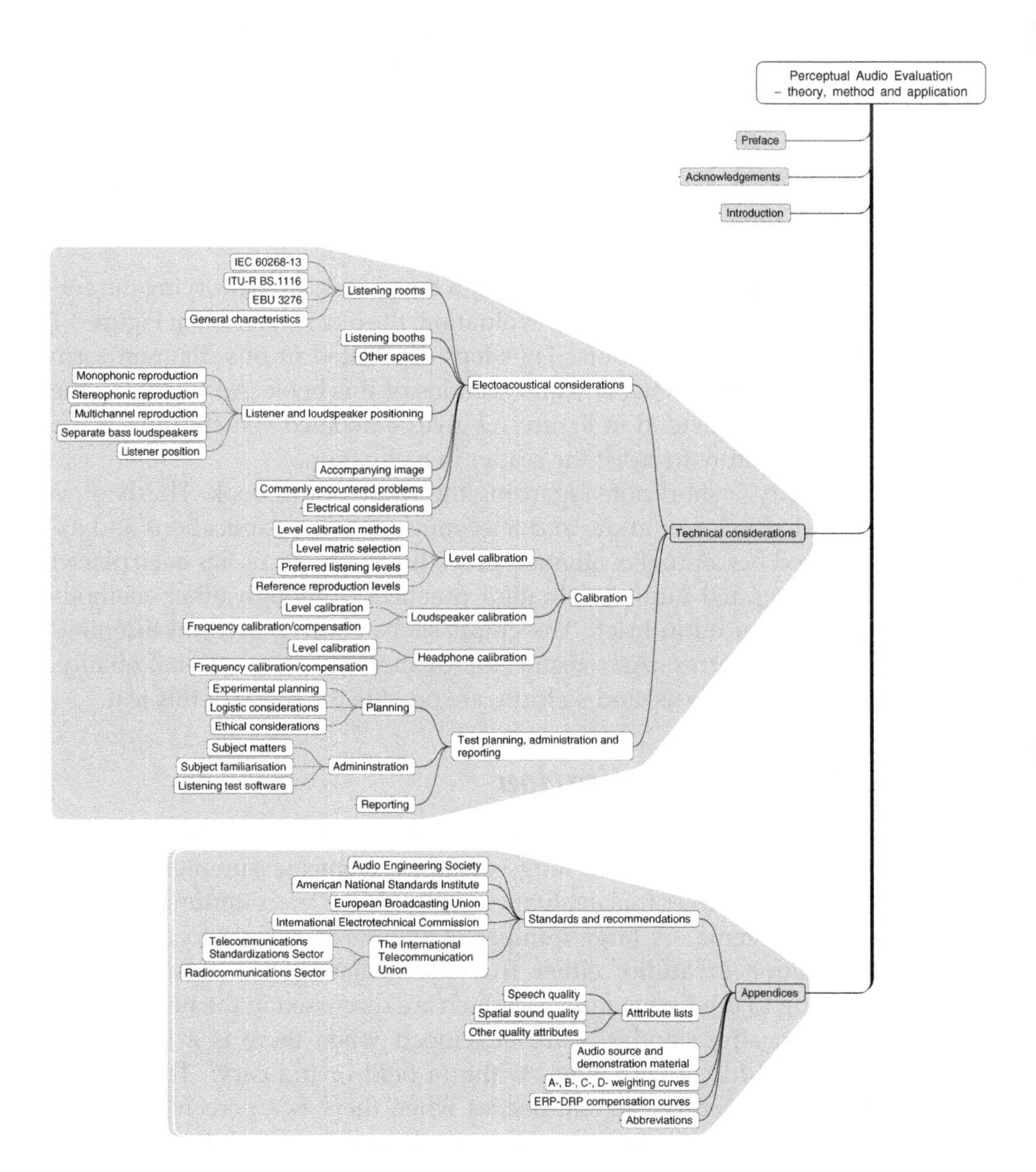

Figure 1.2: Mindmap illustration of the content and structure of this book.

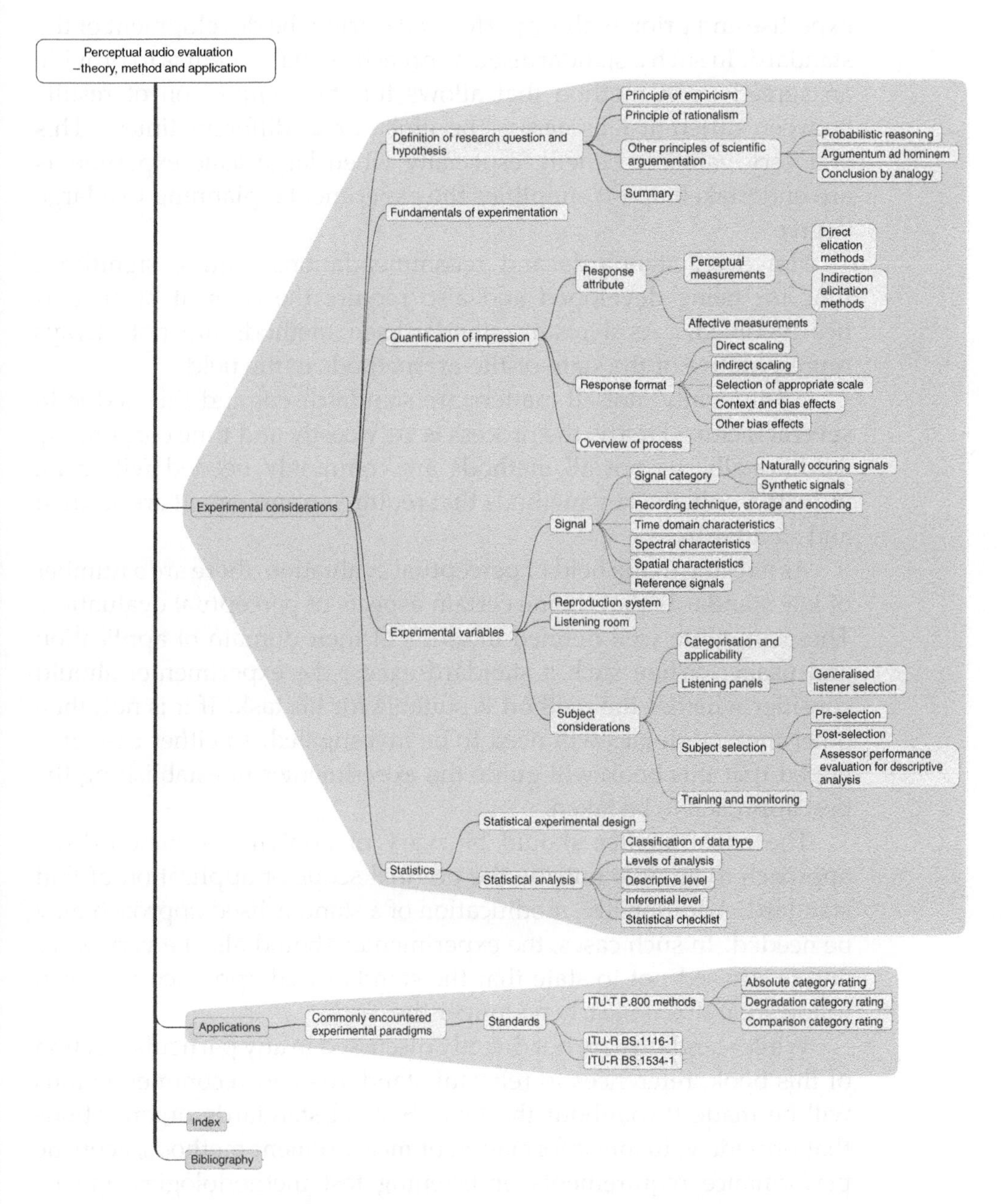

Figure 1.2: Continued.

expertise and prior analysis performed during the development of the standard. In such a standardised approach, a vital benefit is to provide an agreed upon method that allows for the comparison of results between different labs across the globe or at different times. This is a very valuable benefit, especially when large-scale experiments are undertaken and it simplifies the experimental planning to a large degree.

However, standards and recommendations require significant time for being developed and also require the consent of all parties involved. As a result, standardised methods are not always representative of the state-of-the-art methods in the field.

Additionally, not all matters are standardised, and this is due to several factors. Firstly, the process is very costly and time consuming. Additionally, as not all methods are commonly needed within an industry, only the key methods that require an agreement are studied and standardised.

As a result, in the field of perceptual evaluation, there are a number of key standards that define certain aspects of perceptual evaluation. These are quite well defined in terms of their domain of application and usage. When such a standard exists, the experimenter should consider whether the method is suitable for his task. If it is not, then other methodologies will need to be investigated. In either case, it is hoped that this book will guide the experimenter in establishing the best approach to be taken.

The experimenter should be wary of applying a standardised approach to an area outside the original scope of application of that standard. Additionally, modification of a standardised approach may be needed. In such cases, the experimenter should also be very wary and is advised not to state that the standardised approach has been followed, if this is clearly no longer the case.

While standards are not directly discussed in any particular section of this book, references to relevant standards and recommendations will be made throughout the text. Several standards organisations that provide valuable information of measurement methods, acoustic performance requirements or listening test methodologies will be referenced to. An example of some important standards from the International Telecommunications Union are presented in Figures 1.3 and 1.4. Appendix A is provided to summarise a number of key standards that relate directly to perceptual audio evaluation methods. This appendix aims to provide short guidance to the interested reader

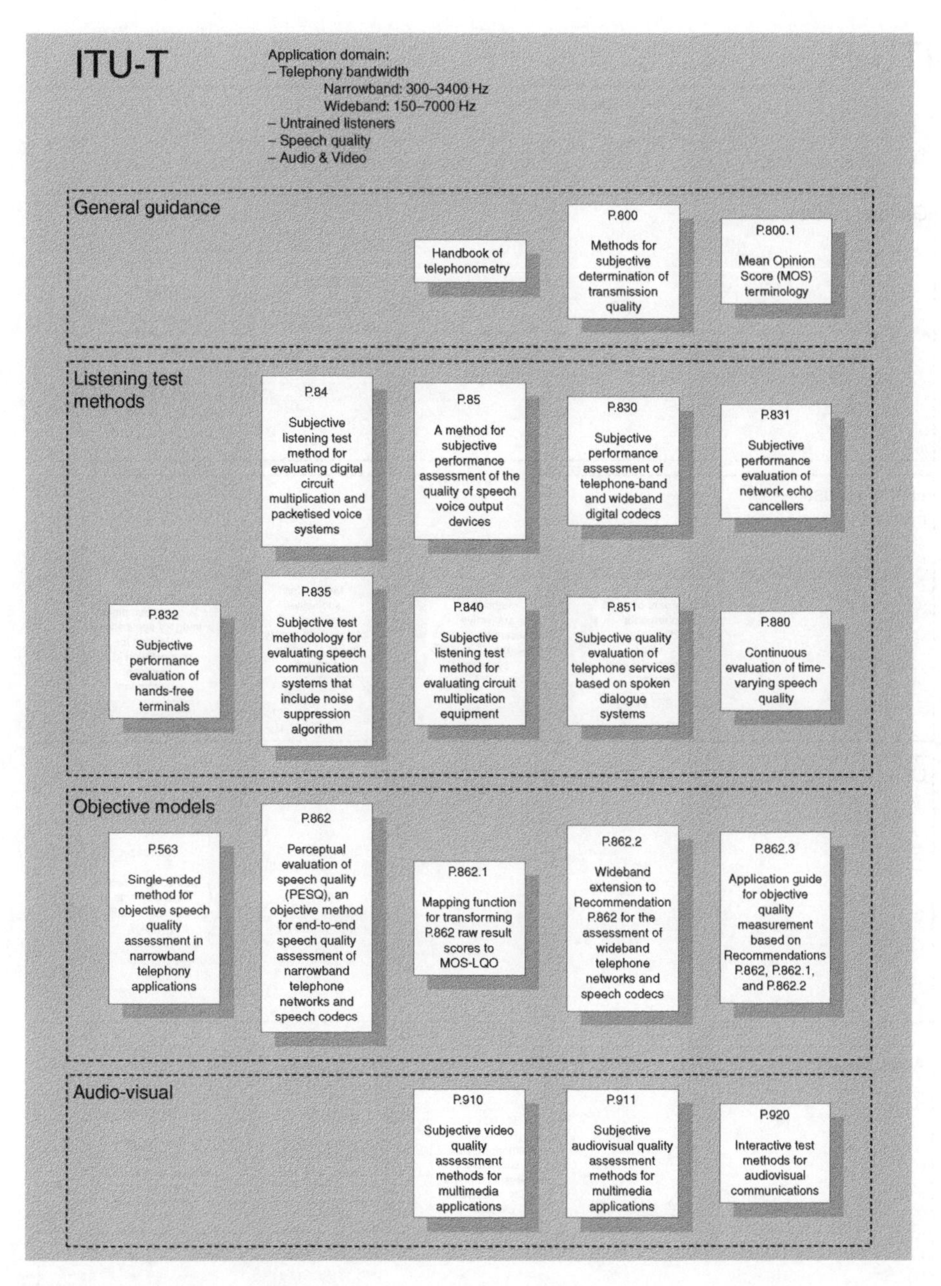

Figure 1.3: Summary overview of key ITU-T recommendations relating to perceptual audio evaluation.

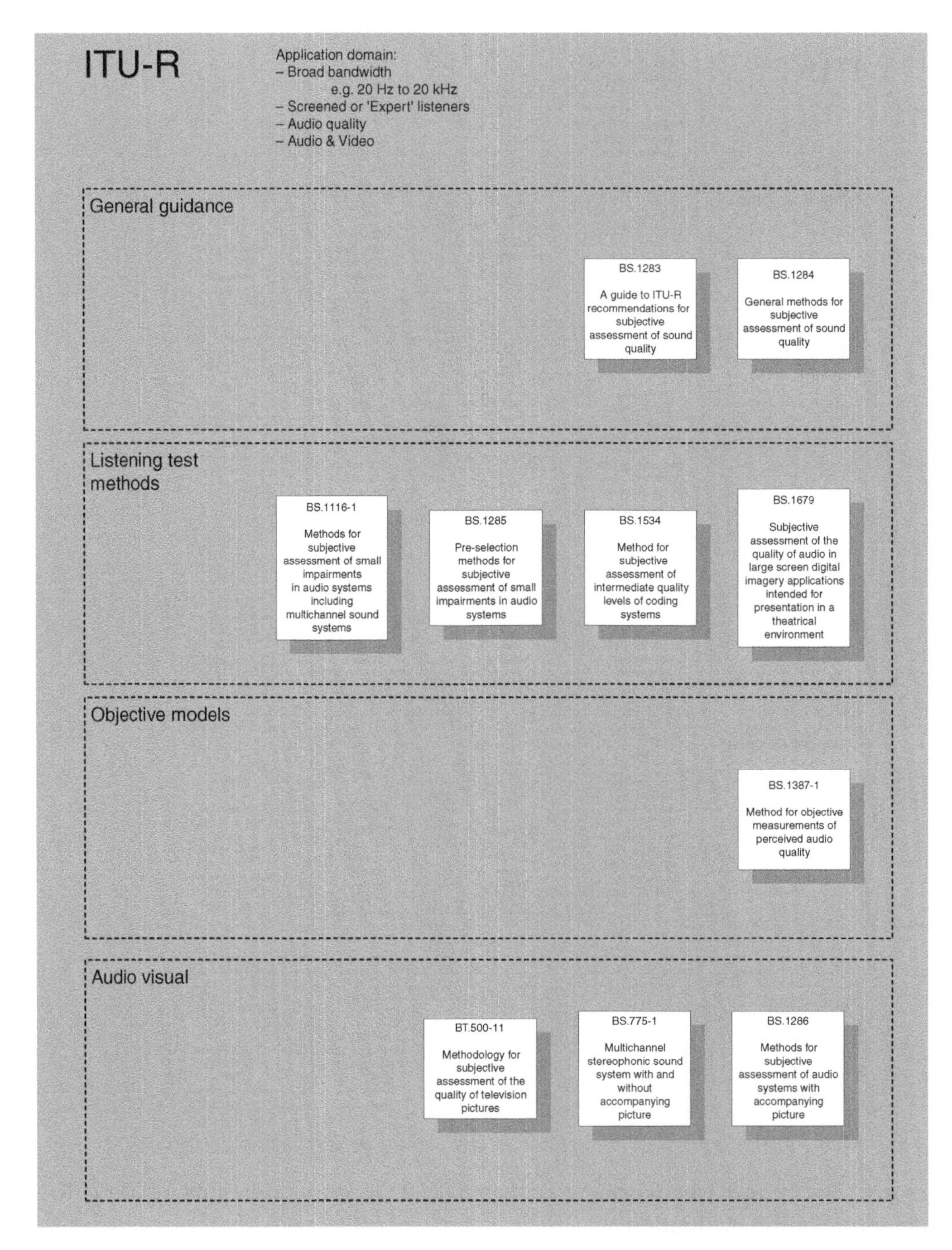

Figure 1.4: Summary overview of key ITU-R recommendations relating to perceptual audio evaluation.

regarding the scope of each standard or recommendation, prior to its acquisition.

1.3 Role of predictive models

One of the ultimate aims in the field of perceptual evaluation is to eliminate listening tests forever and replace them with predictive models that can accurately estimate human evaluation regarding the quality of audio signal. Development of such predictive models has been ongoing for many years. The present-day models typically comprise of a model of the human auditory system followed by a cognitive model to estimate the subjects scoring. Such models, nowadays always exist in the form of a computer program that analyses an input sound file, are very fast and easy to access and, as a result, are highly desirable.

Two categories of predictive models exist, namely, those that aim to predict a particular perceptual attribute, such as loudness, and those that aim to quantify overall performance, such as speech listening quality. In recent years, with the development of speech and audio codecs, there has been a desire to evaluate the performance of codecs and associated devices and this has led to the development and standardisation of numerous predictive models associated with speech and audio quality.

The benefits of such predictive models are quite clear and significant. As illustrated in this book, performing perceptual evaluation through listening tests is a time consuming and involved task, requiring significant knowledge of a number of different disciplines. The use of predictive models allows for the experimenter to obtain an estimate of the outcome of a listening test within a very short time, that is, real time to a few minutes typically and without the need to master a number of disciplines. This time and cost benefit is of great importance in addition to the near instantaneous access to such models, which require only rudimentary skill to operate.

The drawbacks of such models are, nonetheless, sufficiently important to heed. As with all tools, there are both correct and incorrect modes of usage. Predictive audio quality models normally have a particular domain of application beyond which their prediction accuracy is not known. Usage beyond the scope of application is risky and may provide misleading results, and thus the use of models in this manner is advised against. For example, models developed primarily for the evaluation of narrowband speech have been trained extensively with different kinds of speech stimuli, codecs and other relevant stimuli. The applications of such models to audio codecs with music is not necessarily appropriate and may well lead to misleading prediction of the perceived audio quality.

Models that aim to study a unique attribute of human perception, such as loudness, can be found, for example, in [188, 305, 307, 473, 477]. The output of such models, for example, in the form of overall specific loudness, aim to predict the impression of a subject performing a scaling of stimulus loudness.

Several perceptual models have been developed for the prediction of overall audio quality characteristics. With regard to speech and telecommunication applications, they have been developed to predict the results of speech listening quality test as performed using an ITU-T recommendation P.800 absolute category rating (ACR) test paradigm. The perceptual evaluation of speech quality (PESQ) model has been developed for narrowband[1] telephony speech and provides high prediction accuracy in this application. This so-called intrusive model requires a reference signal. Its characteristics are described in ITU-T recommendation P.862 [236] and P.862.1 [242] as well as by its developers in Rix *et al.* [363] and Beerends *et al.* [41]. Recently, a wideband[2] telephony speech version of PESQ has been standardised in ITU-T recommendation P.862.3 [246].

The correct mode of application and the associated limitations are important to pay attention to and they are described in ITU-T recommendation P.862.2 [245] as well as in documents such as [289]. The experimenter who applies such models without taking heed of the application guides risks making prediction errors and drawing scientific conclusions that are ill-founded.

The so-called non-intrusive model for predicting the subjective quality of narrowband telephony applications requires no reference signal. Such models, as standardised in ITU-T recommendation P.563 [243], are of great interest in the monitoring of speech quality in live telephony network. However, owing to their non-intrusive nature, that is, lacking a reference signal, their performance and prediction accuracy are inferior to those of their intrusive counterparts, for example, PESQ.

[1] Narrowband telephony refers to a bandwidth of 300 Hz to 3.4 kHz.
[2] Wideband telephony refers to a bandwidth of 150 Hz to 7000 Hz [230].

For applications in low bit-rate audio codecs, the perceptual evaluation of audio quality (PEAQ) model has been standardised in ITU-R recommendation BS.1387 [213]. Details of this model and its applicability are discussed in [249, 416, 427].

As the reader can see, a number of models exist for predicting the results of listening tests in different and specific application domains. Often, within the scope of the target application, the prediction accuracy of such models is excellent. However, once beyond the scope of application, the prediction accuracy becomes unknown. Presently, there exists no unified predictive model that is able to cover all aspects of audio perception. Additionally, predictive models have not yet been developed for all aspects of audio perceptual evaluation. For example, spatial sound perception models are still evolving. As a result, it will still require some time before such predictive models can replace listening tests completely.

Until that time, listening tests will still be an essential part of perceptual audio evaluation, including the development and verification of such predictive models.

PART I

Experimental considerations

CHAPTER

2

Definition of research question and hypothesis

THIS chapter discusses the formulation of research questions and the principles behind establishing a testable statement based on a premise or hypothesis and a number of initial conditions .

The formulation of the hypothesis, the initial conditions and the testable statement is perhaps the most important part of any scientific experiment. Firstly because they determine the scientific quality of the experimental results, secondly because the truthfulness of all or some of the initial conditions are determined by the design of the experimental set-up and thirdly because the statistical analysis of the observed data depends on the form of the testable statement.

The research question is a broad formulation of the problem that is to be investigated. An example would be the research question that

Perceptual Audio Evaluation – Theory, Method and Application Søren Bech and Nick Zacharov

was formulated for the Archimedes project[3]: 'Why is the perceived sound quality of a loudspeaker different in different domestic sized rooms and in different positions in the same room?'.

This question involves a large number of physical (see discussion in Chapter 5) and psycho-acoustical (see discussion in Chapter 4) variables that need to be carefully considered in order to obtain meaningful experimental results. The next logical step would thus be to divide this general question into a number of experimentally manageable questions where subsets of these variables are examined in more detail. However, the first immediate question one could ask is if it has been proven that loudspeakers actually do sound different in different rooms and in different positions.

Some researchers would argue that their practical experiences have shown this to be the case and thus conclude that it must be a general problem for all loudspeakers in all rooms. Others would argue that though experience indicates that this might be true it is not possible to draw conclusions for all loudspeakers and all rooms on the basis of a few observations and therefore a more rigourous validation is needed before a major research project is started.

The above dilemma, empirical versus rational knowledge, is one of the basic questions when discussing how scientific knowledge may be established. The following discussion will be an introduction to the problem and provide some guidance on how to tackle this and other related questions.

The purpose of this chapter is not to give a complete overview of the history of scientific argumentation, but to introduce the concepts and to provide an understanding of the general principles in formulating a research hypothesis.

The discussion will be based on two main references, Fjelland and Gjengedal [123] and Johansson and Lynoee [248], so only additional references will be noted in the following text.

Scientific knowledge is traditionally characterised by three fundamental aspects:

It is theoretical Scientific knowledge is theoretical in the sense that science does not solve practical problems per se. This is not to

[3]The Archimedes project was a five-year collaborative research effort funded under the European research scheme Eureka. The partners were the Acoustics Laboratory (now Acoustic Technology) of The Technical University of Denmark, KEF Electronics of England, and Bang & Olufsen of Denmark. Results of the project have been published in, for example, Bech [29–31, 34]

say that scientific questions cannot originate from practical problems or that scientific knowledge cannot be used for practical purposes, but if the work is aimed at solving a particular practical situation and the gained knowledge cannot be generalised, then it is not considered science in the traditional sense.

It is the truth The basic difference between knowledge and belief is that knowledge can only be the truth whereas belief can be either true or false. In addition, it is often required that knowledge must be documented using established scientific rules and measures.

It is systematical The systematical aspect is based on the fact that new knowledge must be based on systemised knowledge from previous experiments.

These requirements have led to formal systems representative of how scientific knowledge is established – the so-called axiomatic or deductive systems. Platon (427–347 B.C.) was one of the first to develop such a system and later Galileo Galilei (1564–1642) and Isaac Newton (1642–1727) used such a system to organise the development of their new theories. An axiomatic system is based on a set of *definitions* upon which it is possible to establish *axioms* or proven *hypotheses*, and based on the definitions and the axioms it is possible to establish *theories*. The knowledge represented by the axioms or hypotheses must be true knowledge for the developed theories to be the building blocks for new knowledge. However, the definition of *true* knowledge and how it is proven to be true have been debated since Galilei, who used rational arguments (using common sense) in some cases and empiricism (using his senses to gather experience) in others. This fundamental division between how scientific knowledge is created is represented by two major principles of scientific argumentation. One is the *principle of empiricism*, which claims that only observations and experiments can determine which theories or hypotheses are true. The other, the *principle of rationalism*, claims that deduction using common sense is the correct way to obtain true knowledge.

Most scientists, especially those working in areas involving both exact sciences (for example mathematics) and social sciences (for example psychology), will use a mixture of the two principles, depending on the experimental situation. Perceptual experiments such as evaluation of reproduced sound quality typically involves this mixture, so a short description of both principles will be given in the following text.

2.1 Principle of empiricism

The principle of empiricism is based on the fact that it is only observations and experiments that form the basis for the generalisation of observations: reasoned from the particular to the general. This is expressed in the principle of induction: '(the premiss:) If a certain observation A is observed together with another observation B and further that A has never been found without B then it follows: (the conclusion) the more often A and B are observed together, the greater is the likelihood that they will appear together in another situation where one of them is observed'.

An argument based on induction for the premiss of the Archimedes project, that the same loudspeaker sounds different in different positions in the same room, would be as follows:

Premiss 1: It has been observed through a number of listening tests, including a range of different loudspeaker types, that the same loudspeaker positioned in different positions (observation A) in the same room sounds different (observation B).

Conclusion: It is true for *all* loudspeaker types that the same loudspeaker will sound different in *all* positions in the same room.

According to this approach, there would be no reason to start a series of listening tests to prove the assumption, but instead the experimenter should start a series of experiments to examine various reasons for this observed difference between positions.

The problem, according to critics of the principle of empiricism, is that the conclusion can be flawed but the premiss can still be true. This violates another fundamental law in scientific argumentation: The law of contradiction, which states that a statement and its negation cannot be true at the same time.

When empiricism is the foundation for generalisation, an often applied research strategy is to test the observed principle or law under a range of different conditions and, if it is found to be valid in all the tested situations, it is accepted as a general law. Examples of laws based on empiricism can be found both in physics where Boyle's law states that the relation between the pressure p and the volume V of a gas is always constant ($p \times V = \text{constant}$), and in psychophysics where Stevens, [403, 406] using a direct magnitude estimation procedure, found that the relationship between sensation S and stimulus intensity I can be described by a power law ($S = \text{constant} \times (\frac{I}{I_0})^k$), where I_0 is a reference intensity and the exponent k depends on the modality and the stimulus. The ITU-R recommendation BS.1387 [213], which is intended for prediction of the perceived impairment in sound quality of a low bit-rate codec as a function of bit rate, is also an example of a 'law' or relationship that has been verified for a large number of situations and has been accepted as an international standard.

The principle of empiricism has often been criticised, especially the fact that the main focus is on confirmative results rather than deviating results, which according to the critics is where scientific progress is often made. One of the main opponents of the empirical principle was Popper (see, for example, [350]), who in 1934 published the book 'The logic of scientific discovery'. Popper argues that, no matter how many supportive observations there are, there will always be a finite probability that there are cases that do not support the previous observations and such cases will violate the law of contradiction as discussed earlier.

Popper's alternative is the so-called 'hypothetico-deductive' method, which is discussed in the next section.

2.2 *Principle of rationalism*

Popper [350] suggested the 'hypothetico-deductive' as an alternative (in fact, Popper argued that it is the only method) to the principle of empiricism or inductive method discussed in the previous section.

Before discussing the hypothetico-deductive method, it is useful to examine the principle behind deductive arguments. A deductive argument is based on a number of premisses and a conclusion. A simple deductive argument could be as follows:

Premiss 1: When it rains the street will be wet.

Premiss 2: It rains.

Conclusion: The street will be wet.

The deductive principle means that if the premisses are true the conclusion must be true. The opposite, the premisses were true and the conclusion false, would violate the law of contradiction.

The 'conclusion' is typically of a form that can be tested in an experiment (in the example above, you would look out of the window) and if the conclusion can be proven to be false (the street is not wet) it follows that one of the premisses must be false (probably that it is not raining). However, if the conclusion is found to be true (the street is wet), it does not finally prove that the premisses are true (the street cleaning truck could just have passed by and sprinkled water on the street). This principle is called the asymmetry between falsification and verification, and Popper argued that only falsification leads to true knowledge and progress: A falsification means that the experimenter must re-think or reformulate the premisses, which is equivalent to scientific progress.

The hypothetico-deductive method is based on a set of general observations that lead to the formulation of a hypothesis (see premiss 1 in the example above). The hypothesis together with a set of initial conditions (premiss 2) leads, based on a deductive argument, to a testable statement (the conclusion). The hypothetico-deductive method thus reasons from the general to the particular, which is the opposite of the inductive method discussed above.

Because the test statement is based on deductive arguments, it follows that if the initial conditions are true but the test statement is false then the hypothesis must be false. If the test statement was found to be true it could be assumed, but not formally verified, that the hypothesis was true and this would not lead to any real scientific progress according to Popper. So the experimenter's primary aim is, if the hypothetico-deductive method is used, to formulate one or several hypotheses that together with the initial conditions lead to a testable statement that must be rejected.

If the hypothetico-deductive method is used to test the assumption of the Archimedes project, the rationale would be as follows: The experience shows that for some loudspeaker types the same loudspeaker will sound different in different positions in the same room. However, it has also been found that the sound quality of certain other types are less influenced by their positions. So, in order to follow the principle of falsification, the focus is on the loudspeaker types that are not influenced by their positions in the room:

Hypothesis (premiss 1): Physically identical loudspeakers positioned in different positions in the same room sound similar.

Initial conditions (premiss 2): Two physically identical loudspeakers in two different positions in the same room are compared in a listening test and the results of the test are true representations of the perceived sound quality of the individual loudspeakers.

Testable statement (conclusion): The perceived sound quality of the two loudspeakers will be identical.

This strategy was used in the Archimedes project, and the results have been reported in Bech [29], where the test statement was proven statistically false for some of the tested loudspeakers, and true for others. This was the experimental basis for concluding that the hypothesis was false for at least some of the combinations of loudspeaker types and positions, and it was decided to perform further experiments to examine the influence of position on sound quality.

Now the basic assumption of the deductive principle is that if the premisses are true the conclusion must be true, and further that if the conclusion is false and the initial condition is true the hypothesis is false. The principle also states that the testable statement should be formulated with the purpose of falsification. The truthfulness of the initial condition thus becomes very important and this is often related to the quality of the experimental part of the investigation. The above example from the Archimedes project is a good illustration of why this can be quite complicated and difficult to achieve in the real experimental situation.

The initial condition was that the results of the listening tests represent the true magnitude of the perceived sound quality of the tested loudspeakers. However, this will depend on a number of variables such as the selected programmes (musical genre and number of excerpts), physical properties of the loudspeaker, number and type of subjects, physical identity of the two loudspeakers, and so on. So the truthfulness of the initial condition is related not to the truthfulness of a single condition but instead to a **sum** of conditions.

The scientific quality of the inferences of an experiment thus relies very much on the truthfulness of each individual part constituting the initial condition. The truthfulness of these individual parts is determined by the variables that define the experimental conditions. These variables are identified and discussed in the remainder of this book.

2.3 *Other principles of scientific argumentation*

The principles of empiricism and rationalism discussed above are, in most scientific papers, the two basic principles for establishing a testable statement and for the subsequent discussion of the results. However, there are a number of other relevant types of scientific argumentation that are especially important when considering practical experiments often used in, for example, engineering in contrast to theoretical disciplines like mathematics.

2.3.1 *Probabilistic reasoning*

Probabilistic reasoning is used when the truthfulness of the hypothesis is based on a limited number of observations from a given population. This introduces a finite probability that a rejected hypothesis is not representative or correct for the whole population. This is a very typical situation in medical experiments where only a limited number of persons can be tested and likewise in listening tests where in addition to a limited number of subjects only a limited number of programmes, listening rooms, loudspeaker positions, and so on, can be tested. This makes the probabilistic reasoning approach the most typical for assessments of sound quality using subjects.

Probabilistic hypotheses can be tested using either a deductive principle or an inductive principle, but in addition a probabilistic hypothesis arguments can be

- mathematical.
- objective.
- subjective.

The classical mathematical principle is always deductive, and theoretical probabilistic theory follows the standard mathematical rules. However, when probability calculations are used on an observed data set rather than theory, the truth of the hypothesis becomes testable. However, the deductive line of reasoning can still be applied and a typical example could be as follows:

Hypothesis (premiss 1): For a standard dice, the relative frequency of observing any one number in the range 1–6 is $\frac{1}{6}$.

Initial condition (premiss 2): A standard dice is thrown n times.

Testable statement (conclusion): Approximately $\frac{1}{6}$ of the n observations will be 'one'.

The hypothesis is now based on an observed probability and the truthfulness is now finite depending on the number of observations in the observed sample. A hypothesis originating from a calculated probability (based on a number of samples) is called an 'objective' hypothesis.

A 'subjective' hypothesis refers to the fact that some probabilities, although being objectively determined (based on a number of samples), have other conditions attached to them that must be fulfilled before the objective probability is relevant. The evaluation of these other conditions then becomes the subjective part. An example would be that it is possible to calculate the objective probability that an otologically normal person will hear a given pure tone at a given frequency and sound pressure level, but there might be other conditions (for example, the person is tired or has a cold) that could influence the ability to hear and thus need to be evaluated by the experimenter before that subject is used in the listening test where the objective probability is used.

The argument including the dice is strictly deductive even if it includes an objective hypothesis; however, it is often the case that the objective hypothesis is not tested and instead the hypothesis is based on an inductive argument:

Observation (premiss 1): Approximately $\frac{1}{6}$ of the numbers observed for this dice were 'one'.

Conclusion: The probability of getting a 'one' with this dice is $\frac{1}{6}$.

It is important to note that if the hypothesis is based on an inductive argument then it is impossible for the argument that includes the hypothesis to be anything but inductive.

The above discussion hopefully illustrates that in some cases it is very difficult to determine if an argument is deductive or inductive. In situations where the hypothesis is based on a limited sample, which is the case in statistics, the argument will always be inductive as the premisses describes the sample, but often (but not always) the conclusion refers to the whole population.

This means that the design of most listening tests, theoretically speaking, will be following the inductive principle, but it is still important to apply the hypothetico-deductive principle when designing the experiment to ensure the truthfulness of the initial conditions. However, when the general validity of the conclusions are discussed, it is important to consider and keep in mind the principle the hypothesis was based on. This requires a careful assessment of the statistical assumptions and the subsequent analysis. This is discussed in more detail in Section 6.2.

2.3.2 *Argumentum ad hominem*

There is another principle called 'argumentum ad hominem'. It refers to the principle of using personal statements from authorities within the field as proven facts without checking them formally. This type of arguments should be avoided, in the authors' view, as they carry no scientific weight in themselves. This is not to say that the authorities being referred to are not producing scientifically valid results, but then these should be referenced, and not personal, statements. This line of reasoning is often applied in Hi-Fi magazines, where the reviewers will have or gain status as authorities in their areas and quite often the reader will use the arguments of the reviewer for buying a particular product.

2.3.3 *Conclusion by analogy*

Yet another principle is 'conclusions by analogy', which means that the conclusion is assumed to be valid because it is similar to the premisses. An example would be the doctor who concluded that cold water should be used to treat burnt skin. He drew that conclusion by analogy from the boiled egg he used to have every morning. One morning the egg was more hard boiled than usual and he found the reason to be that his wife had not rinsed the egg in cold water as she used to. He concluded that cold water could stop the process of coagulation in protein and hence that cold water could stop the coagulation process when used on burnt skin.

This line of reasoning is rarely used (i.e. the authors have never seen it being applied) in auditory evaluations, but the principle should not be excluded as not being useful.

2.4 Summary

The above sections have discussed the major principle for establishing a hypothesis and based on a set of initial conditions, a testable statement. The most often applied principle in audio evaluations, based on the experience of the authors, is that of probabilistic reasoning. This, however, is also one of the most complicated as the line between the principles of empiricism and rationalism is not always crystal clear in such experiments. However, it is still the authors' view that rationalism and the hypothetico-deductive principle should be aimed for although it is clear that this is not trivial and requires careful consideration of a large number of factors. Especially, experimental planning and control are important to ensure the truthfulness of the initial conditions. The next chapter will discuss the fundamentals of experimentation and how this relates to the hypothesis and testable statement.

2.4 Summary

CHAPTER

3

Fundamentals of experimentation

This chapter discusses a number of intellectual and practical steps that reflect the scientific practice normally applied when conducting perceptual evaluations of audio systems. The list reflects the current practice within the audio engineering community; however, the reader should be aware that other scientific disciplines could have developed different traditions for perceptual evaluations, for example, in the field of psychology or food science.

The scientific practice, illustrated here by a practical 'to do' list plus the more detailed descriptions provided throughout this book, is what Kuhn [264] termed the 'paradigm' of the field. The paradigm reflects the current set of established rules for scientific work, defined through practical experience, theory and instrumentation. The paradigm

Perceptual Audio Evaluation – Theory, Method and Application Søren Bech and Nick Zacharov

is constantly changing[4]. For example, at the time of writing, a new paradigm was being developed for evaluation of the spatial properties of sound fields (see for example, Choisel and Wickelmaier [89], Lorho [274], Berg [52], Rumsey [367], Zacharov and Koivuniemi [465] and Bech [35]). The reader is thus advised to consult the literature for the latest updates to the paradigm.

The wish or need for conducting a perceptual evaluation of a loudspeaker, low bit-rate codec, amplifier, telephone or any other audio device is usually based on either a need for research or benchmarking. Benchmark experiments are typically conducted in industrial settings and the purpose of such experiments could include the following:

- Comparing (rank ordering) different products with respect to perceptual audio quality
- Comparing different versions (for example, different drive units in a loudspeaker) of the same product
- Qualifying or selecting a candidate algorithm, for example, a speech codec.

In addition, research experiments may be motivated through the need to:

- Verify predicted results from perceptual models
- Characterise the performance of a given type of system, for example, a noise suppressor, in order to further the understanding of the perception of noise
- Collect subjective data for the development of a perceptual model
- Probe the range of a parameter to establish its impact on a perceptual parameter, for example, audibility of a signal in noise.

The process begins when the need for a perceptual evaluation is acknowledged and ends when reporting the conclusions. The complete process is often long and tedious and includes many complicated questions, which can easily distract the experimenter's focus from the overall goal. To aid the experimenter, the following 'to do' list includes the major milestones in the process and the remainder of the book discusses these milestones in more detail. A graphical representation of this process is presented in Figure 3.1. The reader is encouraged to study this list prior to designing and conducting any experiments.

[4]The interested reader should consult Kuhn [264] for a very interesting discussion and theory on how such changes occur.

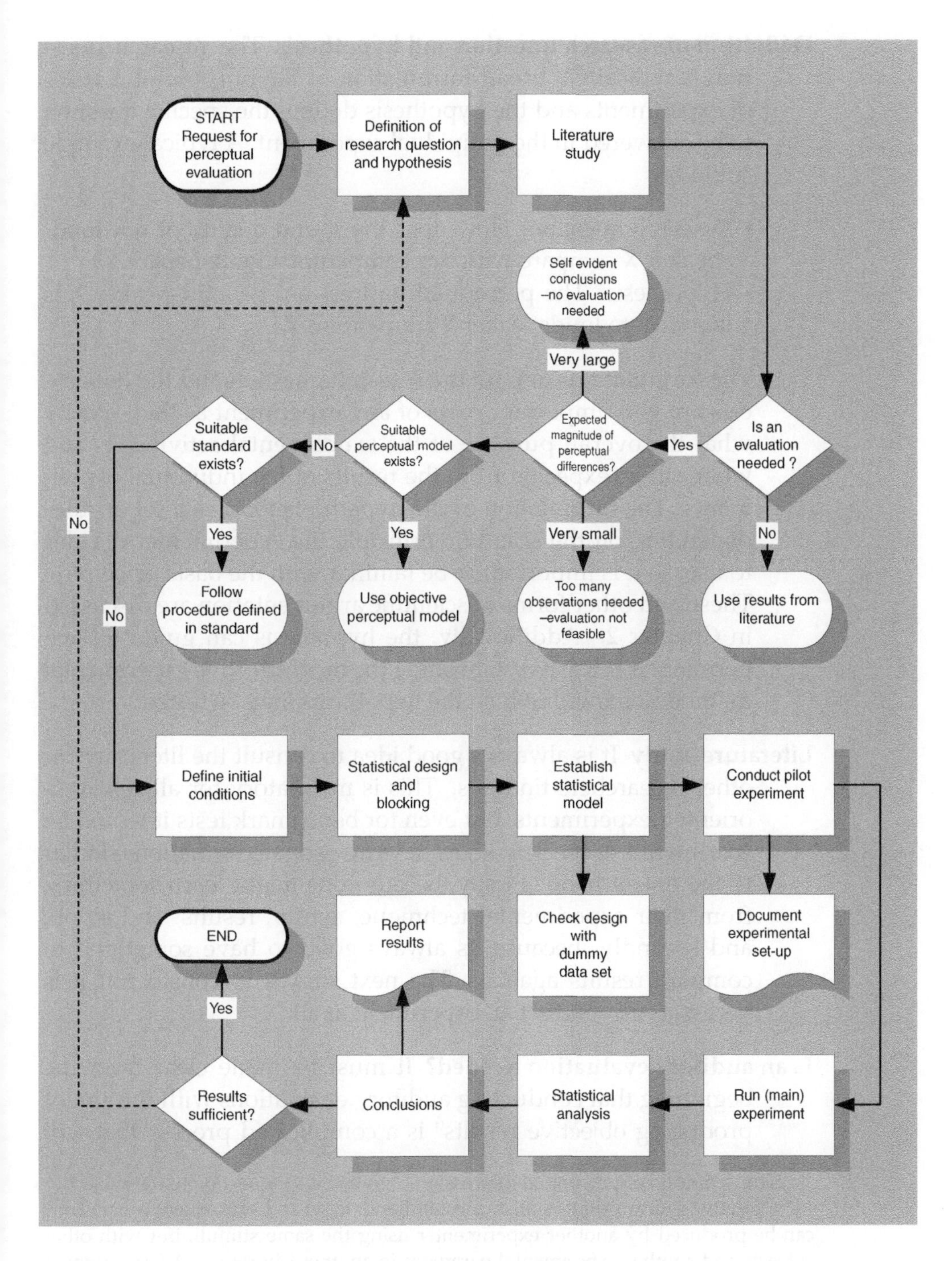

Figure 3.1: Process diagram for the preparation of a perceptual evaluation.

Definition of research question and hypothesis The research question is typically a broad formulation of the purpose of a series of experiments and the hypothesis defines the specific question to be answered in the individual experiment. A typical example could be:

- Research question: How does the sound quality of our loudspeaker X compare with our competitor's loudspeaker Y?
- Hypothesis: The perceptual impression[5] of loudspeaker X is identical to loudspeaker Y from brand Z.

The formulations of both the research question and the hypothesis are very important parts of any experiment as they specify what the overall purpose of the experimental activities is and what can be expected from the results of the individual experiments. The formulation of the hypothesis follows a set of rules depending on the scientific principle the experimenter decides to use, so it is important to be familiar with the basic principles. These and other forms of scientific argumentation are discussed in Chapter 2. Additionally, the hypothesis can guide the experimenter towards establishing the most effective experimental methodology with which the hypothesis may be tested.

Literature study It is always a good idea to consult the literature for other researchers findings. This is mandatory for all research-oriented experiments, but even for benchmark tests it would be worthwhile to check if others have investigated situations similar to the one at hand. Firstly, because one might learn something from their experimental technique, set-up, results, and so on, and secondly, because its always good to have something to compare results against. The next step is to consider if it is necessary to conduct an experiment at all.

Is an auditory evaluation needed? It must be made clear from the beginning that conducting auditory evaluations with the aim of producing objective results[6] is a complicated process that will

[5]The definition of perceptual impression is given by the question to the subject.

[6]Objectivity means that a statistically similar data set and subsequent conclusions can be produced by another experimenter using the same stimuli, but with other subjects and another experimental paradigm in another laboratory. A less stringent criteria is that the data should be reproducible. This means that the same data within the stated confidence interval should be produced by another experimenter using

require time, effort and money. So, it is worthwhile to seriously consider if it is necessary to conduct an evaluation after all. All possible sources of information should be consulted such as the literature as noted above and the expected magnitude of the perceptual differences as discussed in the next section. It should be remembered that an auditory evaluation can always be performed and a result will always be produced. However, as this book hopefully will show, if the experiments are performed incorrectly, the results will be meaningless and the effort wasted.

What is the expected magnitude of the perceptual differences? The magnitude of the perceptual differences is an important factor to consider before starting the experiment. If the differences are large, it might be obvious to everyone what the outcome of the test will be, and if so, how much verification of this is needed. On the other hand, the differences might be so small that it is clear that a disproportionately large number of subjects or repetitions will be needed to establish the statistical confidence needed. This type of analysis is important for both research and benchmark tests as resources are nearly always limited.

Is a suitable perceptual model available? A number of perceptual models have been developed in recent years. These models can be used to predict the outcome of a perceptual evaluation for a number of well-defined conditions. The models can be grouped into those that aim at predicting the impression for a particular attribute like loudness and those that predict the overall quality similar to the mean opinion score (MOS) as defined by ITU-R recommendation BS.1116-1 [207]. Much expensive experimental time can be saved if such a model can be applied. The current models available are discussed further in Section 1.3.

Does a suitable standard exist? A large number of standards have been developed for the perceptual evaluation of audio. These standards have been developed by experts in the field and they represent–at the time of publication–an agreed upon approach that will provide objective results. The experimenter should always examine if one of the available standards can be applied as this will ensure a reliable outcome of the experiments and

a similar set-up, including the same stimuli and experimental paradigm, but with different subjects at another place and at another time.

further enable a comparison with other published results that have applied the same standard. The role of standards are further discussed in Section 1.2 and the currently available standards are described and discussed in Appendix A.

Definition of initial conditions The initial condition together with the hypothesis and a deductive argument defines the testable statement. The testable statement is the question the experiment is designed to answer. This principle is discussed further in Chapter 2.

The initial condition belonging to the hypothesis listed above would be that the perceptual impressions of the two loudspeakers can be objectively measured via a listening test. So the testable statement would be that the mean values of the perceptual impression of the two loudspeakers, as reported by the subjects, are not statistically significantly different.

The initial condition is thus defined by those variables that will determine if the outcome of the listening test is a true and objective representation of the perceptual impression of the two loudspeakers. These variables are divided into two groups:

- Independent variables
- Dependent variables.

The dependent variable is the answer provided by the subject, and the objectivity of the answer depends on a number of variables such as:

- Definition of the question to the subject.
- Response format including choice of scaling method and associated bias effects.

The experimenter must thus be very careful when formulating the question to the subject and when defining how the subject should provide the answer. Chapter 4 includes a thorough discussion of these two issues.

The independent variables are those that can be controlled directly by the experimenter. The independent variables include the following:

- The test signal – typically in the form of the programme sections, their types, duration, spectral properties, degree of stationarity, and so on
- The reproduction system, including all components in the reproduction chain
- The listening room and its acoustical properties
- Acoustical and electrical calibration of the experimental set-up
- The selection of subjects and their characteristics
- The planning and administration of the test.

Chapter 5 and Part II discuss the influence and possible control of these variables.

The results of the listening test are thus objective measures of the perceptual impression if the dependent and independent variables have been chosen and controlled as defined by the experimenter. The next step in the process is thus to make the statistical design and model for the analysis of the results.

Statistical experimental design The statistical design includes determining the number of subjects and replications needed to achieve a certain resolution in the analysis. This can be based on previous knowledge; however, if this is not available and educated guesses are used instead, it is always a good idea to run a small-scale pilot experiment. The statistical design must also ensure that the experiment is correctly administered for a proper control of variables potentially influencing the perceptual responses. This could be the order of presentation of the programmes, number and position (in time) of the breaks, length of sessions (number of presentations per session), and so on. The statistical experimental design is discussed in Chapter 6. The next step is to establish and test the statistical model that will be used for an analysis of the results.

Establish the statistical model The statistical model is a mathematical model used to describe and statistically analyse the data. The statistical model includes the controlled independent variables and the analysis determines if the influence of the independent variables is above chance and thus contributes to the variation in the dependent variable.

Once the statistical model has been established, it is always a good idea to test it on a set of dummy data (for example, random data entered in the response matrix) before the experiment is conducted. This is especially important for highly complex designs where it can be of paramount importance that a certain interaction is testable, whereas other interactions cannot be tested, due to the experimental design, for example, through blocking. In such cases, it needs to be ensured that the planned design provides the desired information. The issues of statistical design and analysis of the data is discussed in Chapter 6.

Prior to the main experiment, it is always a good idea to do a final check in the form of a small pilot experiment.

Pilot experiment A small-scale pilot experiment can only be highly recommended. It serves a number of purposes:

- Final check of the experimental set-up, including any calibration issues
- Check of the suitability of the range of the stimuli
- Re-check on whether the experiment is obvious or not
- Confirmation that the experimental hypothesis can be tested
- Final check on time estimates for conducting the experiment
- Confirmation or recalculation of number of subjects and replications corresponding to the needed resolution of statistical analysis
- Check of the model for statistical analysis.

The next step will be to conduct the experiment and collect the data. However, before doing so, it is highly recommended to make a detailed documentation of the experimental set-up.

Documentation of experimental set-up This should be a trivial exercise, but it is often neglected because the experimenter at this stage is quite eager to run subjects and get real data to work with. However, experience shows that it is worthwhile to take the time to fully document the set-up at this stage. Firstly, because this will be the final version of the set-up and the version that must be described in detail in the experimental report or scientific paper, to allow for replication or continuation of the experiment. Secondly, because if something goes wrong during the experiment (for example, a colleague borrows an important component while travelling to a conference and accidently changes the calibration), it is paramount to be able to reestablish an identical set-up, and lastly because it is still possible to discover errors (logical or functional) that have not shown up during the pilot tests.

At this last stage before actually running the experiment, it is important to realise how much time has already been invested and it will often be realised that the extra time used at this last stage is very minor in relation to the total, but may save the whole experiment.

After the data has been successfully collected, the statistical analysis follows.

Statistical analysis The statistical analysis of the collected data should at this point be a matter of routine as the model has already been defined and tested. The standard statistical assumptions should of course be checked and outliers analysed and explained (if possible), but otherwise there should be no surprises. The statistical analysis is discussed in Chapter 6.

The next step is to discuss the obtained results in relation to other published results, to limitations imposed by number of variables and their levels, and whether extrapolation to a larger population is possible. On the basis of this discussion, final conclusions can be drawn.

Conclusions and reformulation of hypothesis The final conclusion will be an acceptance or rejection of the investigated hypothesis. If the hypothesis is rejected, the next logical step will be to redefine the hypothesis and start all over again, if justified. This closed loop circuit will be typical in the research setting, but in a more applied setting, the final conclusion is often the end of the project. The final step both in research and applied settings is to write the report or journal paper.

Reporting Writing a good report or a journal paper is an art in itself and for most authors it will be quite a time consuming exercise. However, it should be realised that if the work is not documented properly, nobody except the author will be able to gain from it and a lot of effort will have been completely wasted. A number of journals (for example, the Journal of the Acoustical Society of America) have very good recommendations for writing style, content, and so on, and it is highly recommended to consult those for detailed instructions. Extensive guidance on this topic can be found from either [15] or [430]. International standards also often have recommendations regarding content of the test report. For example, the ITU-R recommendation BS.1116-1 [207] recommends that the test report should, as a minimum, include the following items:

- The specification and selection of subjects and programmes
- The physical details of the listening environment and equipment, including the room dimensions and acoustic characteristics, the transducer types and placements and electrical equipment specification
- The experimental design, training, instructions, experimental sequences, test procedures and data generation
- The processing of data, including the details of descriptive and analytic inferential statistics
- The detailed basis for all the conclusions that are drawn.

Further details of this topic are provided in Chapter 9.

CHAPTER

4

Quantification of impression

THIS chapter discusses the variables that will have an influence on the definition of the question to the subject and the subsequent quantification and reporting of the answer.

In the previous chapter it was stated that the definition of the research question, the hypothesis, the initial condition and the testable statement are important parts of the design of an experiment. The definition of the question to the subject and subsequent quantification of the impression are central to all of these because it is assumed that there is a direct relationship between the testable statement and the answer the subject provides. The answer the subject provides represents the so-called dependent variable.

Perceptual Audio Evaluation – Theory, Method and Application Søren Bech and Nick Zacharov

There are two central issues to consider when discussing the dependent variable:

- Definition of the response attribute. This represents the definition of the question to be answered by the subject.
- Definition of the response format. This represents the definition of how the subject quantifies the auditory impression.

These issues will be discussed in the following two sections.

4.1 Response attribute

The type and definition of the question to the subject depend on the purpose of the experiment. The typical question in most research-oriented experiments is related to a specific attribute like loudness (the perceived magnitude of a sound), pitch (the perceived attribute associated with the physical variable of repetition rate of periodicity of the waveform), perceived spatial position of the sound event, perceived duration of the sound, or timbre. These questions are focused on a specific attribute of the sound, and often the stimuli are specifically designed to (hopefully) excite the attribute of interest. This often has the consequence that the stimuli are (very) synthetic and thus quite dissimilar to the natural sounds. However, although the sounds are designed to evoke a single attribute, they might evoke other attributes and it is therefore still important to ask the subject to be analytical and assess only the attribute in question. This requires a very detailed and specific formulation and explanation of the attribute in order to fully ensure that the subject understands what is being asked.

If more complex stimuli, like music, are used instead of designed signals, it means that several attributes are likely to be excited, and it thus becomes relevant to first identify the relevant attributes, then to define and formulate the description of the individual attributes and finally evaluate individual attributes in separate experiments.

In more applied experiments, like assessment of annoyance of noise around an airport or a broadcaster's interest in the general sound quality of lossy low bit-rate codecs[7], it is still interesting to investigate

[7]Codec is a generic term for compressor/decompressor which can be applied to audio signals for efficiency storage and transmission. A lossless codec can recover the original signal exactly, whereas a lossy codec can create only an approximation of the original.

specific attribute–related questions, but in addition such experiments often include questions related to acceptance, (un)pleasantness, annoyance, degree of liking, and so on. For such overall impressions, the subject is asked to use a more integrative frame of mind and take all aspects of the sound into consideration when assessing the sound. This includes personal preferences, the context, personal history, expectations, and so on.

This division between a specific analytical and an integrative frame of mind can be illustrated by the so-called filter model shown in Figure 4.1. This model represents a principle used by, for example, Plomp [346] when studying various aspects of tone sensation, Stone and Sidel [408] in sensory analysis of food, Nijenhuis [325], Bech [39] and Yendrikhovskij [456] in studies of image quality and Bech [35] in studies of spatial characteristics of sound. Pedersen and Fog [338] introduced an early version of the filter representation and this has been modified by the authors to the version shown in Figure 4.1.

The starting point is the existence of an acoustically complex stimulus–a sound field impinging on the ears of a subject–which is then transformed by the physiology and neurology of the hearing system into neural energy. The acoustic stimulus is characterised by *physical measurements* of variables (Φ) like sound pressure level, frequency, and so on.

It is now assumed that an auditory event is formed in the mind of the listener. This event is composed of a number of individual auditory attributes (Ψ) where each attribute represents a specific impression, for example, loudness, of the sound. For each of these attributes, the stimulus gives rise to a sensorial strength S depending on the magnitude of the physical stimulus and the characteristics of the hearing system. It is also assumed that it is possible to quantify the sensorial strength using standard psychometric procedures. The auditory event is characterised by the *perceptual measurements* of relevant attributes.

The next step in the process is the formation of an overall impression of the sound based on a combination of the individual attributes and the so-called cognitive factors. The cognitive factors include, in a somewhat simplified version, the expectations of the listener, the emotional state, the previous experience with this type of stimuli, the context, and so on. This combined impression forms the basis for an assessment of, for example, the degree of liking or degree of annoyance of the sound. The combined impression is characterised by *affective measurements* or hedonic tests.

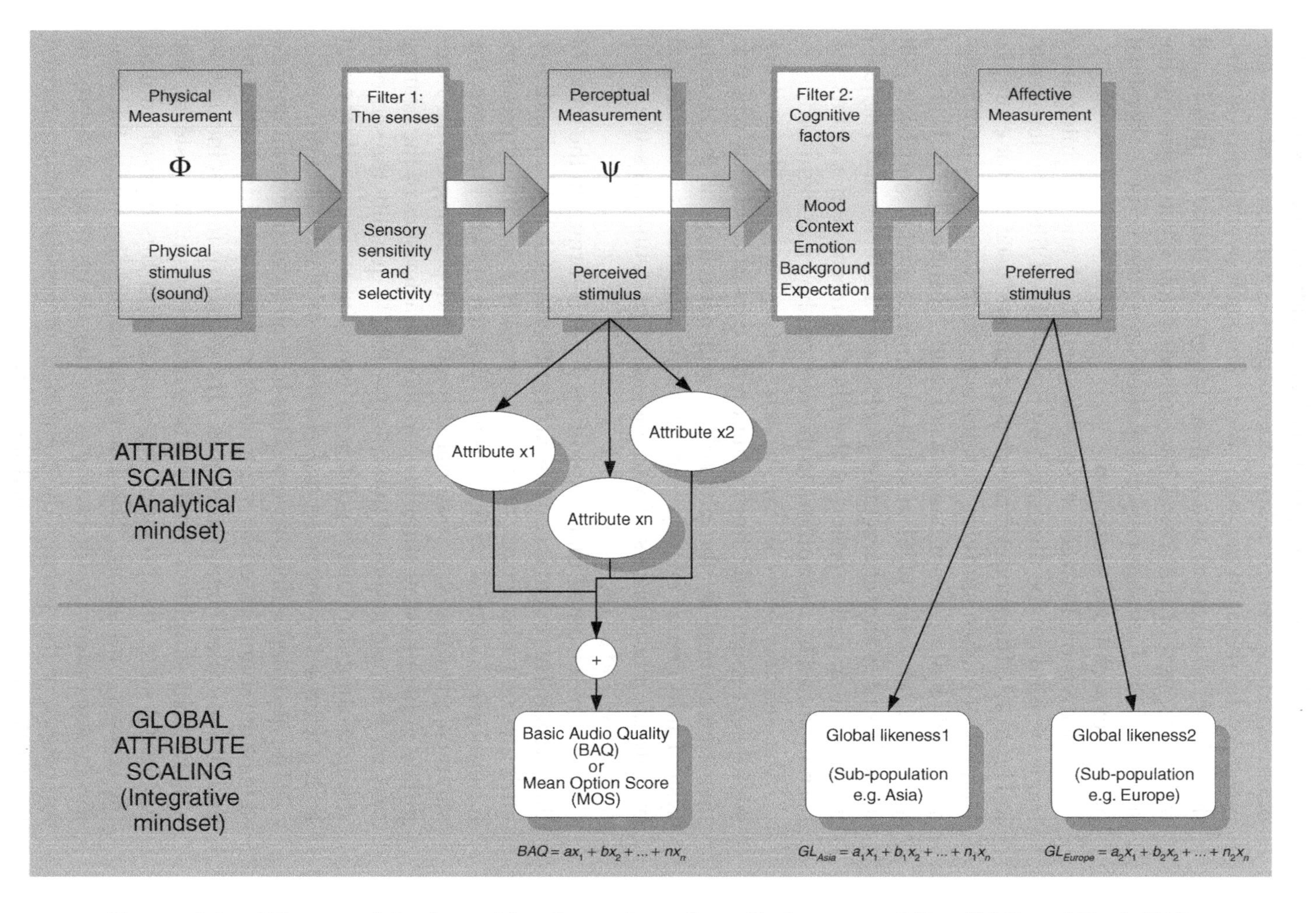

Figure 4.1: Filter model adapted by the authors from Pedersen and Fog [338] and Bech [35].

The identification and definition of the response attribute depends on which of the two types, perceptual or affective measurements, is relevant. The following sections will discuss this for each type of measurement.

4.1.1 *Perceptual measurements*

It is generally accepted that the perceptual measurement of sound quality, using complex stimuli, is a multidimensional problem that includes a number of individual auditory attributes (see, for example, Gabrielsson [132]). It is also accepted that it is possible to identify and elicit these attributes. For example, Wilkins [449] developed a set of 19 attributes for the description of the acoustics of concert halls by interviewing producers, musicians, and so on, and later Bünning and Wilkens [77] used 12 attributes for the evaluation of loudspeaker sound reproduction. Gabrielsson *et al.* [132, 136, 137] developed eight independent verbal descriptors to explore the perceptual space of sound quality for reproduction by loudspeakers, headphones and hearing aids. Staffeldt [398] used a list of 35 attributes including both timbre and spatial aspects for assessment of stereophonic loudspeaker reproduction. Staffeldt's list of descriptors was based on previous experiments, experience and interviews. Staffeldt [399] used nine attributes selected from previous works and also those of Toole [422] and Gabrielsson and Lindstrøm [135].

The above examples document that it is possible to elicit and use individual attributes of sound quality in perceptual evaluations. The examples cited used a mixture of interviews, personal experiences, multidimensional scaling and multivariate techniques like principle components analysis (PCA) to determine the independent attributes.

Food quality is another perceptual domain that can be characterised by a multidimensional approach and food scientists have also developed a number of different techniques to identify the attributes characterising food quality.

Thus, a number of different approaches exist with respect to identifying the individual attributes of multidimensional percepts and there seems to be no universally correct way of doing this. The following will be an introduction to the various possibilities. Two main approaches exist: Direct elicitation and Indirect elicitation.

Direct elicitation methods These methods are based on the assumption that there is a close relationship between a given sensation and the verbal descriptors used by the subject to describe the sensation. It is also assumed that the verbal descriptors can be elicited and a number of procedures have been developed for the same. These are grouped by two main principles:

- Consensus vocabulary techniques: These techniques rely on a common language developed and agreed upon by a group of highly trained subjects. A number of methods have been developed in the area of sensory analysis of food. Lawless and Heymann [268] (Chapter 10) have a good overview of the different techniques. Zacharov and Koivuniemi [465] and Zacharov and Lorho [468] are examples of how some of these techniques have been applied in audio.
- Individual vocabulary techniques: These techniques use the vocabularies developed by the individual subject and a set of principal components representing the common attributes is then identified using statistical procedures. Again, Lawless and Heymann [268] (Chapter 10) provides a good overview with respect to the food industry and Lorho [275] has examples of their use within audio.

Indirect elicitation methods These methods try to separate sensation and verbalisation, as verbalisation, it is argued, depends strongly on the 'size' and availability of suitable terms in the subject's lexicon. Secondly, it is questioned whether the assumed relationship between what is being said and what is being perceived actually exists. Methods include multidimensional scaling (MDS) (see, for example, Shifmann [376] and Borg and Groenen [61]), drawing (see Ford *et al.* [125–128]) and perceptual structure analysis (PSA), a method recently introduced in audio by Choisel and Wickelmaier [89] and Wickelmaier and Ellermeier [447]. Mason *et al.* [283] have provided a detailed discussion of the various aspects of verbal and non-verbal elicitation techniques in the subjective assessment of spatial sound reproduction.

Direct elicitation methods

These methods are based on the assumption that a verbal descriptor elicited from a subject is representative of a sensation. A number of techniques have been developed to elicit a list of words either for a group of subjects, the so-called consensus vocabulary techniques, or for individual subjects, the so-called individual vocabulary techniques. There are a number of characteristics that are common to both types of techniques and which the elicited attributes[8] should possess. Piggott [344] and Lawless and Heymann [268] list the following characteristics in order of importance:

- They should discriminate between the stimuli.
- They should have little or no overlap with others terms.
- They should be related to concepts that influence naive (consumers) preference decisions.
- They should relate to physical measures defining the stimuli.
- They should use singular rather than combinations of several terms or those holistic in nature (clean, fresh etc.).
- They should be precise and reliable.
- They should be capable of generating consensus among the subjects.
- They should be unambiguous to all subjects.
- They should be specifiable by a reference that is easy to obtain.
- They should have communication value and should not use jargon.
- They should relate to reality.

An important point that is not completely clear from the above list is that the attributes elicited should be descriptive and not attitudinal in nature. Attitudinal attributes are by definition highly subjective and related to preference (for example, good vs bad) and such attributes are used for affective and not perceptual measurements. The reader should also note that the following discussion, although aimed at discussing procedures for elicitation of attributes in some cases, will include issues related to scaling of the attributes. The reason is that these two issues cannot be separated for some of the methods discussed.

The so-called consensus vocabulary techniques are characterised by a process in which a group of subjects develops and agrees on the attributes and their meaning. Typical of such techniques are the so-called descriptive analysis techniques Flavour Profile® (see Meilgaard, Civille and Carr [291]), Quantitative Descriptive Analysis® (QDA) (see Stone and Sidel [408]), Texture Profile® (see Brandt *et al.* [65])

[8]Note that in the following text the term 'attribute' will be used; however, the terms 'verbal descriptor', 'word', and 'attribute' are used interchangeably in the literature.

and Sensory Spectrum® (see Civille and Liska [85]). Lawless and Heymann [268] give a good overview of the four methods.

In 1987, one of the authors was faced with the task of developing a method for assessing the image quality of televisions. An international research project, Adonis[9] was established and an extensive literature study was conducted to review possible methods. The results showed that the descriptive analysis (DA) was a good basis and the rapid perceptual image description (RaPID) method [39] was developed. A DA method was also applied in the Medusa project[10] aimed at investigating multichannel sound reproduction. The method was applied in investigations of perceived spatial sound quality and the results have been reported in [259, 465–467].

The attribute elicitation processes used by Flavour Profile®, QDA®, Texture Profile® and the Sensory Spectrum® method differ in various ways, but in order to illustrate the general descriptive analysis process, the QDA method will be used as an example.

The following description represents a typical QDA case. The reader should note that the list of attributes generated using the following and any other procedure will be language specific. Studies by Teunissen [413] have shown that it is not possible to translate specific words to other languages without the risk of losing important information. The consequence of this is that the related rating scale will be distorted.

The standard elicitation process includes six phases and the process includes a group of selected and trained listeners (see Chapter 5.4):

Phase one The first phase of the process involves presenting the whole group of subjects with a representative group of stimuli. To ensure that the developed list of attributes is meaningful, the experiment must include a relevant stimulus set such that many (all) perceptual characteristics are represented for the stimuli (products) under consideration. This initial selection of attributes is thus quite large and typically includes hundreds of words.

---

[9]The Adonis project was a five-year joint research project funded under European EUREKA scheme. The partners were the Institute for Perception Research (IPO)(NL), Philips Sound and Vision BG-TV/TV lab. (NL), Philips Components B.V. (NL), and Bang & Olufsen a/s (DK).

[10]The Medusa project was a three year joint research project funded under the European EUREKA scheme. The partners were British Broadcast Corporation (BBC) (UK), University of Surrey (Institute of Sound Recording) (UK), Genelec (FIN), Nokia Research Centre (FIN), and Bang & Olufsen a/s (DK).

As a preparation for generating this list, subjects could be asked to generate a preliminary list, before they participate in the first session. This list will be based on their experience (recollection) of the scenario at hand. This, however, requires that the subjects are somewhat familiar and/or have received a detailed explanation to the stimuli that will be presented. One benefit of making this preliminary list is that the subjects are forced–on their own–to think carefully of the task and further that the influence of group pressure is minimised. Martin and Bech [282] describe the generation of such a preliminary list.

The first phase typically lasts a couple of two-hour sessions, depending on the complexity of the stimuli. It is important to note that the panel leader[11] should interfere as little as possible during the discussion process; however, the subjects should be encouraged to avoid attributes with an attitudinal nature, such as good, bad, and so on.

Phase two The second phase includes removing duplicate attributes and grouping them to ensure a comprehensive description using a minimum of words. The whole group is present during the discussions and all stimuli are also available. This part could last five to six sessions and it is the task of the panel leader to ensure that continuity is maintained throughout the process.

Phase three The third phase includes further discussion and reduction in the number of attributes. A group of stimuli are selected to represent good examples of the selected attributes.

Phase four The fourth phase includes the introduction of simple test stimuli that activate all attributes and represent a wide range of intensities. These stimuli are used to introduce the concept of scaling, and the subjects will discuss the use of the scales and define the end-point terms for each scale. For example, the scale suggested by Stone and Sidel [408] for standard QDA experiments is a 15-cm horizontal line with end-point markers offset by 1.5 cm from each end. The intensity increases from left to right. The subject inserts a vertical line at the point that best reflects the relative intensity of the attribute in question. The distance from the left end of the scale then represents the numerical score.

[11] A panel leader is a person who is responsible for recruiting, training and supervising a panel. Panel leaders are often employed to facilitate the elicitation process.

A number of different scales could be suggested; however, it is important that the subjects participate in the definition of the scale and especially in defining the end-point terms. The activities should include evaluation of a large number of stimuli to increase the subject's consistency and familiarity with the selected terms. The process of eliminating or changing terms is continued during this phase.

Phase five The fifth phase introduces stimuli with smaller differences and repeat the scaling exercise introduced in phase four. Repetitions are included so that the consistency of the subject's responses can be monitored. The results are used to check the response system, examine the consistency of the panel and of the individual subjects, and establish the base-line performance for future tests. One point of interest is the question of when to terminate the panel training. Stone and Sidel [408] list the following points for examination:

- Are individuals consistent from trial to trial?
- Is the level of panel performance consistent?
- Are individual terms providing useful information?
- Are responses to product differences consistent across all subjects?

These questions can be answered via standard statistical analysis tests, as discussed in depth in Section 5.4.3. Specific values for acceptable performance in food tests can be found in Stone and Sidel [408], and Bech [27] presents results based on listening tests. It is noted that lack of consistency of either a subject or the panel is often attributable to the subject or panel itself. However, other possibilities are lack of control of other experimental variables and lack of agreement within or between subjects on certain terms. Experiences from numerous viewing experiments have shown that if there is a large between-subject variance for a given attribute, it is often indicative of a multidimensional attribute that should be divided into additional attributes.

Phase six The last phase is to conduct a number of tests with test conditions as in the real tests.

The number of attributes elicited depends on the difference between the stimuli, but a typical number is between 8 and 15. The attributes for spatial sound reproduction that were elicited in an

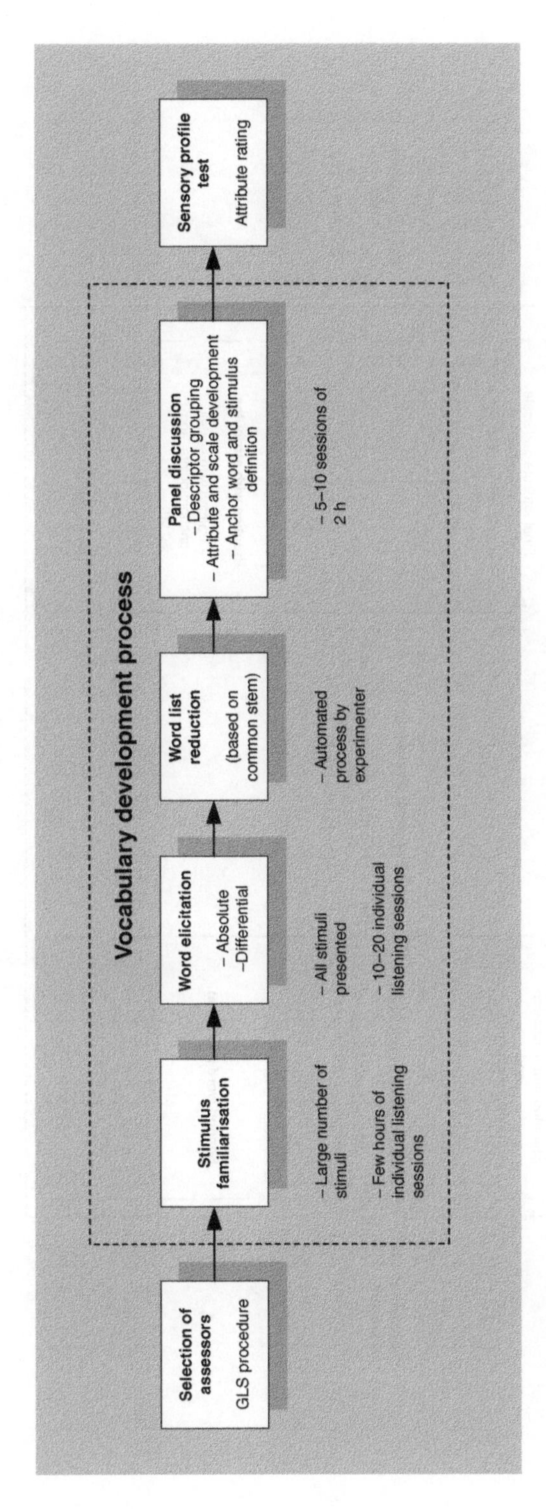

Figure 4.2: Example of implementation of the six phases in a descriptive analysis process as implemented by Zacharov and Lorho [468].

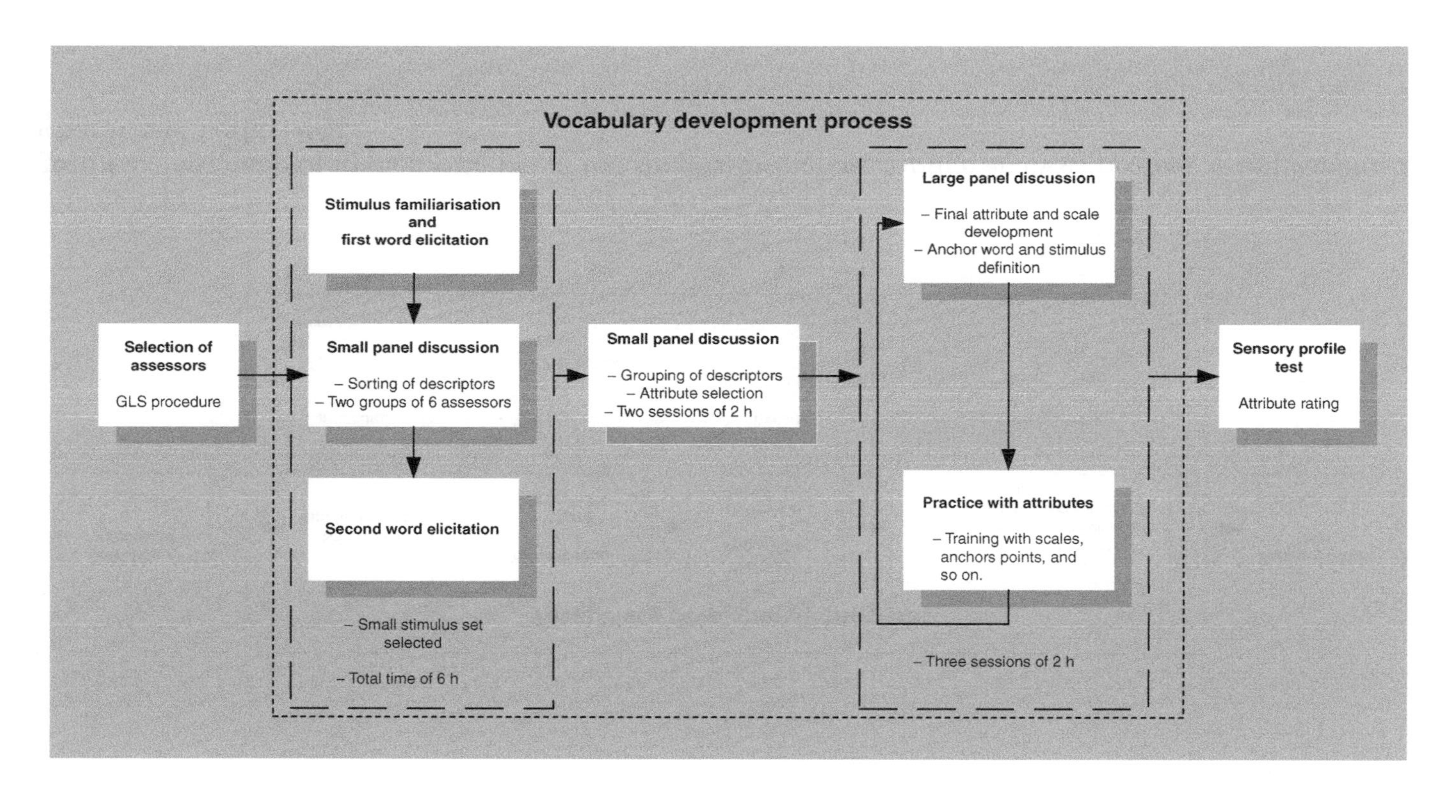

Figure 4.3: Time optimised attribute elicitation process as proposed by Zacharov and Lorho [468].

experiment conducted as a part of the Medusa project (see Koivuniemi and Zacharov [259]) are shown in Tables B.3 and B.4.

The attributes of speech quality have been investigated by Matilla [286] and the resulting list of attributes is shown in Tables B.1 and B.2. Appendix B includes additional examples of attribute lists for sound reproduction in headphones and timbres of reproduced sound.

Another good example of the development of attributes using descriptive analysis is the wine aroma wheel developed by Noble [326], shown in Figure 4.4. The attributes of wine aroma are arranged such that the most basic overall attributes are in the inner circles and the outer circles include the more specific attributes.

It is important to note that the development of attributes is a continuous process. In the Adonis project, an initial list of attributes was defined for the image quality of static images on cathode ray tubes (CRT) displays. This list was the starting point when moving pictures, liquid crystal displays (LCD) and plasma displays were introduced, and during the weekly training session subjects added/deleted or redefined attributes depending on the stimuli presented. Currently, the introduction of digital-based TV platforms has brought in another set of artifacts that leads to the elicitation of new attributes.

When the elicitation process is finished and the listening tests have been conducted, it should be considered whether the team should be permanent or whether a new team should be formed for each test. The experiences at Bang & Olufsen (B&O), which include establishing permanent listening and viewing teams, strongly suggest that permanent teams should be established if possible. A permanent team ensures a high degree of consistency over the years and makes it possible to develop lists of terms that encompass a wide range of situations. Maintaining a permanent team, however, requires a weekly training programme and regular visits to 'real' events such as concerts, theatre shows, exhibitions, and so on, to maintain the 'absolute' calibration of the team members. It has also proven worthwhile to equip the team members with the latest technological developments in the field (loudspeakers, televisions, and so on, of both B&O and other brands).

The six phases can be implemented in different ways, and an example is given by Zacharov and Lorho [468] (Figure 4.2) together with the approximate time spent on each phase. Note that some of the above phases have been combined and that phase one has been split into two where the second part includes elicitation on an individual basis. It is noted that the individual elicitation process and the group

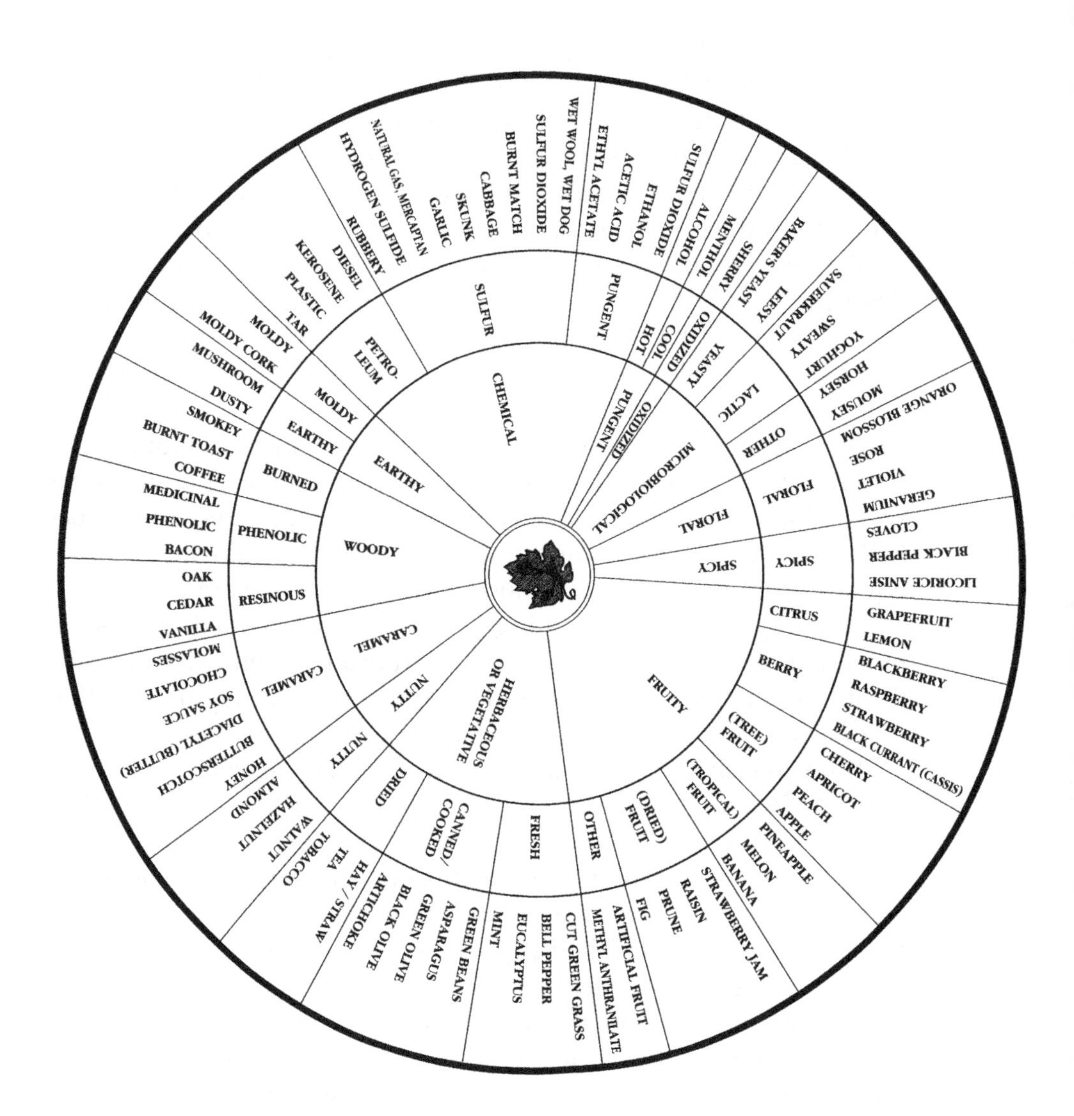

Figure 4.4: The Wine Aroma Wheel represents the attributes of wine aroma. The inner circle represents the more overall attributes while outer circles include the more specific attributes. For a more specific introduction to the wheel, the reader is referred to `www.winearomawheel.com` and the specific attributes are discussed in Noble [326]. The wine aroma wheel is copyright A. C. Noble 1990. Reproduced by permission of A. C. Noble.

discussions are quite time consuming, and in an effort to save time another approach was introduced. The new method is illustrated in Figure 4.3 and it included replacing the individual elicitation part with elicitation in smaller groups which further improved efficiency compared to larger group discussions.

The descriptive analysis techniques discussed and described above are consensus vocabulary techniques where the group, by consensus, defines the attributes and their scales. This however, includes very time consuming group discussions, as illustrated by the example of Zacharov and Lorho [468]. In industrial and applied settings, time is always a constraint and in an effort to improve the methods in this respect a number of so-called individual vocabulary techniques have been developed.

The individual vocabulary techniques avoid the group discussions and the continuous training and instead let each subject develop and use his/her individual lists of attributes. The 'group' language is then extracted by using advanced statistical procedures. These methods include free-choice profiling (FCP) [266, 278, 451, 452], repertory grid technique (RGT) [50, 251] and most recently flash profile (FP) [103, 380].

The following will be a brief description of the basic principles of each method and the interested reader is referred to the original papers as listed above for more detailed discussions.

Free-choice profiling The FCP technique was proposed by British sensory scientists in the 1980s [266, 278, 451, 452] and it differs from the standard DA in two major ways.

Firstly, it requires the subjects to develop their own idiosyncratic lists of attributes. The attributes are understood only by the individual subject and no training is given before the subjects evaluate the stimuli. The subjects also develop their own end-point descriptors for the applied scales. The lack of training and group discussions means that the FCP is much faster than the standard DA procedure.

The second major difference is in the statistical analysis of the data. The general aim of most experiments is to form the conclusions based on some form of consensus of the groups of subjects involved. The standard DA techniques achieve this via group discussions and training; however, for FCP data based on the individual attribute's spaces the consensus averaging must be done in the subsequent data analysis. The statistical method called generalised procrustes analysis (GPA) [107, 266, 374] was applied to achieve this goal. The method takes the data from each subject and mathematically creates a common group solution by standardising the scores from each subject, adjusting them for the subject's different usage of the scales and finally applying rotation, where the subject's inconsistencies in applying the attributes are compensated for. The interpretations of the dimensions of the common solution is performed by plotting the individual subject spaces in the consensus space. This process, however, is also one the of the more tricky points in the method and often presents the experimenter with a difficult task. Both Lawless and Heymann [268] (Chapter 17) and Dijksterhuis [107] provide good discussions of the procrustes analysis process.

The FCP method assumes that all subjects are equally capable of producing attributes and a corresponding scale with end-point descriptors and further of employing the attributes consistently. This is not always the case and especially the ability to produce attributes can differ considerably between subjects. This was the primary motivation behind employing a screening test for verbal fluency as a part of the selection procedure for subjects to participate in DA experiments (see Wickelmaier and Choisel [446] and Martin and Bech [282] for further details). Another potential problem of the FCP method is that there is no structured process to employ for subjects who are not confident with the task. The Repertory Grid Technique (RGT) offers such a structured process as discussed below.

Repertory grid technique The repertory grid technique (RGT) was developed by Kelly [251] and revised for audio tests by Berg [48] and Berg and Rumsey [50].

The original RGT elicitation process is based on the presentation of triplets of stimuli to the subject. The subject is asked to identify the stimulus that differs the most from the other two and then to identify, by means of words, in which way the selected stimulus differs from the other two and further to describe in which way the two similar stimuli are similar. The obtained words or expressions are likely (but not always) to represent the bipolar structure of an attribute with the words representing the end-point descriptors. Subsequently, the subject is asked to rate all stimuli on a scale with the words representing the poles of the

Attribute	Stimulus 1	Stimulus 2	Stimulus 3	Stimulus N	Attribute
Inside head	2	5	1	3	In front of head
Smaller	4	5	9	4	Larger
Unpleasant	1	2	7	2	Pleasant
Artificial	4	8	8	5	Natural
Metallic	1	1	4	5	Live
Less realistic	5	4	9	1	Realistic

Table 4.1: Example of a stimulus–attribute matrix for one subject using the repertory grid technique. The entries are ratings on an arbitrary scale.

scale. All possible pairs of triplets are presented to the subjects and for each pair of words identified, all stimuli are rated on the corresponding scale. The following possibilities for rating procedure have been suggested (see Berg and Rumsey [51], Berg [49], and Choisel and Wickelmaier [89]):

- Dichotomisation. The subject is asked to select which of the two poles best describes the stimulus.
- Rating using a category scale. The subject is asked to rate how the poles match the stimulus using, for example, a five-point scale with 1 indicating the best match and 5 the worst.
- Rating using a continuous scale. The word pair represents the end point on a continuous scale and the subject is asked to indicate, by means of a marking on the scale, where a given stimulus is relative to the end points. This could be done with a line for each stimulus (Choisel and Wickelmaier [89]) or by placing all stimuli on the same line (Berg [49]).
- Ranking. The subject is asked to indicate for a given word pair which stimulus matches the left-hand word best (assuming the poles to be on the left- and right-hand sides of a horizontal line). This is repeated with the remaining stimuli until all stimuli have been selected. The first chosen stimulus is given the grade 1, the second grade 2, and so on. This ranking procedure is followed for all word pairs.

An example of a stimulus–attribute matrix for one subject is shown in Table 4.1. Triplets based on N stimuli have been evaluated and the elicited attributes are shown. The entries are the ratings (arbitrary scale values) for each stimulus for each set of attributes.

Berg [49] has developed a software tool (OPAQUE) to facilitate the process of elicitation of the pairs of word, scaling of the stimuli, analysis and presentation of the result.

The RGT procedure using presentation of triplets does not specifically ask the subject to produce bipolar attributes, and further, there is a potential risk that attributes describing the two similar stimuli in the triplet could be missed. Both of these potential problems could be solved by using paired stimuli and asking the subject to describe the difference between the two pairs with a pair of opposite attributes. Choisel and Wickelmaier [89] compared the triplet and the paired comparisons (PC) based methods and found that the PC method produced attributes that were easier to interpret as end points of a scale.

The attributes elicited for each subject should ideally be representing independent impressions of the stimuli, but this is generally not the case and some form of reduction procedure is often applied. A fewer and independent number of attributes per subject will further facilitate the identification of a common attribute space based on all subjects.

The reduction in the number of attributes can either be by classification of elicited attributes according to their semantic category and type or by an analysis based on the numerical ratings of the elicited attributes.

The semantic classification of the elicited attributes should serve both to reduce the number of attributes and to verify that the attributes are of a descriptive, and not attitudinal, nature. Guastavino and Katz [154] present an example of reduction of number of attributes by classification into semantic categories. Berg and Rumsey [53] employed a so-called verbal protocol analysis (VPA) as described by Samoylenko *et al.* [372] for classification of the attributes into descriptive and attitudinal categories.

The reduction in the number of attributes can also be performed using the obtained ratings of the stimuli. Several standard statistical methods exist such as principle components analysis (PCA), factor analysis (FA), multidimensional scaling (MDS) and cluster analysis. Examples of employing cluster analysis can be found in Berg [49] and in Choisel and Wickelmaier [89].

Flash profile Flash profile (FP) was developed by Sieffermann [380] to reduce the time needed for attribute elicitation and panel training in the standard DA procedure. The procedure includes four phases:

Phase one. Phase one basically applies a free-choice profiling procedure where all subjects generate individual lists of attributes sufficient to discriminate between the stimuli and subsequently rank them. The important point is the focus on the attribute's capability to discriminate between and rank order the stimuli and not to fully describe them as is the norm for the standard DA. Another important point is that the subjects should be experienced sensory evaluation experts, which means that they should be able to understand the task and generate perceptive, and not attitudinal, attributes.

Phase two. In phase two, all the subjects are handed a list of pooled attributes based on the lists from all the subjects. The pooling is done by the experimenter and the subjects are asked to update their individual lists with any new attributes. After the subjects have familiarised themselves with their updated lists, they proceed to rank order the stimuli. All stimuli are available to the subject and ties[12] are allowed. The rank ordering of the stimuli forces the subjects to focus on and use discriminative attributes.

Phases three and four. These phases are repetitions of the rank ordering part of phase two.

The main difference between the FCP procedure and the FP is that the FP procedure forces the subject to focus on the discriminative attributes by allowing all stimuli to be compared during the elicitation process. Further, the FP procedure specifies that experienced sensory evaluation experts be used instead of naïve subjects, and this ensures that descriptive, and not attitudinal, attributes are generated.

Delarue and Sieffermann [103] compared the standard DA and the FP procedures for two groups of dairy products (yoghurt and cheese). They found that the two procedures gave identical

[12] A tie means that two stimuli can have the same rank.

sensory profiles for the yoghurt products but not for the cheese products and they were not able to explain the differences for the cheese products. They also found that the products were better discriminated using the FP method, but less well described (more attributes but difficult to interpret) compared to the DA method. They concluded that the FP method should be considered a supplement to the standard DA, especially in situations where a preliminary quick impression of the attributes involved for a given set of stimuli is needed.

Lorho [275] used a slightly modified FP method to characterise the spatial properties of spatial enhancement systems for reproduction of sound through stereo headphones. He included a familiarisation step before phase one and a second training step before phase three. He found that while the standard DA procedure applied for a similar task required 20 and 60 hours to develop the attributes, the present version of FP only needed 3 hours.

The above discussion has been based on the assumption that a direct relationship exists between the verbal attributes, elicited using one of the described processes, and the auditory sensation. This, however, rests on a number of assumptions and is quite complicated (if possible) to verify. This has lead to a search for other methods for communicating the sensation without the use of verbal descriptors. These so-called indirect methods of elicitation are discussed in the next section.

Indirect elicitation methods

The direct elicitation techniques described earlier assume that it is meaningful to assign and use words as descriptors of sound percepts. However, experience shows that some auditory concepts can be very difficult to describe verbally and other non-verbal elicitation methods for verification of the elicited attributes are thus needed. Mason *et al.* [283] has a good discussion of the pros and cons of verbal and non-verbal elicitation techniques. In this book, non-verbal elicitation is described as the information communicated from one person to another in addition to the verbal communication. This includes hands and arm movements such as when the person points in a given direction to indicate the origin of the sound or when the explanation of something is accompanied by a sketch, and so on.

The need for non-verbal elicitation techniques is a result of the ambiguity of the language. Language has a symbolic purpose whereby things are identified and catalogued; however, the interpretation of the symbolic meaning of a word is determined by the culture, education, tradition, and so on. So there are plenty of possibilities for misunderstandings between the subject and the experimenter. Language can also be used to communicate emotions; however, this situation only increases the degree of ambiguity of the message. Generally, it is known that the need for other communication tools in addition to language increases with an increase in the complexity of the experience.

The use of non-verbal elicitation in psycho-acoustic experiments exploring auditory localisation is not new. Blauert [56] describes a number of experiments where the subject was asked to point his finger in the direction of the sound source, position an arrow on a circular diagram in the direction of the sound source, turn the head in the direction of the sound source, and so on.

Theories for how humans describe and communicate perceived events (see Mason *et al.* [283] for an introduction with emphasis on auditory perception) also include the use of the visual-motor space in addition to a verbal response. This includes the responses humans can use to communicate by means of their visual system and the subsequent (or simultaneously) use of physical gestures, for example, pointing using their arms/hands or making drawings.

The elicitation of response attributes by indirect methods can thus be divided into two main areas:

- Elicitation using body gestures like eye or arm movements to report the sensation
- Elicitation without labelling the sensation.

Elicitation using body gestures Within the disciplines of acoustics and audio engineering, the most often used methods are pointing techniques and drawings.

Pointing techniques have often been applied in investigations of simple spatial properties, such as the perceived position of a single sound source radiating a simple signal. The techniques include asking the subject to point towards the sound event using a hand-held stick or a more advanced device such as a laser pointer (to provide visual feedback to the subject) in combination with a magnetic tracking system measuring the position of the laser pointer. Choisel [88] has a description and experimental verification of a new highly accurate laser pointing system. He also discusses the benefits and drawbacks of other pointing methods. Blauert [56] also includes descriptions of various pointing techniques used for the examination of spatial impressions.

The use of a pointing system (or drawings) introduces a very basic problem in relation to interpretation of the results: what is the reference point in the coordinate system the subject is using to report the position of the percept?

The general notion is that humans use an egocentric system for orientation within a spatial environment. Several definitions of egocentricity exist; however, the one favoured by Mason *et al.* [283] is that objects are positioned with respect to the body or some part of the body with no reference to an external point or place. The slightly modified version of an 'ego-centre' is that an external centre can be defined (fixed with a reference to the body) to which objects can be referenced as up, down, to the left, to the right, and straight ahead. Accepting the 'ego-centre' theory introduces the next step, that is, to define an intuitive way of communicating the position of the object with a reference to the 'ego-centre'. Mason *et al.* [283] discuss a number of possibilities, one of them being to draw on a piece of paper or a screen in front of the subject. The obvious problem here is the translation of a three-dimensional space into two dimensions, which will limit the intuitiveness of the methods and thereby increase the variance of the responses. The conclusion of the review by Mason *et al.* is that there are no obvious solutions to the conversion process, so each situation will have to be evaluated individually for the optimum choice of response.

Investigations using more complex stimuli such as real music, including several instruments, have often employed drawing techniques to describe the perceived spatial properties of the sound field in, for example, concert halls or the reproduced sound in the environment of a car. The need for using a minimum of two dimensions for the reporting is due to the fact that the sound sources in these situations, in addition to perceived main positions of, for example, the instruments, often include a spatial extension or envelopment of the sound field.

Ford *et al.* [125] report on an experiment where a non-verbal elicitation technique was used to examine the influence of loudspeaker and listener locations on the perceived spatial impression. They investigated the combinations of three different loudspeakers and three listener positions. Two of the listener positions represented a typical domestic listening set-up and the third the optimum set-up for stereo reproduction. They used five sections of standard programme material selected to contain clear spatial information. Three subjects participated and they were asked to draw their impressions of the spatial elements that they noticed. They did not receive any special instructions, but they were all students of music at the university. The paper provided for their drawings included certain landmarks: the three listener positions and the acoustically transparent curtain in front of the loudspeaker. The subjects were not trained, but they had all participated in a RGT experiment prior to the graphical experiment. Their results were analysed using three different techniques. First, the individual responses were visually inspected to detect any trends per subject, secondly, the drawings were measured to identify any common trends between subjects, and lastly, density plots were established to further examine the general trend for all the subjects.

The inspection of the individual responses showed clearly that subjects adopt different styles of drawing and emphasise on different spatial aspects of the reproduction. A representative example is shown in Figure 4.5. It was found that one source of these large differences between responses was caused by the–too high–complexity of the programme material.

It is worthwhile noting that if it is assumed that the drawings in Figure 4.5 represent the true mental images of the spatial sound reproduction, they clearly illustrate the problem encountered using, for example, a DA technique for identification and

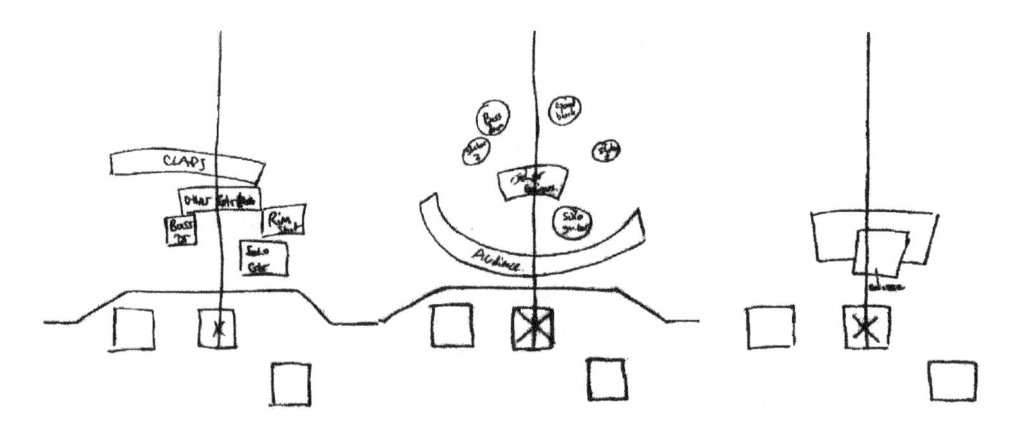

Figure 4.5: Drawings made by three subjects positioned in the centre position (indicated by x), listening to the same programme section. The acoustically transparent curtain in front of the listener is indicated on the two left most drawings. From [125]. Reproduced by permission of N. Ford.

scaling of the primary attributes: forcing the subjects to use the same attributes to scale the impression would clearly introduce some form of distortion in their response.

The individual drawings were then quantified by measuring 'total image width', 'total image depth' and 'image skew' (see [125] for a definition of the physical measures). The results for 'image width' were overall consistent with general experiences such as a decrease in image width when the listening position is off-centre and an increase in image width with increasing source width (only for the centre position). The 'image width' results also indicated rather large differences between the drawings for repeated presentations of the same programme. The results for 'image depth' were non-conclusive. This could be explained by the fact that the subjects, in the RGT part of the experiment, did not mention depth impression, indicating that image depth is not a primary attribute in this experiment. The results for 'image skew' were also consistent with the general experience that an off-centre listening position increases the image skew.

The next step in the analysis procedure was to create a density plot where all drawings were combined per stimulus. The results indicated that for a standard stereophonic set-up

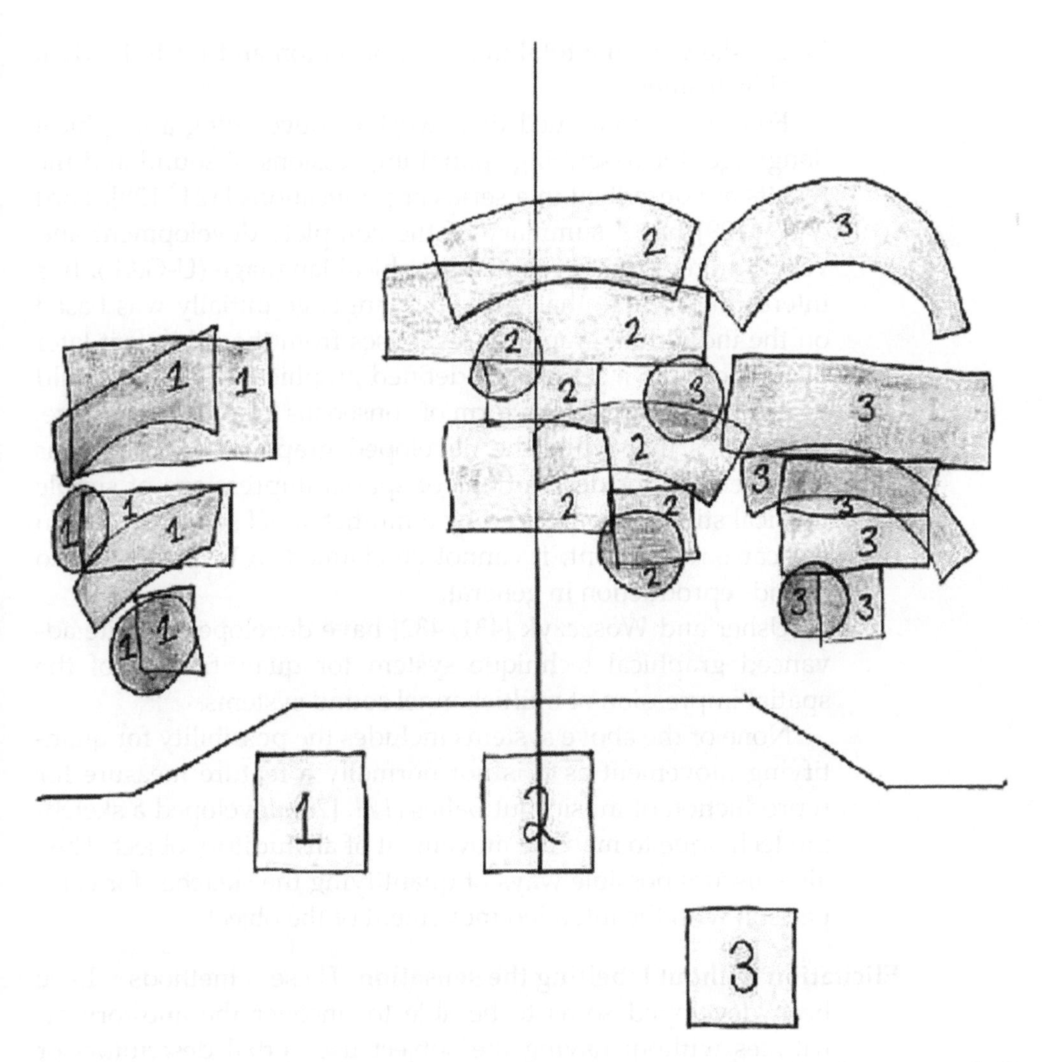

Figure 4.6: Total image plot for a jazz programme based on the drawings of all the subjects. The stereophonic set-up had a 120° angle between the loudspeakers and the subjects were seated in three positions, indicated by the squares 1–3. The number in the images indicates the position from which the image was depicted. From [125]. Reproduced by permission of N. Ford.

(60° between loudspeakers), image skew is not greatly influenced as a function of listening position. However, if the angle between the loudspeakers is increased to 120°, image skew becomes much more sensitive to the listener position as shown in Figure 4.6. This type of plot proved to be very useful for evaluations of image skew for the total image information and for individual focal instruments.

Ford *et al.* continued their work of developing a graphical 'language' for describing spatial impressions of sound and the results are described in a series of publications [124–128]. Ford *et al.* [129] has a summary of the complete development and verification of the developed graphical language (U-GAL). It is interesting to note that while the language initially was based on the individual graphical responses from the subjects it later developed into a set of well-defined graphical icons that could be used by all subjects–a form of consensus graphical language. Ford notes that while the developed graphical language has been verified for descriptions of spatial impressions of simple musical stimuli reproduced by a multichannel sound system in the car environment, it cannot be claimed to be applicable to sound reproduction in general.

Usher and Woszczyk [431, 432] have developed a quite advanced graphical technique system for quantification of the spatial impression of multichannel sound systems.

None of the above systems includes the possibility for quantifying movement as it is not normally a feature measure for reproduction of music, but Ballas *et al.* [20] developed a sketching technique to map the movement of an auditory object. They also discuss possible ways of quantifying the sketches for comparison with the intended movement of the object.

Elicitation without labelling the sensation These methods have been developed so as to be able to uncover the auditory attributes without having the subject use verbal descriptors or even being aware of the existence of multiple attributes.

A well-known method is based on the so-called multidimensional scaling technique (MDS) (Carroll [82] and Borg and Groenen [61]). The subject's task in this type of experiment is to indicate the perceived dissimilarity between pairs of stimuli. The subject is presented with all possible stimuli pairs and dissimilarity matrices are generated for each subject. These are analysed by standard MDS routines such as the INDSCAL (Carroll and Chang [83]). The basic assumption is that it is possible to represent the inter-point distances, representing the perceived dissimilarities between the n stimuli, by fewer than the $n - 1$ dimensions ideally required. The dimensions of this reduced space is assumed to represent the main auditory attributes responsible for the perceived dissimilarities. The method, however, does not include simple methods for determining the perceptual meaning of the attributes. Grey [150] used MDS to uncover the dimensions of musical timbre and Choisel and Wickelmaier [89] describe an MDS analysis to uncover the dimensions of spatial sound perception. Martens and Zacharov [281] used MDS to uncover the spatial attributes of processed speech.

Recently, Choisel and Wickelmaier [89] and Wickelmaier and Ellermeier [447] introduced the so-called perceptual structure analysis (PSA) for indirect elicitations of attributes. The subject is presented with triads of stimuli and the question posed is 'Do sounds A and B share a common feature that sound C does not have?' The subject must answer 'Yes' or 'No'. The subject is presented with all possible triplets and based on the responses it is possible to extract the auditory features underlying the responses. The method is new to psycho-acoustics, so only few results have been reported but interesting results have been produced by Choisel and Wickelmaier [89] when extracting the auditory features of multichannel sound reproduction and Wickelmaier and Ellermeier [447] when uncovering the features of narrowband noises and complex tones. The reader is also referred to these papers for a more detailed discussion of the underlying mathematical theory.

4.1.2 *Affective measurements*

The purpose of affective measurements is to objectively quantify an 'overall' or 'general' impression of the sound stimulus. This assumes that the subject is in some form of integrative state of mind where the influence of the impression for the individual attributes, the context, the mood, the expectations, the previous experience, traditions and so on, are all combined into one single impression that establishes the basis for some form of action of the listener. The action can be to buy a product, to complain because of being annoyed, to feel happy, and so on. These types of measurements are typically conducted using the so-called naïve subjects where no formal training or selection criteria are applied. It is possible to use trained subjects, but great care should be taken in such cases as the 'integration' of the various impressions could be quite different compared to the corresponding results when naïve subjects are used.

The question to the subject should reflect or force the subject to apply an integrative state of mind. However, despite the more general approach the subject needs to take in this type of experiments, it is still important to carefully consider how the question should be formulated. In audio, the typical question would be to evaluate the 'fidelity' of the reproduction. This concept has been defined in many ways. Typical examples would be 'the degree of naturalness' or 'similarity to a live performance' and it is evident that the two definitions could differ with respect to the degree of integration, context, expectations, and so on.

The ITU-R standard for assessment of low bit-rate codecs [207] asks the subjects to evaluate basic audio quality (BAQ), which is defined as 'this single, global attribute is used to judge any and all detected differences between the reference and the object'. It is assumed that the subject integrates the impairment caused by all the different types of artifacts–represented by an attribute for each of them–as illustrated in Figure 4.1. The same principle applies for evaluation of the mean opinion score (MOS) as defined in the ITU-T standard [238], where the subject is asked to provide 'listening-quality opinion' on a 5-point scale ranging from 'bad' to 'excellent'.

The various forms of the question to the subjects have been the object of quite a lot of research related to sensory analysis of food. However, the results and conclusions are equally applicable to the auditory impression, so most of the following discussions are related to food and again Lawless and Heymann [268] is an excellent reference.

A number of different test types are typically used and each of them has specific questions associated with it:

- Preference tests
- Acceptance tests
- Appropriateness tests.

Preference tests Preference tests are conducted in two typical forms: paired preference tests with two stimuli presented to the subject at a time and ranked preference tests with multiple stimuli presented. In both cases, the question to the subject is to make a choice or a ranking based on preference. In most tests, the subject is forced to make a choice, but in some cases ties (i.e., no decision) are allowed. The preference is not explained further to the subject, so it is entirely up to the subject to define what the preference decision is based on.

Acceptance tests In acceptance tests, the question to the subject is to report the degree of liking. This is typically done using some form of hedonic scale like the 9-point scale (d) shown in Figure 4.7. Instead of asking for the degree of liking, other forms of acceptance tests use the 'just-right' or 'just-about-right' versions. Then the question to the subject is to evaluate the desirability of a given product or the amount of a given property of the product in relation to the general desirability of the product. The scales typically include the categories 'not enough', 'just right', and 'too much'.

Appropriateness tests The appropriateness tests combine the degree-of-liking question with a context relation. The question could be 'How much do you like the sound quality of the loudspeaker if you were to use it in the kitchen when cooking?' This type of question tries to disentangle the fact the degree-of-liking could be (is) dependent on the context: a given reproduction of a voice in a mobile phone context could lead to a fairly high degree-of-liking, but a much lower liking would result if the same reproduction was evaluated in an audiophile context.

The type of questions listed above are usually asked as single questions, but nothing prevents a combination, for example, in a questionnaire-based experiment.

When the question to the subject has been formulated, either in the form of a specific attribute to be evaluated or as an acceptance test, the next step is to decide on the format that the subject should use for responding. This complex issue is discussed in the following section.

4.2 *Response format*

The definition of the response format includes specifying a scale and a scaling method that the subject can apply when reporting the auditory impression.

A straightforward scaling process would be to ask the subject to report, using his own words, what he perceives. However, the problem with 'results' from such an experiment is that they would not comply with the basic requirements for scientific data:

Objectivity Personal descriptions of impressions in the form of 'stories' using daily language do not represent objective statements or data (see footnote 6).

Quantification Personal descriptions are not in quantified form such that the resolution can be checked and statistical analysis can be applied.

Communication Personal description cannot be exchanged in a clear and well-specified format between users (scientists, clients, the public).

Scientific generalisation Personal statements are specific to the person providing them and thus they cannot be generalised to a larger population.

To fulfil these requirements, it is necessary to specify a standardised procedure for reporting the impressions. This involves using a standardised scaling paradigm that possesses the following characteristics [328]:

- The procedure or the rules for assigning numbers to the impressions is clearly stated and has been tested and verified by many scientists under many circumstances.
- It is practical.
- It is easy to implement and use by experimenters.
- The results obtained using the procedure do not depend on the individual experimenter.

The development of scaling procedures has followed two main lines–direct scaling and indirect scaling. In a direct scaling technique, the subject is asked to directly convert the sensation into a sensory magnitude and to report it on a scale. These techniques are often used for perceptual measurements. Stevens [401, 404] was one of the founding fathers of direct scaling techniques.

The indirect scaling techniques measure the sensation through the subject's ability to discriminate between the stimuli. The indirect scaling techniques originate from Thurstone [418, 419], who developed a technique (the law of comparative judgements) for construction of a scale based on minimum information (simple preference decisions in a paired comparisons experiment). These techniques are often used for affective measurements.

The intention of this section is not to present a complete overview of scaling theory and techniques but to give the reader an introduction to the major issues and to present references for further study. Ekman and Sjöberg [115], Gescheider [143], and Nunnally and Bernstein [328] give excellent overviews of the many theories and questions related to scaling of human responses.

4.2.1 Direct scaling

There are two basic aspects that need to be considered when applying a direct scaling technique. The first is the properties of the numbers that the subjects have assigned to the stimuli. The second aspect is the method that has been used to obtain the numbers. These two aspects are not independent as certain methods tend to dictate the properties of the numbers.

Four different categories of measurement scales are normally considered in accordance with the suggestions of Stevens (see Nunnally and Bernstein [328] for a thorough discussion and further references). The scales reflect four basic ways of reporting a sensation. The definitions of the four scale types as given by Stevens are shown in Table 4.2.

Brief explanations of the categories are as follows:

Nominal The basic operation is to decide if two objects are equivalent or not. The objects can be labelled by number, names or by any property that is relevant for distinguishing between the items being considered. Categorising loudspeakers according to their brand names or model numbers would be representative of applying a nominal scale.

Ordinal The basic operation is to decide if the magnitude of sensation of one object is higher or lower than that of the other object, for the specified attribute. The simplest ordinal scale would be a pass/fail scoring of assigning a 1 to the loudspeaker that is preferred in a pair and 0 to the not-preferred loudspeaker.The outcome of the test would be a rank-ordered list of the loudspeaker according to the specified attribute.

Scale	Basic operation	Permissible transformations	Permissible statistics	Examples
Nominal	Equality versus inequality	Any one-to-one	Number of cases, mode	Telephone numbers, and so on
Ordinal	Greater than versus less than	Monotonically increasing	Median, percentiles, order statistics	Hardness of minerals, class rank
Interval	Equality of interval or differences	General linear $x' = b \times x + a$	Arithmetic mean, variance, Pearson correlation	Temperature (Celcius) (conventional test scores)
Ratio	Equality of ratios	Multiplicative (similarity) $x' = b \times x$	Geometric mean	Temperature (Kelvin)

Table 4.2: Steven's level of measurement, basic defining operations, permissible transformations, example of permissible statistics, and examples. From [328]. J. Nunnally and I. Bernstein, Psychometric Theory, 3rd ed., 1994, McGraw Hill. Reproduced with permission of The McGraw Hill Companies.

Interval The basic operation is to specify a unit of measurement such that it is possible to specify how much more an object is greater or less than another not-equivalent object. The outcome would be a rank-ordered list of items where the distances between the items are specified. The absolute magnitude of the scale values are not known as no true zero can be defined. An example would be the 5-point scale recommended by the ITU [207] for quantification of differences between low bit-rate audio codecs.

Ratio The basic operation is the same as for an interval scale, but now the absolute value of the unit of measurement is known (i.e. a true zero has been defined). Choisel and Wickelmaier [90] present results using a ratio scale for preference ratings of different sound reproduction modes.

Stevens suggested that the type and degree of statistical analysis was linked to the scale type as indicated in Table 4.2 in the column 'permissible statistics'. Data based on nominal and ordinal scales should be analysed using methods based on, for example, the binomial distribution (frequencies for nominal scales and ranking for ordinal measurements). Interval- and ratio-based data could be analysed using quantitative analysis methods based on an assumption of a normal distribution (t-test and ANOVA). The suggestions of Stevens gave rise to much debate with (typically) statisticians in one camp claiming that 'numbers do not know where they come from' [138] and therefore any set of data could be analysed as long as the statistical assumptions (distribution etc.) were satisfied and psychologists in the other (the so-called representational position [328]) camp maintaining that the properties of the scale should be documented before analysing the data. The debate is still continuing as no conclusive argument has been produced by either camp.

A more pragmatic view that is preferred by the authors and recommended, for example, by the ITU-R Recommendation BS.1116-1 [207] and Lawless and Heymann [268] is that data based on an–assumed–interval scale is analysed by quantitative statistics unless severe violations of the scale properties or statistical assumptions are present, in which case it is re-analysed using categorial based methods. The two sets of results can then be compared and valid conclusions made.

It has often been debated as to how many categories a scale must include in order to be treated as an interval scale and not a categorial scale. Nunnally and Bernstein [328] argue in Chapter 4 of their book that a scale can be considered continuous if it provides 11 or more categories. The often-applied 10-point fidelity scale with one decimal place [184] and the ITU-R preferred 5-point scale [203] with one decimal place could thus be considered continuous and the results analysed using quantitative statistical procedures.

A more detailed discussion of statistical analysis of data including a description of the various statistical procedures available for categorial and quantitative data is found in Section 6.2.

The scaling procedure is the process that defines how the subject should assign the numbers from the scale to the stimuli. Stevens suggested two direct scaling methods for the construction of scales: partition scaling and ratio scaling.

Partition scaling In these methods, the subject is required to partition the sensory continuum into equal intervals. Two methods, equisection and category scaling, have been developed for this purpose.

In the equisection method, the subject is presented with a group of stimuli and asked to select a limited number of stimuli that produce equidistant sensations on the attribute of interest. The end points of the continuum are defined to the subjects by the lowest and highest stimuli in the set.

In category scaling, the subject is asked to assign a category (a number or label) to each of the stimuli in the set presented. The categories or labels are determined by the experimenter. This scaling method is often employed for audio evaluations.

Ratio scaling These methods are divided into four groups: ratio production, ratio estimation, magnitude estimation and magnitude production.

In ratio production, the task of the subject is to adjust the magnitude of a variable stimulus to be a prescribed ratio of a reference stimulus. The Sone scale for loudness has been constructed using this method. This method requires that the stimulus is continuously variable over the sensation's range of interest. This is often not the case in audio experiments where fixed solutions are compared to each other.

In ratio estimation, the task of the subject is to report the ratio between two stimuli for a given attribute. A range of stimuli values is available and all possible combinations are evaluated by the subject.

In magnitude estimation, the task of the subjects is to assign a number representing the sensory magnitude of the stimulus. It can be employed using a standard stimulus having a prescribed value and the subjects must then assign a value to the new stimulus that reflects how much greater (or smaller) the new stimulus is relative to the reference. It is also possible to let the subject assign any value to the first stimulus and then judge subsequent stimuli against the first.

In magnitude production, the task of the subject is to adjust the auditory sensation of a variable stimulus to be at a pre-specified magnitude.

A number of different implementations of various category scaling methods have been used in audio. Figure 4.7 includes an overview of the scales that are recommended by the various standards. It is strongly recommended that one of the standardised scales be applied whenever possible. It is not an easy nor trivial task to establish and document a new scale, so unless it is the specific purpose of the study to develop a new scale or scaling method, one should always try to accommodate one of the standardised methods.

However, there is a continuous need for new scaling methods, for example, the interest in assessment of spatial audio has resulted in proposals for new scaling methods, as discussed below.

The introduction of the 'surround' channels in addition to the two standard (stereophonic) front channels increases the possibilities for conveying an impression of auditory space to the listener. Subjective assessment methods that can quantify these aspects of sound reproduction are thus required and consequently there is a need for scaling procedures. A joint research project 'Medusa' (see footnote 10) was established in 1998 to investigate some of the problems related to multichannel sound reproduction, including that of quantifying the spatial impression of sound. A part of this effort was focused on the identification of the spatial attributes of sound and another on the development of a graphical assessment language (GAL) [125–128]. The GAL procedure requires the subjects to make drawings of their impressions of the spatial attributes [125] included in the presentation or for specific attributes such as 'instrument width', 'ensemble width' and 'location' [128]. The perceived images are then converted to

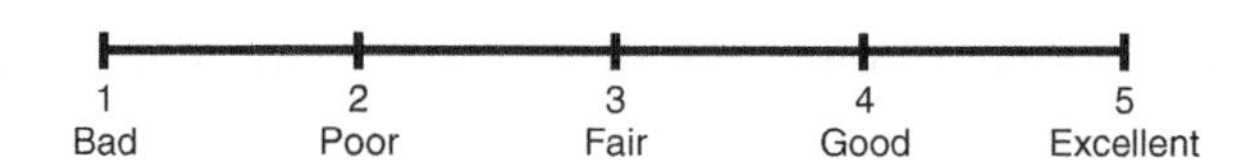

(a) Absolute category rating (ACR) scale [224]

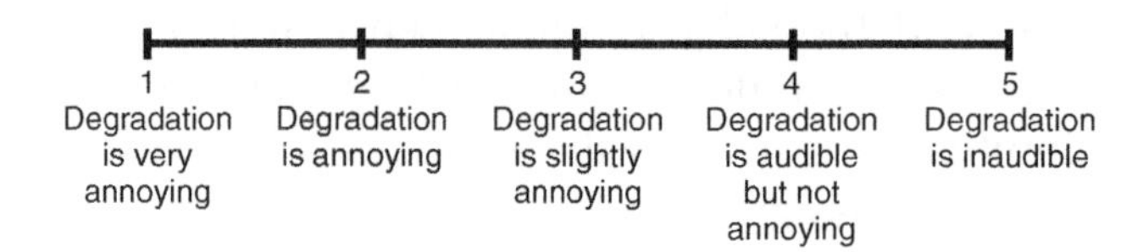

(b) Degradation category rating (DCR) scale [224]

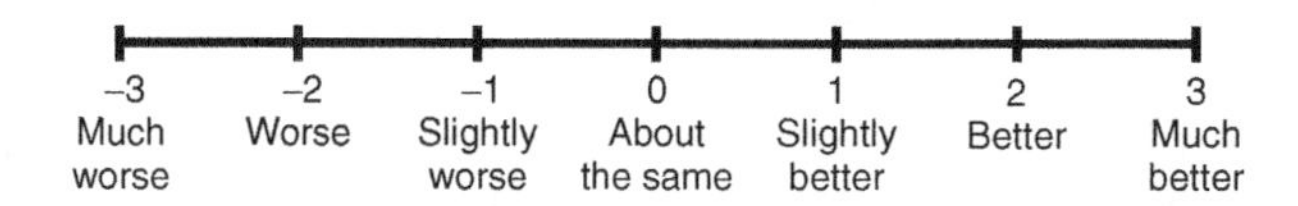

(c) Comparison category rating (CCR) scale [224]

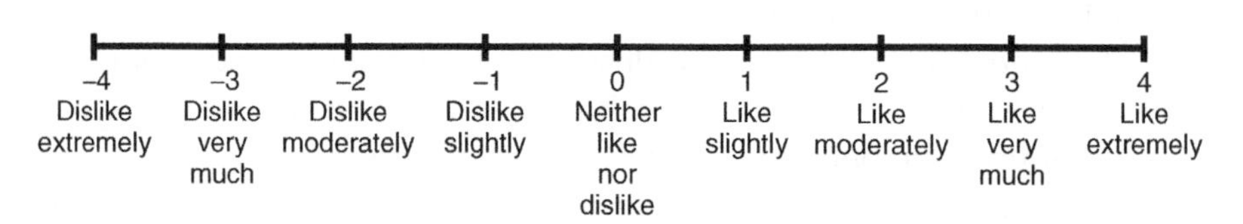

(d) 9-point hedonic categorical scale [342]

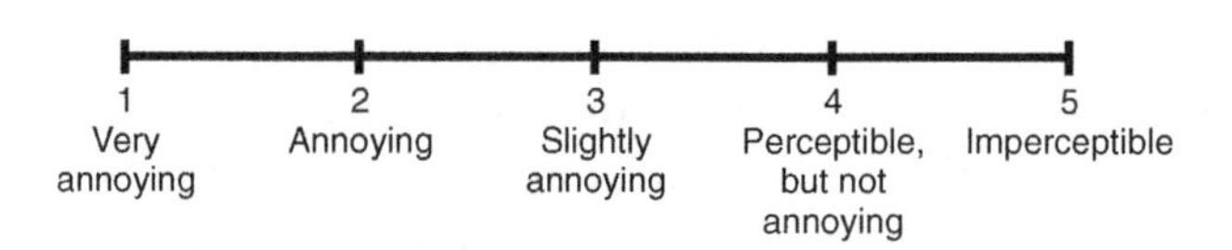

(e) ITU-R 5-point continuous impairment scale [203]

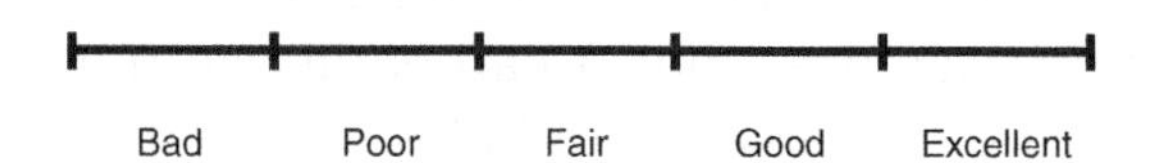

(f) ITU-T P.1534 continuous scale [217]

Figure 4.7: Commonly employed rating scales. The subjects are instructed to use scales a–d as categorical scales and scales e–f as continuous scales.

numerical data by measuring the drawings. The GAL procedure is an interesting approach that could supplement the traditional direct scaling method and is discussed in detail in Section 4.1.1. The elicitation of attributes is further discussed in Section 4.1.

One of the main problems in scaling spatial impressions is that the auditory image is not a specific point in space (it can be, but often it's not), which introduces the problem of indicating an area in space. Usher and Woszczyk [431] have examined this problem and suggested a new 2D technique where the subject indicates both an area of the image and the certainty ('hotness') by which it is perceived to be in that area. An interesting post-processing technique of the data is suggested that clearly illustrates the 2D extent of the perceived images.

Choisel [88] has devised and verified a novel pointing technique for standard localisation experiments that involves using a laser pointer and a tracking device that records the position of the laser. The new technique has been compared to a standard procedure using a 2D tablet to indicate the position of the image. The new technique has been shown to have a higher absolute accuracy and lower variance of repeated ratings and, for the subjects, it is more natural to use. The verification experiments have further highlighted a number of interesting bias problems with such pointing devices.

Pernaux *et al.* [341] studied scaling methods for 3D sound localisation experiments. They compared a standard 2D reporting technique with two more elaborate 3D pointing techniques (including 3D visual feedback) and found that the most elaborate (3D finger pointing and 3D individual visual feedback) was more accurate (less errors for front–back and up–down positions) and much faster than the other two.

Kelly and Tew [252] employed a matching paradigm where the positions of target and the variable sound source must be matched. The response data is thus the position of the variable sound source.

4.2.2 *Indirect scaling*

The indirect scaling methods are based on the assumption that the number of times a stimulus is judged different (greater or smaller) from another stimulus is directly related to the degree by which the two sensations are different. The basic measure is the probability that the two stimuli are considered to be different. The estimated probability must be based on the average of many presentations over many subjects to be reliable.

The strength of indirect scaling is that it avoids the complicated problem of asking subjects to assign a number to an impression. Instead, it relies on the variability of their responses as a measure of the sensory differences. This, however, has a number of experimental consequences such as the need for a large number of data points for each scale value, which means that the method has a lower efficiency than for example direct scaling methods. A second problem is that it is assumed that the distribution of responses are overlapping for the selected stimuli. This means that if the discrimination probability approaches 100 %, then the models fails. Consequently, the model works only for smaller sensory differences. A third problem is when, if at all, it is reasonable to use variability as a unit of measure. The experimental consequence of this is that all factors that could contribute to the variability of the data must be very tightly controlled.

Two basic methods are often used for indirect scaling: difference threshold (DL)[13] or paired comparisons (PC).

The difference threshold is the amount of change in the stimulus ($\Delta\phi$) needed to produce a just noticeable difference (JND) in the sensation. The DL scaling method is based on the assumption proposed by Fencher, which states that all JNDs are equal psychological increments in sensation magnitude, independent of the size of DL. This is the basis for Fencher's law ($\psi = k \times \log \phi$, where ψ is the sensation magnitude, ϕ is the stimulus intensity and k is a constant that depends on the sensory modality and attribute assessed) (see Gescheider [144] for further details). The idea behind the DL derived scales is that the DLs are determined experimentally as a function of stimulus intensity by successively increasing the stimulus intensity by one JND. First, the stimulus intensity corresponding to the absolute threshold is determined, which is equal to zero sensation. Then, using one of the methods discussed below, the DL value corresponding to one JND above the threshold is determined. This DL value is added to the threshold stimulus intensity to obtain the stimulus intensity that corresponds to a sensation magnitude of one unit. Starting from this stimulus level, the JND level is determined and the DL value recorded. This value is added to the starting stimulus intensity and the new level corresponds to a sensation level of two JNDs. This procedure is

[13] DL originates from the German 'Differenz Limen'

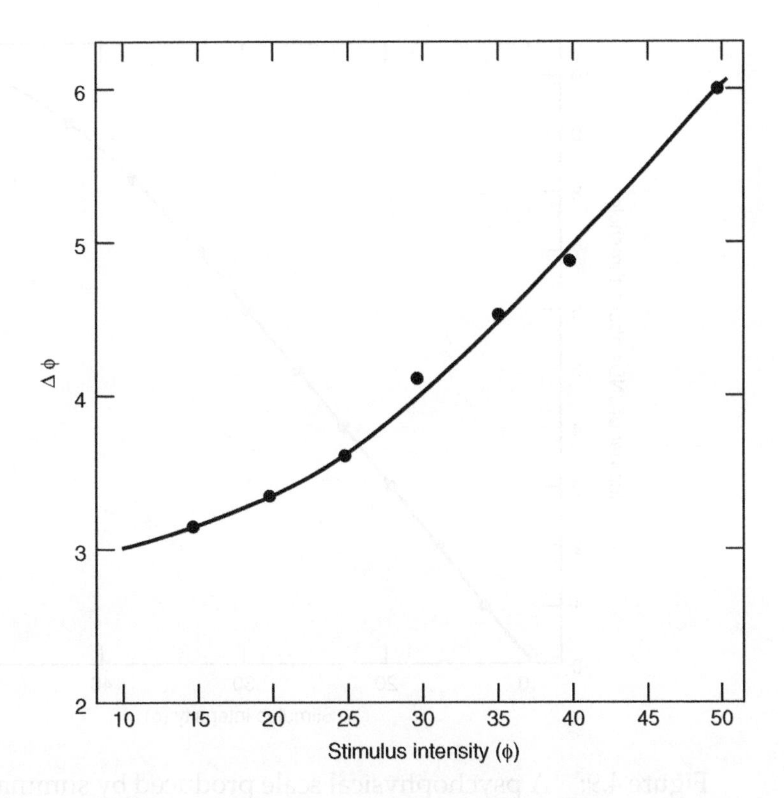

Figure 4.8: The results of a hypothetical experiment in which DL ($\Delta\phi$) was determined for seven values of stimulus intensity. The curve is extrapolated down to an absolute threshold of 10 units. Gescheider, GA, Psychophysics: Methods, theory and application, 2nd ed., 1984, Lawrence Erlbaum Assoc. Inc,. Reproduced with permission of Lawrence Erlbaum Associates Publishers [144].

continued until the sensation's range of interest is explored. On the basis of the results, the corresponding values of stimulus intensity and DL values are recorded as shown in Figure 4.8, and using the curve, the corresponding vales of JND and stimulus intensity can be determined and plotted as shown in Figure 4.9.

The DL method can use any of the classical scaling methods like the method of constant stimuli, the method of limits, and the method of adjustment for determining the JNDs.

Constant stimuli The method of constant stimuli requires the subject to indicate which of two stimuli has the greater sensation. The pair of stimuli includes a standard or reference with a fixed level

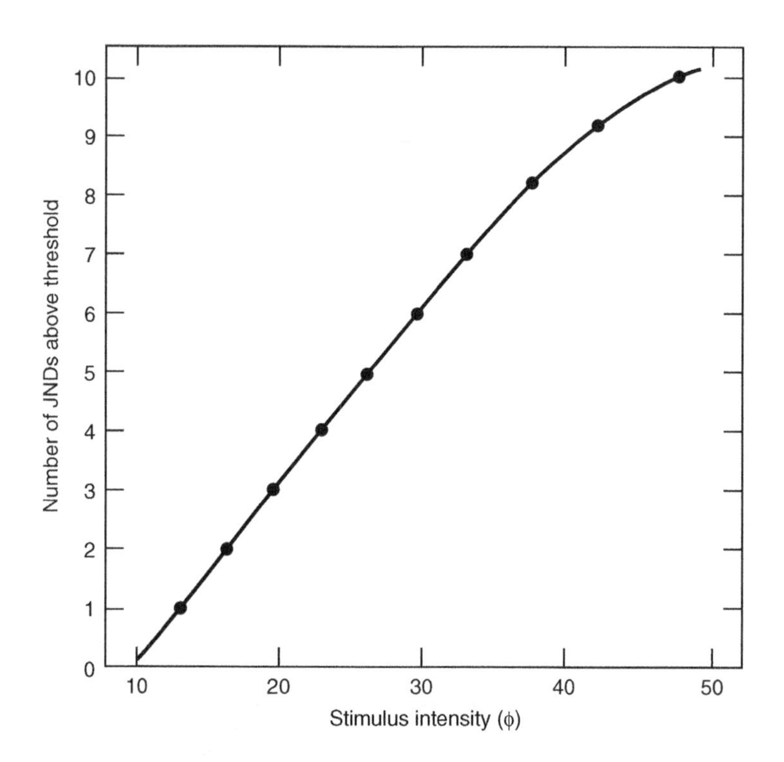

Figure 4.9: A psychophysical scale produced by summation of JNDs above the absolute threshold. The scale is derived from the $\Delta\phi$ function of Figure 4.8. Gescheider, GA, Psychophysics: Methods, theory and application, 2nd ed., 1984, Lawrence Erlbaum Assoc. Inc,. Reproduced with permission of Lawrence Erlbaum Associates Publishers [144].

and the variable stimulus has a variable level (randomly). A range of levels is compared to the reference and, based on the judgements, a probability for 'greater than' can be calculated as a function of the stimulus level. The DL is then often calculated as the average of the differences in stimulus level between 50 and 75 % probability of 'greater than' (upper DL) and 50 and 25 % (lower DL).

Method of limits This method is also based on comparisons of a standard and variable stimuli. The level of the variable stimuli is changed in small successive steps from either above ('greater than') or below ('less than') and the subject is asked to respond if the variable stimuli is 'greater than', 'equal to' or 'less than' the reference stimuli. Two points can be determined on the basis of the responses: upper limen and lower limen defined by a shift from 'greater than' to 'equal to' and from 'less than' to 'equal to'. The DL is thus half the difference between the two limen values.

Method of adjustment In this method, the subject is asked to adjust the level of a variable stimulus until it is equal to that of a reference stimulus. The experiment is repeated a number of times and the result is a distribution of adjusted levels clustering around the level of the reference stimulus. The DL is defined as the standard deviation of the distribution of adjusted levels.

The other standard indirect scaling method is the paired comparison. The task of the subject in this method could be to indicate if the two stimuli are different or not or which of the two stimuli is preferred. All possible pairs of stimuli combinations are judged by the subject and the result is a matrix of probabilities indicating if pairs of stimuli are different or which one is preferred. The conversion of this matrix into scale values is done using Thurstone's 'law of comparative judgments'. This law [418] is the basis for modern discriminant models and is based on a number of important assumptions concerning the process that takes place in the subject's mind when presented with a stimulus(i).

The basic assumption is that each stimulus will lead to a distribution of responses, on the attribute continuum that is assumed to exist in the subjects mind, if presented a repeated number of times. The distribution is considered to be normal for each stimulus with a mean value r_j and a standard deviation σ_j, where r_j is assumed to represent the 'scale' value that the experimenter is interested in obtaining. A number of additional assumptions can be formulated and the interested reader is referred to, for example, Nunnally and Bernstein [328] for a detailed discussion. The simplest form (Case V) relates the distance, in scale units, between two stimuli to the probability of preferring one to the other:

$$\bar{r}_j - \bar{r}_k = z_{jk} \tag{4.1}$$

where $\bar{r}_j$ is the estimate of the scale position of stimulus j, $\bar{r}_k$ is the estimate of the scale position of stimulus k, and z_{jk} is the normalised z score.

The estimation of scale values is done by converting the observed matrix of probabilities for all stimuli pairs to z scores using a unit normal distribution. The average z score is calculated for each column (stimuli) and this value corresponds to the deviation from the average stimulus in the set. The final scale value for each stimulus can then be transformed by normalising with the lowest scale value such that all scales values are positive.

The conversion of the probabilities using the law of comparative judgement, as described above, is based on a very important assumption, which is often only stated implicitly or is not stated at all: the unidimensionality of the attribute continuum that is assumed to exist in the mind of the subject. Unidimensionality means that all the perceived differences between the stimuli can be represented on a single scale. This further means that all the calculated differences (distances in z scores) between the stimuli can be represented by a single scale: if stimulus A is preferred to stimulus B and stimulus B is preferred to stimulus C, then it follows that A should be preferred to stimulus C. This property of the data is termed transitivity and it can be checked quite easily (see, for example, [116] or [90]). The dimensionality of the matrix can also be checked using, for example, MDS analysis. The assumptions behind Case V (equal variances for the two discriminal processes) can be checked by using a chi-square test on an arcsine transformation of the reconstructed matrix of proportions. The reader is referred to Engeldrum [117] for a description of the procedure.

The Bradley–Terry–Luce (BTL) model is another method that can be used to statistically test the validity of the assumption of a unidimensional scale and further to construct a ratio scale based on preference probabilities. The model has not been applied often within psycho-acoustics; however, recently Ellermeier and Zimmer [116] used it to model 'unpleasantness' of environmental sounds and Choisel and Wickelmaier [90] used it to scale preferences for different sound reproduction systems. Wickelmaier and Schmid [448] presents a Matlab function for the estimation of scale values for a number of versions of the BTL model.

4.2.3 *Selection of an appropriate scaling procedure*

The first step in defining the scaling procedure is to select between direct scaling and indirect scaling, as discussed in Sections 4.2.1 and 4.2.2. The next step is to decide on the scaling method (e.g. partition or ratio scaling) and the final step is to select an appropriate implementation of the method (see e.g. Figure 4.7).

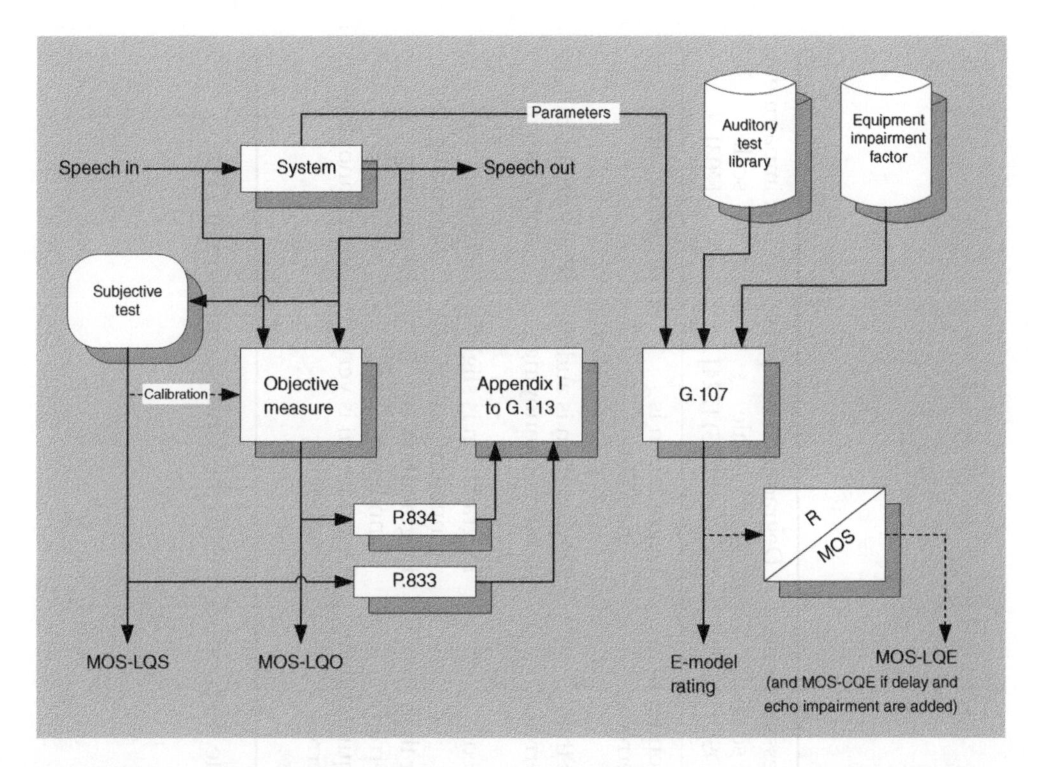

Figure 4.10: Relationship between some ITU-T MOS qualifiers [238]. Published with permission from ITU.

The reader should note, when studying Figure 4.7, that different terminologies are associated with MOS scales derived for different sources, such as listening tests or voice quality metrics, which are discussed in Chapter 1.3. A clear terminology for differentiation of MOS data is laid out in [238] and an outline summary is provided in Figure 4.10. This terminology makes a clear distinction between MOS qualifiers derived from listening tests (MOS-LQS[14]) and those from objective measurements systems (MOS-LQO[15]) (Table 4.3).

[14]Where MOS-LQS is defined in [238] as 'The score has been collected in a laboratory test by calculating the arithmetic mean value of subjective judgments on a 5-point ACR quality scale, as it is defined in ITU-T Rec. P.800 [224]. Subjective tests carried out according to ITU-T Rec. P.830 [226] give results in terms of MOS-LQS.'

[15]Where MOS-LQO is defined in [238] as 'The score is calculated by means of an objective model which aims at predicting the quality for a listening-only test situation. Objective measurements made using the model given in ITU-T Rec. P.862 [236] give results in terms of MOS-LQO.'

Score	Listening quality scale (MOS) [203, 224]	Listening effort scale (MOS_{le}) [224]	Loudness preference scale (MOS_{lp}) [224]	Degradation category scale (DMOS) [224]	ITU-R impairment scale [203]
5	Excellent	Complete relaxation possible; no effort required	Much louder than preferred	Degradation is inaudible	Imperceptible
4	Good	Attention necessary; no appreciable effort required	Louder than preferred	Degradation is audible but not annoying	Perceptible, but not annoying
3	Fair	Moderate effort required	Preferred	Degradation is slightly annoying	Slightly annoying
2	Poor	Considerable effort required	Quieter than preferred	Degradation is annoying	Annoying
1	Bad	No meaning understood with any feasible effort	Much quieter than preferred	Degradation is very annoying	Very annoying

Table 4.3: A selection of a 5-point grading scale. Reproduced with the kind permission of ITU.

The mention three-step process for selection of scaling procedure might seem quite simple, but it has been the object of much controversy amongst psychologists. It is not within the scope of this book to provide a detailed discussion of this process. However, the interested reader is encouraged to consult the literature (for example, Nunnally and Bernstein [328]) as much can be learned–also on the practical implementation level–from the discussions.

Engeldrum [117] (Chapter 12) suggests the use of three factors for selection of the best method for scaling of image quality which are also useful for scaling of sound quality:

Degree of confusion in the stimulus set This factor is related to the perceived differences between the stimuli. Clearly, different stimuli have low confusion and equal or nearly equal stimuli have high confusion. The degree of confusion could typically be measured as a part of a pilot experiment.

Number of stimuli The number of stimuli is directly related to the time the subjects will have to spend judging and thereby to the experimental effort required both per subject and in total.

Effort of the subject The effort required by the subject is related to the time spent per subject for scaling n stimuli.

Lawless and Heymann [268] (Chapter 7) and Stone and Sidel [408] have fine discussions of the selection of scaling procedures for sensory evaluation of food quality, and their arguments are also relevant for audio tests. They list the following requirements and factors to be considered in the selection process of an appropriate scaling method:

It should be meaningful to the subject The scaling process must be meaningful to the subjects. This includes an unambiguous explanation of the scale and what it is intended to measure. If the task is to scale the impression for a specific attribute, then it is very important that the attribute is explained in detail and it must be ensured that subjects understand the instruction.

It should be uncomplicated to use The meaningfulness of the scale and its ease of use are closely linked to the user interface–especially if a screen–based interface is employed. The user interface should clearly define the different steps in the scaling procedure. It is highly recommended to test the user interface in a series of pilot experiments. Experience has shown again and again that what the experimenter thinks is obvious could be complete nonsense to the subjects.

It should possess the ability to differentiate between the stimuli of interest
The scaling process should allow for a differentiation of the stimuli being tested. This requires that the subjects should feel they have enough resolution on the scale–that is, they do not feel that an intermediate scale value is missing between the noted values. Most interval scales used for listening tests (see Figure 4.7 regarding audio scales) have a one decimal resolution or can be considered continuous. However, the use of, for example, anchor labels can limit the apparent resolution. Zielinski *et al.* [471] illustrates how anchor labels on a continuous scale can bias the distribution of scores. This problem is further discussed in Section 4.2.4. The latest version of ITU-R Recommendation BS.1116 [207] recommends that anchor words be avoided. However, omitting the known and traditional anchor words could present difficulties in communication of the results, especially if there is a long tradition in the field. In general, it is highly recommended that the distribution of the ratings on the scale in a series of pilot experiments be examined to ensure that the distribution is free of severe anomalies.

It should be unbiased Bias effects present a serious problem and it is discussed in more detail in the next Section 4.2.4. If no experience has been gained with the suggested scale, it is highly recommended that a series of pilot experiments be conducted to test the properties of the scale.

It should be relevant for the task This would seem to be a trivial requirement, but it is worth considering the issue seriously before any test is conducted. The problem lies in the specification of the attributes and the description of the scale. These are done during the elicitation process (see Section 4.1.1); however, they are relevant only if the same group of subjects is available for the experiments. If this is not the case, then the new subjects need detailed training and instruction in the use of the attributes. The relevance of the attributes–once they are understood by the subjects–should be checked when the stimulus set is defined. This is an important step as the selected stimuli should excite all the attributes–subjects will become confused if some of the attributes they are asked to scale are not present in the stimulus set.

The end-point effects The term 'end-point' effect refers to the problem that arises when subjects are reluctant to use the end points of the scale. The problem is most pronounced when the stimulus set is unknown to the subjects. This causes the subjects to 'save' the extreme ratings to any unknown cases that might need be presented later in the experiment. The resulting effect is that the ratings tend to cluster around the centre of the scale, which in practice is a limitation of the scale range that again could limit the resolution. To limit this problem, all or a selected group of stimuli that illustrates the range of impressions should be presented to the subjects in training or pilot experiments. Another solution is to introduce anchor scale points and associated samples; however, anchors represent another set of problems as discussed above. The use of anchor stimuli is recommended by the ITU-R Recommendation BS.1534 standard [217]. The end-point effects are a sub-class of the bias effects that will be discussed in the next section.

Context effects The term 'context' effects refers to the problem that the impression of one stimulus could (will) be influenced by a range of non-experimental factors such as the absolute range spanned by the whole stimulus set, subject-specific likes/dislikes concerning the selected programme, light conditions (spectral content) in the listening room, and so on (see also Section 4.2.4 for a further discussion of context effects). The practical consequence of context effects is that results cannot be compared across experiments unless the context set is kept constant. A number of solutions have been suggested to minimise the context problem and Lawless and Heyman [268] (Chapter 7) list the following main categories:

- **Randomisation and counterbalancing**. Simple order effects and sequential dependencies will be eliminated by using either truly random presentation orders or by applying a specific scheme (e.g. Latin squares). However, it is important to note that the global level of context (specific experimental setting etc.) will still be present even if random presentation order is used.

- **Stabilisation**. Stabilisation refers to keeping all experimental conditions constant. This is difficult, if not impossible, but a first step would be to introduce a constant reference by which all stimuli are evaluated and then use the difference scores (stimulus standard). This strategy is used in ITU-R Recommendation BS.1116 [207]. In addition, anchor points for the end points of the scale could be introduced. However, it must emphasised that even when using the strict procedures and attempting to keep the experimental conditions constant as suggested by ITU-R Recommendation BS.1116, it has proven very difficult to, for example, compare results between different laboratories that all strictly follow the standard (see ITU-R Recommendation BS.1387 [213] standard for an example)
- **Calibration**. This refers to calibration of the subjects by specific training programmes such as those specified for descriptive analysis [268]. It still remains to be proven scientifically that subjects are more immune to context effects after having participated in such training programmes, but the experience suggests that training decreases the influence of context.
- **Interpretation**. This refers to the fact that the experimenter is aware of the problems associated with context and uses that knowledge when analysing and interpreting the results. The statistical analysis could especially be used for checking some of the effects.

The availability of statistical analysis methods The data produced by the scaling method should be checked for compliance with the statistical procedures that are intended for the analysis. The standard checks of statistical assumptions are further discussed in Section 6.2.1; however, it is highly recommended that a pre-analysis is conducted on a set of dummy data. This will ensure that the experimental design is complete and allows for the comparisons planned.

4.2.4 Context and bias effects

Context and bias effects are basic issues related to all forms of scaling of subjective impressions. The difference between a context and a bias effect is slightly blurred as context effects often will be the cause of what is termed bias in other circumstances. However, bias effects can also occur independent of context effects.

A typical context effect would be the influence of the presentation order of stimuli. The perceived quality of a loudspeaker will be highly dependent on the quality of the other loudspeakers that have been evaluated before it. If a low-quality loudspeaker is evaluated right after a high-quality loudspeaker, the rating for the low-quality loudspeaker will be lower than that would be obtained had it been preceded by a loudspeaker of similar quality. This effect is often termed *sequential contraction bias*, that is, the preceding stimuli will have an influence on the assessment of the succeeding stimuli. So the general rule for context effects is that humans adapt to the environment that they are presently in. This means that evaluations of impressions will not–as a general rule–be absolute, but will always be a function of the surroundings–that is the context.

A bias effect that is independent of context would be the general conservative nature of subjects to save parts (upper and lower) of the rating range to be assigned to the yet unknown stimuli. Other bias effects that are independent of the context will be discussed below.

Both context and bias effects need to be considered carefully before specifying a scaling method. A detailed discussion of bias is somewhat out of the scope of this book. So the following will be a brief summary largely based on the book by Poulton [351], which the reader is strongly recommended to consult for more detailed information.

Poulton [351] uses the decision tree shown in Figure 4.11 to classify the various bias types and he divides bias effects into three main categories:

Contraction bias This type of bias is caused by the subject's tendency to be conservative so that large differences are underestimated and small differences are overestimated (levels 1 and 2 in Figure 4.11) and thus the response range is different compared to the range covered by the stimuli. The contraction bias will be determined by both the stimulus range and the response range. When judging a given set of stimuli that are either familiar from daily life or learned by training, subjects tend to form a reference magnitude that typically corresponds to the average stimulus. The contraction bias thus predicts that stimuli larger than the

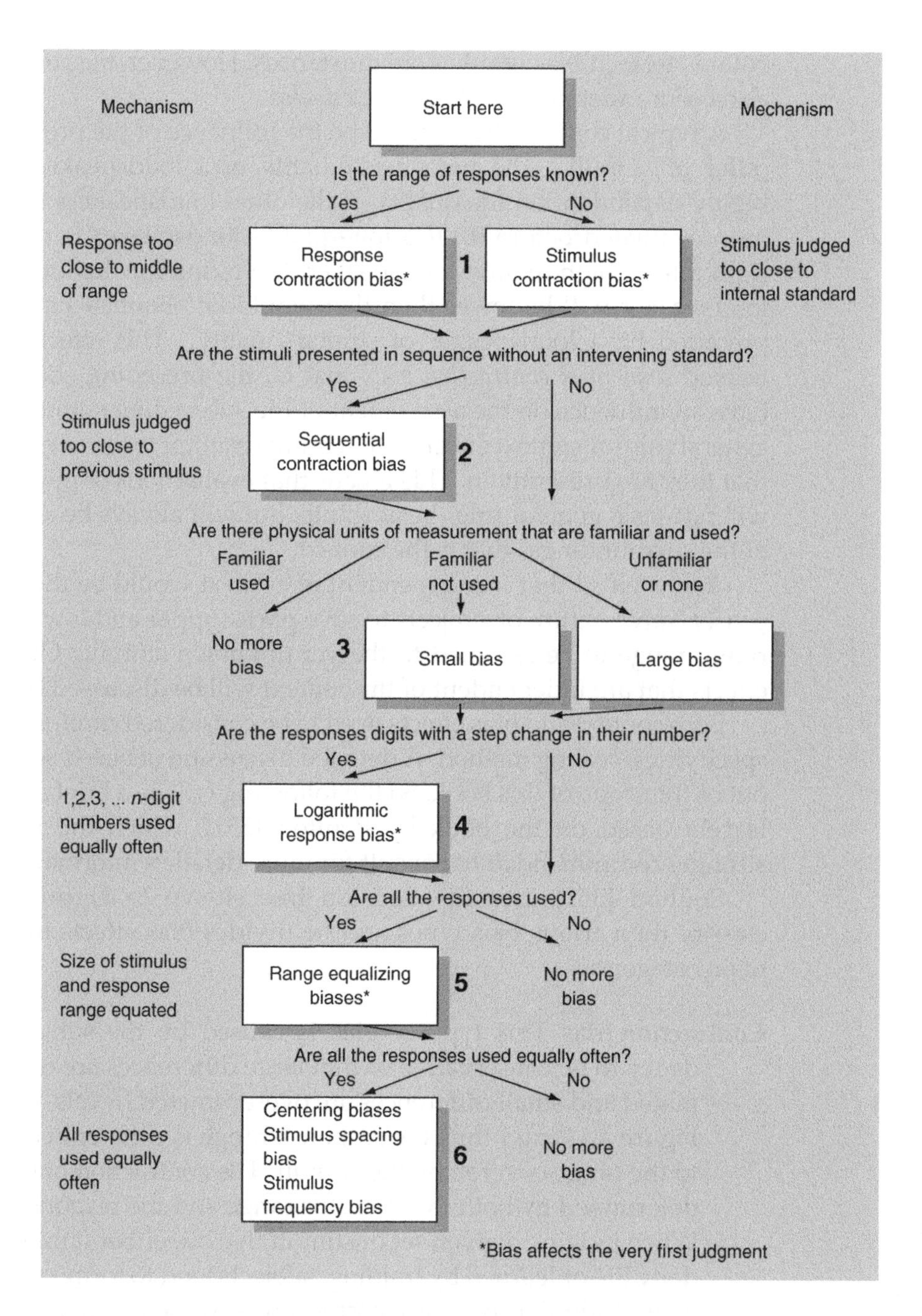

Figure 4.11: Decision tree showing how bias can arise. Poulton, E. C. *Bias in Quantifying Judgements*. Lawrence Erlbaum Associates, 1989. Reproduced with permission of Lawrence Erlbaum Associates Publishers [144].

reference will be underestimated and stimuli smaller than the reference will be overestimated. The same effect will be present on the response scale if the scale has an obvious centre that is familiar to the subjects.

The contraction bias also includes the effects caused by the order of presentation of the stimuli. If stimuli are presented sequentially, then the subjects will use the previous stimuli as a reference and then the effects discussed above will be relevant.

To avoid the stimulus and response contraction bias, Poulton suggests to counterbalance the bias by reversing the stimuli and responses and then average. It would also help to anchor the response range to the stimulus range by anchor stimuli.

The contraction bias caused by the presentation order could be avoided by taking the order effect into the experimental planning by using, for example, balanced Latin squares (see, for example, Box *et al.* [63]) or by counterbalancing the descending and ascending differences between pairs of stimuli.

Bias caused by (un)familiarity with units of magnitude The third level in Poulton's decision tree (see Figure 4.11) is the bias effect that will be present if the magnitudes to be judged are not familiar. An example of an unfamiliar unit would be the degree of annoyance caused by a noisy aircraft. The opposite example would be to ask for a judgement of the size of a mouse. Most subjects will be familiar with the size of a mouse, but evaluation of noise from an aircraft is likely to be unfamiliar to most people.

This type of bias is a difficult problem to tackle in audio evaluations as most of the related magnitudes are unfamiliar to the subjects. For example, if subjects are asked to evaluate the perceived weight of a number of loudspeaker boxes, they will do so using a unit of magnitude (weight) that is familiar to most of them. However, if the same subjects are asked to evaluate the perceived quality of the high-frequency reproduction of the same loudspeakers, they will not be able to do so using a familiar unit of magnitude.

One way of limiting this problem is to establish an extensive training programme for the subjects where they are continuously subjected to the attributes that they will be asked to assess in the listening tests.

Bias caused by unfamiliarity with the mapping of the responses to the stimuli

These biases include logarithmic response, range equalising, centering, stimulus spacing, and stimulus frequency (levels 4–6 in Figure 4.11). An example would be the ratings of a loudspeaker on a quality scale of 0 to 10. The subjects would have to make their own assumptions about the scale values in order to assign them to the reproduction quality and this introduces a number of possible bias effects:

- Logarithmic response bias (level 4 in Figure 4.11). This type of bias will be present if the applied scale forces subjects to use both one- and two-digit (or two and three, etc.) numbers. If the subjects have been using one-digit numbers (1, 2, 3, 4 ...) in a linearly fashion, then they either continue with a linear use of the two-digit number or shift to a logarithmic strategy where they use steps of 10 (10, 20, 30, 40 ...) linearly. The obvious solution to this problem is to restrict the scale to include numbers with no changes in the number of digits.
- Range equalising bias (level 5 in Figure 4.11). The range equalising bias has the effect that subjects will use the same range of responses independent of the range of stimuli. This means that the rating of a given loudspeaker depends on how many of the other loudspeakers in the group are given, for example, higher ratings. Subjects also tend to use the entire given range of responses to map the stimuli independent of the range of the stimuli. This means that the rating of a given loudspeaker depends on the number of categories that are available on the rating scale. The obvious solution to this problem is to have subjects choose their own response ranges and to avoid category ratings. This, however, is not always a possible solution in listening tests paradigms, but it is always a good idea to consider it when deciding on the rating scale.
- Centering bias (level 6 in Figure 4.11). The centering bias will influence both stimuli and responses and the effects are caused by the fact that subjects tend to use all the possibilities for the response equally often. This means that a given response range will always be centred on the midpoint of the stimulus magnitudes independent of the stimulus range.

Similarly, a given stimulus range of magnitudes will always be mapped to the midpoint of the response range independent of the response range. Poulton specifically mentions that the centering bias should not be confused with the contraction bias. The contraction bias influences the judged difference between stimuli, whereas the centering bias influences only the general relationship between the response and the stimulus ranges.

- Stimulus spacing and frequency bias (level 6 in Figure 4.11). The stimulus spacing and frequency bias are both the results of the same effect. The problem is that subjects tend to use all parts of the response scale equally often. Assume that the stimuli in the upper half of the stimulus range are spaced one unit apart and those in the lower half of the range are two units apart. However, owing to the spacing bias, all parts of the response range will be used equally often, which means that the average responses are more equally spaced than they should be according to the spacing of the stimuli.

    The stimulus frequency bias is a result of the fact that subjects tend to respond as if all stimuli are equally frequent. Thus, if the stimulus group includes one group of stimuli that are nearly equal and another with larger differences, it follows from the fact that subjects tend to use all parts of the response range equally often that the closely spaced stimuli will be allocated too large a part of the response scale. Poulton has a number of 'solutions' to avoid the bias affects at level six; however, they appear very difficult (impossible) to implement in the case of listening tests on loudspeakers.

Poulton lists two other bias effects that are worth considering in listening tests on loudspeakers. The first is the so-called local contraction bias, which has so far only been observed for loudness evaluations: the second stimulus in a pair will be judged to be lower in loudness if preceded by a low (absolute level) loudness stimulus, independent of whether the first stimulus is relatively higher or lower compared to the second. The opposite effect is that it will be judged to be relatively higher if it is preceded by a loud stimulus, independent of whether the first stimulus is relatively higher or lower than the second.

This bias effect should be observed when calibrating loudspeakers to have equal loudness independent of the programme. The solution is to use a calibration signal that has a relatively constant loudness level.

The last bias effect is termed transfer bias and it occurs when the same subjects are used for assessments of different attributes and for different conditions of the same attribute. This is a serious problem that will be present if a permanent and trained team of subjects are used for the tests. The solution is of course to not use the same group of subjects; however, this should be weighted against the benefits of having a permanent group. No scientific studies on this issue that are relevant for listening tests are presently known.

4.2.5 *Other bias effects*

Some of the bias effects discussed in this section are based on experience and have not been formally scientifically documented, while others have been formally tested, but do not belong to the standard bias categories as suggested by Poulton.

Visual bias A very important bias effect in listening tests is the effect introduced when subjects can see the loudspeakers that are being tested. Toole [425] has shown that it has a significant effect on the assessments if the loudspeakers being tested are visible to the subject. Thus, the general recommendation is that listening tests on loudspeakers must be conducted 'blind'.

Perceptual oversensitivity If specific attributes are being tested, it must be ensured that the subjects can actually hear those artifacts or are not overly sensitive. It is a well-known fact in sensory analysis (see e.g. Lawless and Heymann [268] (Chapter 2)) that certain subjects cannot taste or smell certain flavours. This can be the result of, for example, specific physiology defects or illnesses. In listening tests, similar effects have been observed as some listeners cannot hear or are oversensitive to certain artifacts in low bit-rate coding systems [261, 353, 379]. Thus, it is important to have such aspects in mind when assessing why either individual subjects or ratings in general are deviating from the expected responses.

Halo bias The Halo bias is another effect that describes the influence that a very positive evaluation of stimulus for one of the rated attributes can have when the other attributes are evaluated for that stimulus. Toole [422] describes how the increased quality of the spatial ratings can significantly influence the general sound quality rating. In the experiment, the same loudspeakers were evaluated for both monophonic and stereophonic reproduction and the increased ratings for the spatial properties in stereophonic mode had a significant influence on the aspects of sound quality that mainly included timbral attributes. This effect also illustrates why it is a problem to use the same fixed anchor loudspeakers across tests. Depending on the other loudspeakers in the group, they could be either superior and thereby introduce a halo effect or they could be inferior and be the subject of a halo effect. In either case, the result is that they would not be functioning as the intended fixed anchors.

Dumping bias This bias is present if the scale is restricted or if an attribute is missing in a descriptive analysis experiment. The basic problem is that if subjects miss an opportunity to scale an impression that they clearly perceive, then it will have an effect on the scaling of the other attributes. The influence is likely to be dependent on the interaction between the attributes, which often depends on the training and experience of the subjects as well as the experimental paradigm. Thus, it is very important to determine the number of attributes required in pilot experiments and further to allow subjects to comment on the missing possibilities–this holds good for the main experiment too. It is important to remember that the experimenter might be biased by focusing on 'the' attributes and thereby forgetting other minor, but for the subjects important, aspects that eventually could lead to dumping effects.

Expectation bias This bias effect is related to the expectations of the subject, for example, if a subject has an expectation that a certain loudspeaker will be included in the tests or if a fixed order of stimuli is used, so the subjects knows (or expects) what the next stimulus will be. Such expectations will influence the assessments; however, it will be very difficult to quantify the magnitude after the experiment. One of the authors has experienced that a subject saw a box carrying the name of a certain loudspeaker when entering the laboratory on the way to the listening room (It had been left there by another

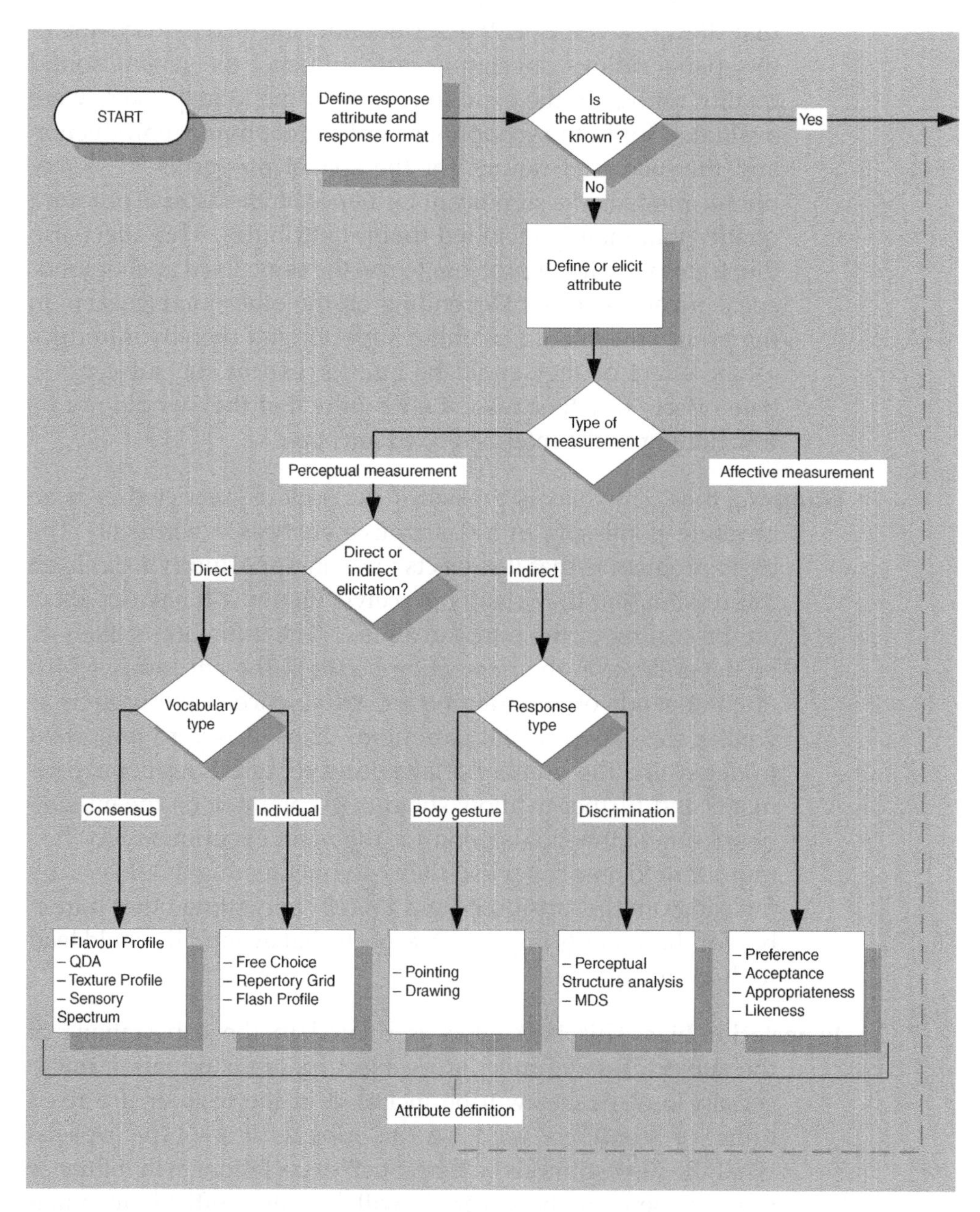

Figure 4.12: Summary of the decision steps included in the process of identifying the response attribute and selecting the appropriate scaling procedure.

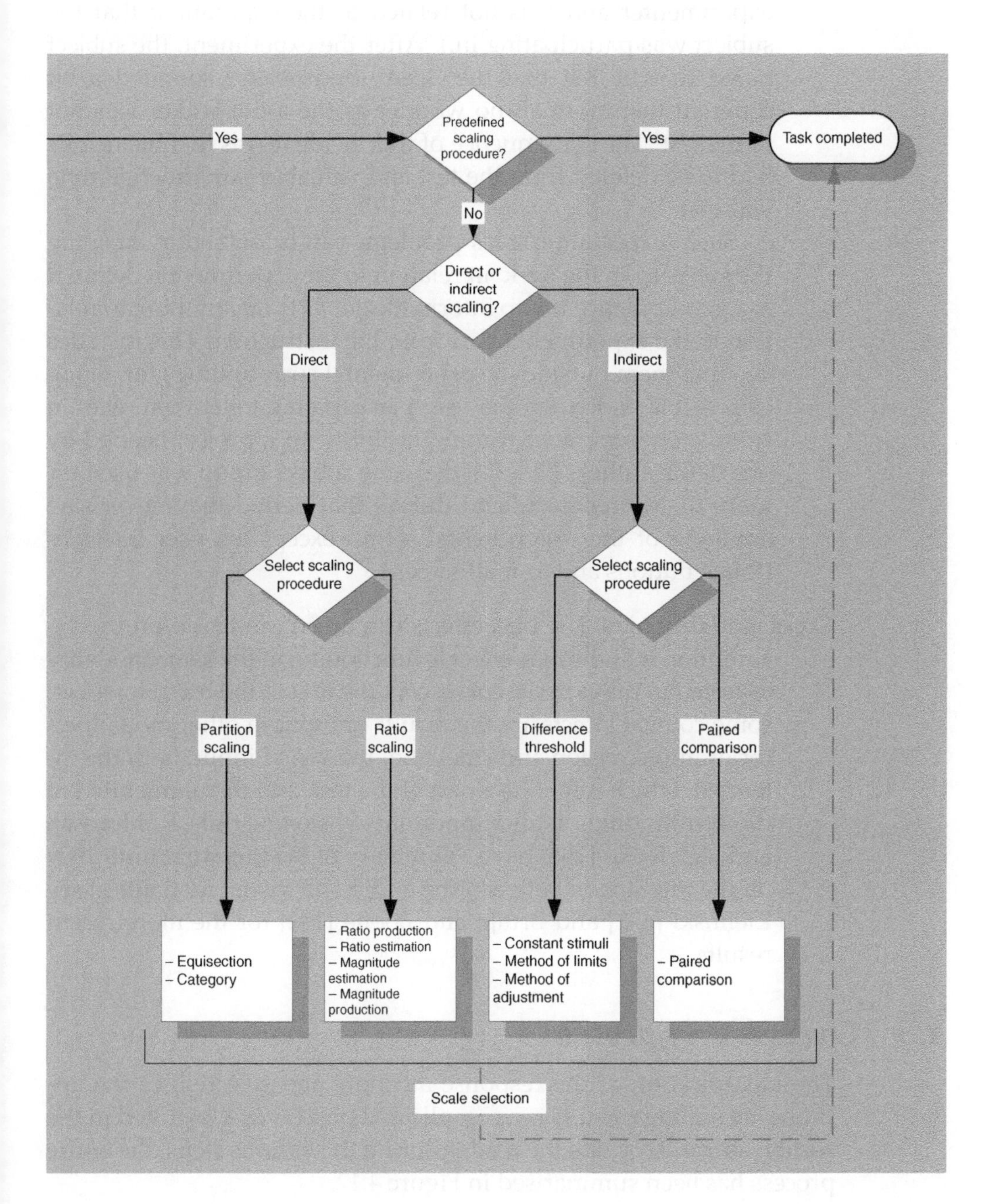

Figure 4.12: Continued.

experimenter and was not related to the experiment that the subject was participating in.). After the experiment, the subject noted that he felt that the 'seen' loudspeaker sounded a bit different than normal–no wonder as the loudspeaker was not in the test! The consequence of this incident was that his results had to be deleted from the test and valuable experimental time was lost.

Such 'expectation' bias problems can be solved by keeping the subjects 'in the dark' in relation to the experimental detail. It is a good practice not to inform the subjects on anything in relation to the experiment apart from the instruction. This includes keeping them outside the experimental surroundings for a long as possible and when they are participating to let them wait in completely separated waiting facilities. In a project headed by one of the authors [29–32], the same subject group was used for more than three years and during that period they never saw anything of the experimental set-up except the user interface (touch buttons and a small screen).

Cross modality bias The bias effects discussed are based on the assumption that the bias effect is functioning in the same modality as tested in the experiment (except the first in the list). However, some context effects are the results of input in other modalities, for example, vision and smell. So, the visual input, both the intended, which might be a part of the test, and the unintended in the surroundings, is also important in listening tests. Kohlrausch and van de Par [258] have a summary of the literature until 1999 on the interaction between the audio and visual modalities; see Zielinski [471] and Bruijn and Boone [102] for the more recent results.

4.3 Overview of process

The identification of the response attribute and selection of the appropriate scaling process is a complicated process as illustrated in the section above. To guide the reader during the various steps, the entire process has been summarised in Figure 4.12.

CHAPTER

5

Experimental variables

An important goal for any perceptual experiment is that the data are objective. Objectivity means that a statistically similar data set and subsequent conclusions can be produced by another experimenter using the same stimuli, using different subjects and another experimental paradigm in another laboratory. A less-stringent criteria is that the data should be reproducible. This means that the same data within the stated confidence interval should be produced by another experimenter using a similar set-up, including the same stimuli and experimental paradigm, but different subjects at another place and another time.

The degree of objectivity and reproducibility will be influenced by a number of variables and of these the known and experimentally verified variables will be discussed in this chapter.

Perceptual Audio Evaluation – Theory, Method and Application Søren Bech and Nick Zacharov

An experimenter usually defines an experiment with the purpose of examining the influence of a number of variables on a given process. The purpose of the experimental design is thus to ensure that the main variation in the data set is caused by these variables and not by other uncontrolled or unknown variables. The fewer unknown variables in the experiment, the higher the expected accuracy and objectivity of the resulting data set is likely to be.

The division between known and unknown variables is reflected in the statistical model used for the analysis of variance (ANOVA) as discussed in Chapter 6. The known variables are factors in the ANOVA model and the unknown variables define the error variance.

The known variables are divided into *dependent* and *independent* variables. The independent variables (discussed in this chapter) are those defined and controlled by the experimenter (the selected loudspeakers, signals, subjects, etc.) and the dependent variables (discussed in the previous chapter) are the data of the experiment–the answers provided by the subjects.

Another important aspect that should be considered for each of the variables is whether it is a random or a fixed variable in the ANOVA model. If the levels of the variable are considered as a representative selection, for example, if the included loudspeakers are representative of all loudspeakers in the world, the variable should be considered random in the ANOVA model. This also means that the statistical conclusions of the experiment with respect to this variable is applicable to the whole population of loudspeakers. If, on the other hand, the levels of the variable are not representative of a greater population, then the variable is fixed and the conclusion of the experiment applies only to this particular selection of levels. The issue of random variables versus fixed variables in the ANOVA model is discussed in more detail in Section 6.2.4.

The following independent variables will be discussed (not in order of importance):

- Signal
- Reproduction system
- Reproduction or listening room
- Subjects.

5.1 Signal

The basic purpose of the signal is to excite the perceptual differences between the devices (for example loudspeakers) under test, for the chosen response attribute(s). For example, if there is no energy in the signal at low frequencies, it is not likely that any differences in the attribute related to the low frequency range will be detected between the devices under test.

Basically, there are no limitations as to what a signal can be: music, speech, and any type of artificial signal generated by the experimenter, as long as the signals stimulate the selected response attribute and are well motivated. Listening tests on loudspeakers or low bit-rate codecs typically use music or speech; however, these signals are often specifically chosen after careful examination in pilot experiments, where it is ensured that they excite the specific response attributes. The choice of the so-called critical programme material is discussed in, for example, ITU-R Recommendation BS.1116 [207] (Section 6) for evaluation of low bit-rate codec systems.

On the basis of the known response attributes (see Section 4.1), it is possible to establish a number of classifiers to aid the grouping and selection of signals:

- Signal category
- Time domain characteristics
- Spectral characteristics
- Spatial characteristics
- Reference signals.

5.1.1 Signal category

The signal category is important when the general validity of the results is planned/considered. For example, if a broadcaster is planning an evaluation of low bit-rate coders, it is important to consider what the primary type of signals are for that particular broadcaster (e.g. speech in case of a mobile telephone network operator) and then subsequently to select signals that represent that primary signal category.

Signals can be divided into many groups depending on the purpose of the classification, and for the purpose of evaluation of audio devices, experience has shown that a division into naturally occurring and synthetic signals will be sufficient for most cases.

Naturally occurring signals

The naturally occurring signal can be divided into three (minimum) groups:

- Music
- Speech
- Noise.

The 'Music' category includes the different musical genres like classical, rock, pop, country, and so on. The problem is that this division cannot be completely specified, but each category is often defined by the number and type (e.g. electrical vs acoustic) of instruments and the spectral range covered by the instruments in the orchestra. The music category is also important to consider in relation to the background and skills of the subjects. For example, if young subjects are preferred (due to lack of hearing deficiencies) they might not be familiar with classical music, but only with the music of their generation. Such bias towards a certain type/genre of music is very difficult to identify and quantify in the data set and it should therefore be ensured that the subjects have no idiosyncrasies towards the music selected for the test.

The 'speech' category includes male/female voices and of course the different languages. The spoken text and its properties are also part of the characteristics of the signal and this aspect has been of great concern in, for example, speech intelligibility tests (see, for example, the ITU-T standards on evaluation of telephone systems [84] and Section 10.1.1). Experience further shows that the vocal characteristics of the speaker are important. Issues like dialect (or lack of), the speaking rate, the ability to maintain a constant sound pressure level (SPL), articulation, and so on, are all important if the subjects are going to listen to the same pieces of speech several times. Experience shows that recordings of professional speakers (e.g. actors, news announcers) should be used whenever possible.

Naturally occurring noise signals can be anything that is normally considered to be a noise signal (excluding music and speech), for example, the noise of cars in traffic and aircraft noise. These types of signals are not frequently encountered as the primary signals in listening tests, but are sometimes used, for example, as background noise signals, and so on.

Synthetic signals

Synthetic signals include any signal that can be constructed either mathematically or by using, for example, random noise generators and some analogue processing. Constructed signals are often employed in basic psycho-acoustic experiments where the purpose is to understand how the auditory system functions. A very important aspect of such signals is that they can be specified mathematically (often) and the properties of the resulting sound field can be precisely specified in physical terms. This enables the experimenter to relate specific physical properties of the signal or the sound field to the dependent variables investigated. Another advantage of using such signals is that they can be defined to excite specific properties of the auditory system.

5.1.2 *Recording technique, storage and encoding*

The recording technique is often an integrated part of the signal, but it need not be so. Berg [50], for example, used five different recording techniques to examine the spatial attributes of reproduced sound and the recording technique was thus an independent variable in the experiment. The general issue of how to record a signal 'correctly', for example, a live performance of an orchestra, is a complicated matter that is somewhat out of the scope of this book. Hugonnet and Walder [172] have an excellent introduction to stereophonic sound recording and Rumsey [366] discusses multichannel sound recording for 5.1 channel reproduction systems. Blauert [56] and Hammershøi and Møller [156] have excellent introductions to the binaural recording technique, and Begault [42] discusses the general issue of 3D sound reproduction.

The signal once recorded or generated has to be stored for later retrieval and reproduction. The storage process often includes some form of compression or bit-rate reduction system to minimise storage space. Many forms of compression schemes (for an overview of such schemes see Brandenburg and Bosi [64] and Bosi *et al.* [62]) introduce audible artifacts, so it is important to ensure that the signals selected do not include artifacts caused by compression as this will otherwise be a confounded variable–the best solution is to use the uncompressed file.

Another part of the storage/retrieval process is the analogue-to-digital (A/D) and digital-to-analogue (D/A) conversion process and it must be ensured that this process does not introduce any artifacts like quantisation noise, and so on, that could influence the quality of the signal. See also Section 7.7.

5.1.3 Time domain characteristics

The structure of the signal as a function of time is important for a number of reasons discussed below.

If a paired comparisons paradigm is employed, it is required that the signal is reasonably stationary to allow the subject to switch between the two presentations to make an estimate of the difference or similarity between the presentations. This requirement is also present when multiple stimuli presentation methods are employed as, for example, in the ITU-R Recommendation BS.1534-1 [217].

The degree of stationarity of the signal has also been shown to have a direct influence on the audibility of the various aspects of the signal. Olive and Toole [335] have shown that the audibility of resonances is different for stationary signals and time varying (impulse) signals. Grey [150] has shown that synchronicity in the attack and decay of upper harmonic is important for the timbre of the signal, so to excite all timbral differences between the devices under test, the signal should include, for example, the attack, the stationary part and the decay part of single instrumental sounds. See also Mattila [286] for a summary of timbre research and the variables that influence timbre perception.

5.1.4 Spectral characteristics

The spectral characteristics are important in particular if specific areas of the frequency range are to be tested. For example, Bech [36] and Zacharov *et al.* [462] found that it is difficult to find suitable signals in the form of standard programme material for experiments examining sound reproduction in the frequency range below 120 Hz. The same applies to experiments concerned with the audibility of spectral components above 20 kHz. The spectral characteristics in general are one of the primary factors determining the timbre characteristics of instruments, so it is important to include a representative selection of instruments covering the frequency range of interest. Benjamin [43, 44] and Chapman [87] has determined the spectral characteristics of different types of programme material.

5.1.5 *Spatial characteristics*

Toole [422] was one of the first to quantify that the spatial quality of reproduction–in this case the difference between monophonic and stereophonic reproduction–can have a significant influence on the overall evaluation of a loudspeaker. The improved quality of spatial reproduction has also been one of the major arguments for introducing the so-called multichannel systems, for example, the 5.1 system as defined by ITU-R [205].

Listening tests aimed at examining the qualities of such systems thus need to be able to quantify their ability to convey the spatial information in the programme material. This requires signals that can excite the spatial attributes of sound reproduction. The problem, however, is that these attributes have only been partly identified, so selection of the appropriate signal is a difficult task. See also Section 4.1.

A simple division of signals with respect to the spatial properties is based on the number of reproduction channels: a single reproduction channel (monophonic reproduction), two channels (stereophonic reproduction), and more than two channels (e.g. 5.1 and other multichannel systems).

High quality single-channel signals in the form of standard programme material are becoming increasingly difficult to obtain. A few examples, however, are still available such as the Archimedes CD [158] and the CD issued by the Audio Engineering Society (AES) [18], which include programme material for evaluation of low bit-rate codec systems. See also the list in Appendix C.

Two-channel signals are still the dominant format in the commercial music market; however, the main problem is in obtaining the technical details of the recording and in quantifying the spatial properties of the material. Most of the available recordings are based on multi-track recordings that are down-mixed to a two-channel track. The spatial properties of this type of material can be assessed only via listening tests, and then we enter what Toole [426] has termed the *circle of confusion*: 'Loudspeakers are evaluated by programme material that has been evaluated by loudspeakers that has been evaluated by programme material that has been .....'. Fortunately, a number of technically well-defined signals have been produced and are commercially available, for example, the Archimedes CD [158].

Signals intended for reproduction via more than two channels have the same problems in relation to reproduction of spatial properties as those discussed above for two-channel signals. Rumsey [367] and Slawek *et al.* [470], in the search for useful signals, introduced a classification system based on the perceptual use of the information in the various channels. Two categories, front or background information, were introduced and proved useful for classification of commercially available signals. A number of physical measures were also introduced to quantify the properties of the signals in each channel and for comparisons between channels and signals. Recently, Neher [318] has developed a stimulus set for training and evaluation of spatial properties of multichannel reproduction.

5.1.6 *Reference signals*

Reference signals are considered in a class of their own because they serve a particular purpose in the experiment–that of being the reference to which other signals in the experiment are being compared. This makes it very important to select a signal with well-specified auditory and physical properties. However, note that all the issues discussed above also apply to a signal that is chosen to act as reference.

A reference signal can serve either as a reference representing one end of the scale (typically the highest possible score on the scale applied) as suggested in the ITU-R standard for testing of high-quality low bit-rate coders [207] or as anchors representing several levels of the attributes being examined as used in the ITU-R for testing of medium quality coders [217].

The term 'absolute reference' is often heard in relation to listening tests, especially if the applied scale is a 'true-to-nature' scale. However, the term 'absolute' in the sense of being 'the original source as heard in the real environment' should be used very carefully: A true absolute reference signal could be a situation where the talker/voice being reproduced by the loudspeaker was placed in the same room and next to the loudspeaker. The real voice and the reproduction system would be alternated. The comparison would thus be between the real (absolute) reproduction and the loudspeaker reproduction including the recording technique, the transmission, the loudspeaker and the interaction between the loudspeaker and the room.

Similar absolute references could be established using, for example, single instruments that could be accommodated in the listening space; however, the spatial extension of the instruments should be considered: it should preferably be a small and relatively well-defined acoustic centre as is the case for the human voice.

The term absolute reference can thus be correctly used only for a very limited number of cases, and it is noted that the reference signal as suggested in ITU-R [207] is intended not as being an absolute reference, but only to serve as the original to be compared with the (assumed) degradation of the signals after having been processed by a low bit-rate coding system.

5.2 Reproduction system

Various parts of the reproductions system can be defined as independent variables in the test. All aspects of the part in question must be documented and considered before the experiment is conducted. For example, Bech [29] found that it was not possible to find two loudspeakers of the same type that were sufficiently similar in terms of frequency response such that any audible differences between them could be excluded. If the reproduction system includes a loudspeaker, it is important to take the interaction between the loudspeaker and the listening room into account. This aspect is further discussed in Chapter 7.

5.3 Listening room

The listening room is another obvious variable in the experiment, and the technical and physical aspects of such rooms are discussed in Section 7.1. However, again it must be ensured that all aspects are included in the ANOVA model: for example, if several rooms and positions of the loudspeakers are used in the experiment, these must be included as variables in the model (see Bech [29] for an example).

5.4 Subject considerations

The experimental factor associated with subjects and their use in listening tests will be discussed in this section. Aspects relating to the subject variable include the following:

- Categorisation and applicability
- Subject (pre- and post-) selection
- Listening panels

- Training and monitoring
- Familiarisation with experiments.

All of these topics will be presented in this chapter, bar familiarisation with experiments, which will be handled in Chapter 9. While ITU-R Recommendation BS.1116-1 [207] provides some introduction to listener expertise and selection, additional guidance on the selection, training and monitoring of subjects can be found from several standards outside the field of audio. Several standards from the field of agricultural food products will be referred to in this respect during the course of this chapter, for example, [192, 194], which provide excellent background information regarding subject matters.

The subject is the central measuring device that supplies the basic data to be analysed, and the more the knowledge about the 'instrument', the easier it is to understand the data that it provides. The main problem with subjects, however, is that they are more difficult to describe objectively, compared to, for example, the listening room or the reproduction loudspeakers.

The characteristics of the subjects are important when assessing the generality of the conclusions of the experiment: are they relevant for the population, in general, or only for a selected group with special listening abilities? They are also important for the planning of the experiment as they will influence factors such as the number of subjects needed, the level of complexity of the task of the subjects, the degree of training, and so on.

The main issue is thus to establish objective measures that can be used to quantify those characteristics on which the selection of relevant subjects can be based. The selection of subjects is typically a two-step process: pre-selection and post-selection.

This section will describe some of the objective measures that have been found useful for characterisation of subjects' listening abilities during these two phases.

5.4.1 *Categorisation and applicability*

When considering subjects, four key topics areas are to be considered relating to their categorisation and applicability for listening tests including the following:

- Population
- Acuity

- Ability
- Logistics.

The topic of population is related to whether or not the subject is a member of the desired sample of the population. Typically, this matter is of relevance in all experiments and is further discussed in Section 5.4.3.

Acuity of a subject relates to the healthiness of the subject's hearing mechanism, and is associated with filter 1 in Figure 4.1. Most commonly, this issue can be evaluated through audiometry, which may or may not be of importance to the experimenter.

The ability of a subject relates to his/her capability to rate stimuli in a reliable and repeatable manner and is related to filter 2 in Figure 4.1. This is typically assessed through an evaluation of a subject's performance during either special pre-selection tests or analysis and post-selection following an experiment. Depending on the type of experiment, number of subjects, and so on, this may or may not be of relevance to the experimenter and will be discussed in the following sections.

Lastly, the practical topic of logistics is indirectly very important to the experimenter. It is essential that subjects are available, prompt and reliable for the purpose of completing an experiment. Selecting reliable and available individuals is thus vital to collecting experimental data in a timely fashion, leading to the completion of the experiment as designed.

The experimenter should review these four topics and establish which are important and relevant for any given experiment.

In several fields of sensory evaluation, including audio, terminology has been developed to describe the differing levels of skills, expertise and suitability for performing perceptual evaluations. However, in the audio industry this terminology is ambiguous and spread across a number of different standards including [84, 207, 224, 231, 235]. The following terms are frequently encountered in audio and provided here with some of their associated meanings.

Untrained subject (according to ITU-T Recommendation P.831 [231])

> 'Untrained[16] subjects are accustomed to daily use of a telephone. However, they are neither experienced in subjective testing nor

[16] Also commonly referred to as *naïve* subjects, though this term is not formally described in the literature.

are they experts in technical implementations of echo cancelers. Ideally, they have no specific knowledge about the device that they will be evaluating'.[17]

Un-trained subject (according to ITU-T Recommendation P.800 annex B [224])

'Subjects chosen at random from the normal telephone using population, with the provisos that:

- they have not been directly involved in work connected with assessment of the performance of telephone circuits, or related work such as speech coding.
- they have not participated in any subjective test whatever for at least the previous six months, and not in any listening-opinion test for at least one year.
- they have never heard the same [sic] sentence lists before'.

Experienced subject (according to ITU-T Recommendation P.831 [231])

'Experienced subjects (for the purpose of echo canceler evaluation) are experienced in subjective testing, but do not include individuals who routinely conduct subjective evaluations. Experienced subjects are able to describe an auditory event in detail and are able to separate different events based on specific impairments. They are able to describe their subjective impressions in detail. However, experienced subjects neither have a background in technical implementations of echo cancelers nor do they have detailed knowledge of the influence of particular echo canceler implementations on subjective quality'.

Expert (according to ITU-T Recommendation P.832 [235]) 'Experts (for the purpose of hands-free terminal evaluation) are experienced in subjective testing. Experts are able to describe an auditory event in detail and are able to separate different events based on specific impairments. They are able to describe their subjective impressions in detail. They have a background in technical implementations of hands-free terminals and do have detailed knowledge of the influence of particular hands-free terminals implementations on subjective quality'.

[17] This standard refers to the subjective evaluation of telephony network echo cancelers.

Expert subjects (according to ITU-R Recommendation BS.1116-1 [207])

> 'Expertise and subject selection based upon the following criterion:
>
> - Subject's expertise in detecting small impairments
> - Pre-screening based upon audiometric performance, previous experience and performance in previous tests
> - Post-screening based upon correct sample identification.

As can be seen from the above list, the terminology employed in the audio industry is limited and somewhat ambiguous.

A well formulated structure of similar terms widely employed in the sensory analysis of agricultural food products is presented in detail in ISO standard 8586-1 [192] and 8586-2 [194]and summarised in Table 5.1. An overview regarding the development of subject expertise is also defined in [194] and outlined in Figure 5.1. The standardised approach presented in these two standards is well structured and rigorous and is almost directly applicable to the field of audio. The authors strongly encourage the audio industry to adopt a similar unambiguous terminology regarding the definition of expertise of subjects.

There are several additional implications associated with the categorisation of subjects that the experimenter should be aware of regarding the determinability of results. Firstly, the meaning of the results differs on the basis of the category of subject employed. When random samples of the population are employed (i.e. untrained or naïve subjects), it may be possible to extrapolate population means.[18] This is one of the reasons for using such methods in speech codec evaluations. However, when employing experts, extrapolation to the general public is not possible [207]. The expert listener's role in this case is to establish whether or not any degradation can be perceived, indicating results of the most sensitive subjects within the normal hearing population.

Additionally, the category of subjects has a significant impact on the number of subjects that should be employed for a listening test. While this aspect will be presented in the following sections, a short guideline regarding the number of subjects typically employed is presented in Table 5.2, according to the common practices in the application of common standards. Bech [27], has shown that a

[18]Assuming a large enough sample of the population has been tested, discussed further in Section 5.4.3 and Chapter 6.

Assessor category	Definition
Assessor	Any person taking part in a sensory test
Naïve assessor	A person who does not meet any particular criterion
Initiated assessor	A person who has already participated in a sensory test
Selected assessor	Assessor chosen for his/her ability to carry out a sensory test
Expert	In the general sense, a person who through knowledge or experience has competence to give an opinion in the field about which he/she is consulted. (Please note that the term *expert* does not provide any indication regarding the qualification or suitability of the individual to perform listening tests.)
Expert assessor	Selected assessor with a high degree of sensory sensitivity and experience in sensory methodology, who is able to make consistent and repeatable sensory assessments of various products
Specialised expert assessor	Expert assessor who has additional experience as a specialist in the product and/or process and/or marketing, and who is able to perform sensory analysis of the product and evaluate or predict effects of variations relating to raw materials, recipes, processing, storage, ageing, and so on.

Table 5.1: Summary of assessor categories employed in sensory analysis, as defined in ISO standard 8586-2 [194], applied to the food industry and recommended for adoption in the field of audio. Reproduced by permission of ISO.[19]

reduction by a factor of up to 7 can be achieved if highly trained subjects are used instead of naïve subjects.

Additional information regarding the influence of the number of subjects on the power of a test is also discussed in Chapter 6.

The experimenter should however bear in mind that if expert subjects are to be employed, this will typically involve the development

[19]The terms and definitions taken from ISO 8586–2:1994 Sensory analysis–General guidance for the selection, training and monitoring of assessors, Part 2: Experts, sections 3.1, 3.1.1, 3.1.2, 3.2, 3.3 (excluding Note 2), 3.3.1 and 3.3.2 (excluding Table 1), as well as the upper part of Figure A.1, are reproduced with permission of the International Organization for Standardization, ISO. This standard can be obtained from any ISO member and from the Web site of the ISO Central Secretariat at the following address: `www.iso.org`. Copyright remains with ISO.

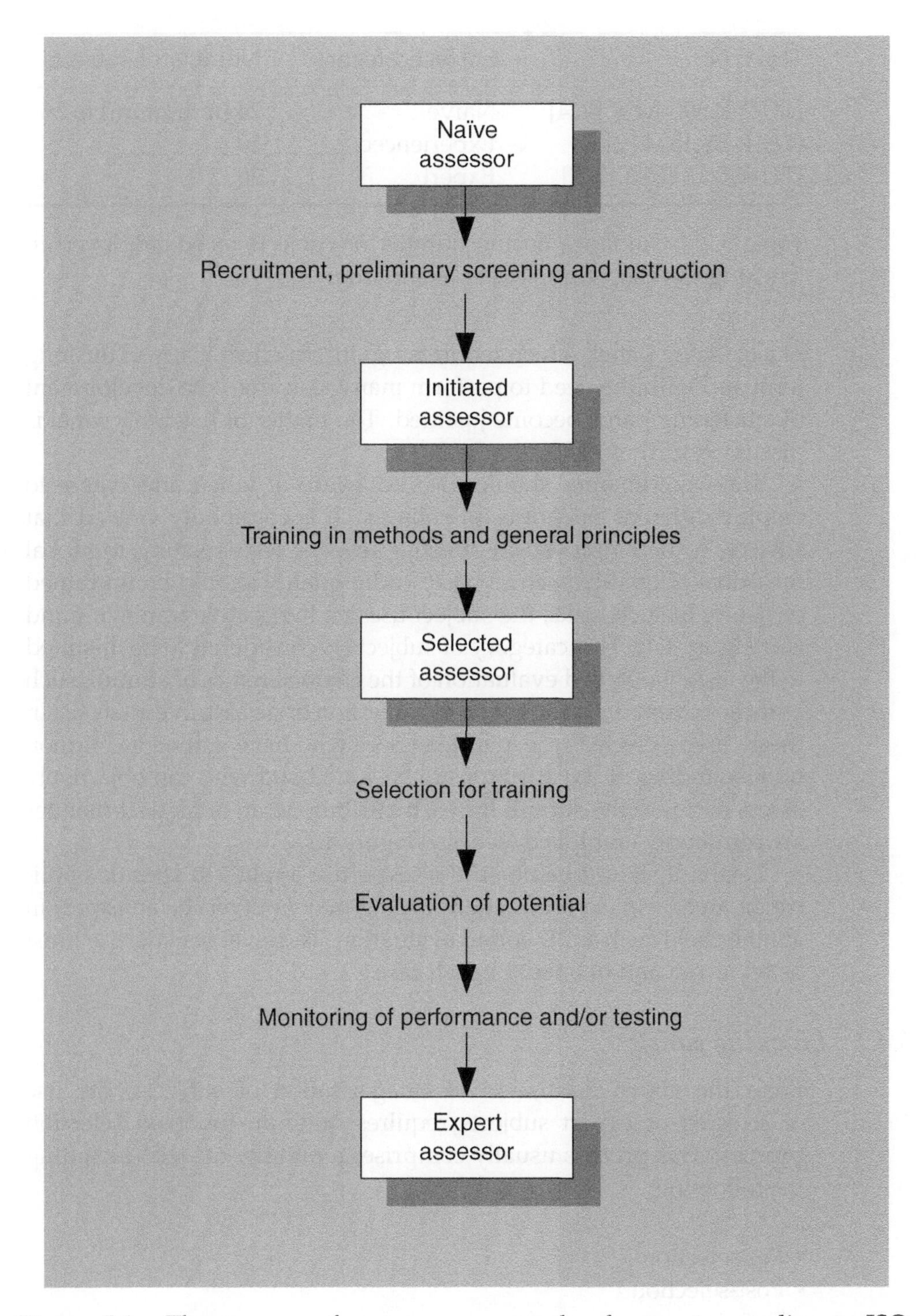

Figure 5.1: The process of sensory assessor development according to ISO 8586-2 [194]. Reproduced by permission of ISO.[19]

Test type	Subject category	Number of subjects
ITU-T P.800 ACR [224]	Naïve	24 (minimum) to 36
ITU-R BS.1534 [217]	Experienced	20
ITU-R BS.1116-1 [207]	Expert	20

Table 5.2: Guidance on the number of subjects to be employed in listening tests according to ITU standards.

of a listening panel, which requires significant effort. Only in the long term and with the need to perform many tests does the development of a listening panel become justified. The matter of listening panel is discussed further in Section 5.4.2.

The experimenter should also be aware of when and where to employ different categories of subjects. It is commonly viewed that subjects to be employed in affective tests or tests relating to global measures of quality such as basic audio quality should be untrained or naïve. In such tasks, the subject uses an integrative frame of mind (see Figure 4.1). This category of subjects is considered to be ill-suited to the task of detailed evaluation of the characteristics of stimuli, such as those required during sensory evaluation or descriptive analysis. In these cases, selected or expert assessors who have gained a common understanding of the attributes to be scaled and who can objectively assess and rate the stimuli for each attribute in an analytical manner are commonly employed (see also Figure 4.1).

Lastly, it should be noted that expertise applies to specific application areas. An expert in audio coding may not (yet) be an expert in another field such as 3D sound evaluation. The level of expertise must be reviewed and defined for each case.

5.4.2 *Listening panels*

From the above discussion of categorisation of subjects, the use of selected or expert subjects requires quite an involved selection process. This process usually comprises a number of steps including the following:

- Pre-selection
- Post-selection
- Training.

Pre-selection may be used in isolation to form a group of subjects. However, to be confident that one has a group of experts, typically all three stages must be applied.

There is also a strong commonality between selecting subjects and establishing a listening panel. Setting up a listening panel typically comprises all three stages, but it also aims to create a (semi) permanent group of subjects that can be employed as subjects in listening tests on a regular basis. If subjects are to be employed in a regular manner, additional logistical matters regarding motivation, availability, and so forth, must also be considered for members of a listening panel.

The true benefit of listening panels becomes apparent when listening tests play a key role within an organisation and many tests are run on a regular basis.

The main benefit of applying additional selection criteria is that fewer subjects are needed to obtain statistically significant results as shown by Bech [27]. Typically 5–15 subjects will be enough to ensure a sufficient resolution in the test. This conclusion was one of the main motivations for Bang & Olufsen to apply a set of additional criteria when establishing a permanent team of subjects as described by Hansen [157]. Other organisations, such as Nokia Research Center, have followed suit in establishing listening panel where regular listening tests are performed.

An overview of the subject selection procedure commonly employed is outlined in Figure 5.2.

The following section will provide an overview of one example, termed generalised listening selection (GLS) [187, 290], for the establishment of a listening panel, though slightly different approaches have been proposed by Hansen [157] or Bech [27] and other researchers.

The remainder of this chapter will provide detailed discussions of the various aspects contributing to the establishment of listening panels and subject assessment. Section 5.4.3 will go into further detail regarding pre-selection procedures and in Section 5.4.3 post-selection will be reviewed. Section 5.4.3 will also provide an overview of the more advanced methods for panel assessment and selection employed in the field of descriptive analysis. Lastly, the topic of training and associated techniques will be discussed in Section 5.4.4.

Generalised listener selection

The generalised listener selection (GLS) procedure has been evolved by Mattila and Zacharov [290] and Isherwood *et al.* [187] as a relatively

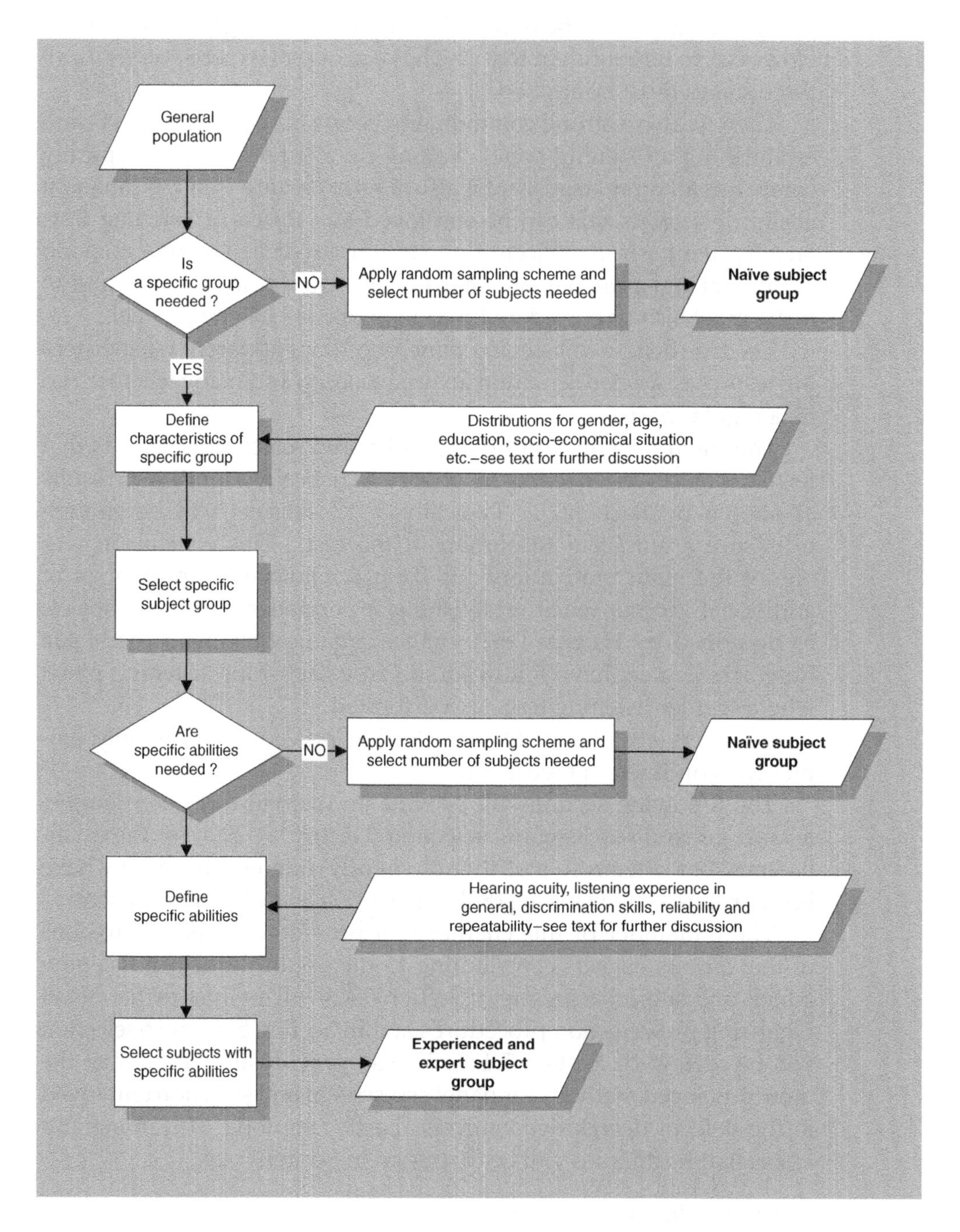

Figure 5.2: Flow diagram for pre-selection of subjects.

effective means of screening subjects for establishing a multidisciplinary listening panel.

The primary aim of such a selection procedure is to establish a group of subjects who can be categorised as selected assessors (see Table 5.1) fulfilling a number of selection criteria. The GLS procedure provides an indication that such subjects exhibit some of the following desired qualities:

- Are members of the desired sample population
- Have normal hearing acuity
- Are sensitive to audio quality characteristics
- Have the ability to repeatedly rate stimuli consistently
- Are available for performing listening tests.

These subjects may then be trained further (see Section 5.4.4), with the aim of developing their expertise.

The GLS procedure follows the outline presented in Figure 5.2 and comprises of three key phases prior to the selection process. The three stages include the following:

- A web-based questionnaire
- Pure tone audiometry
- Screening experiments.

The questionnaire as shown in Figures 5.3 and 5.4 is used to establish whether the subjects belong to the desired sample population.

Pure tone audiometry, which is further discussed in Section 5.4.3, is employed to ensure that the subject's hearing mechanisms are physiologically healthy.

Lastly, a number of screening experiments are applied to test the subject's abilities in performing listening tests. The intention here is to establish whether the subjects are sensitive to different types of audio characteristics and additionally whether they are able to rate these consistently and repeatedly.

The screening task comprises five sub-experiments[20] relating to loudness, speech, music and spatial sound evaluations. Each experiment is identical in structure to simplify the task for the subject and comprises a simple paired comparison task[21] employing the user

[20] The original GLS study included the first three sub-experiments.

[21] The developers of the GLS experiment presently consider that the application of a triangle test method may provide improved analysis for this task as well as a

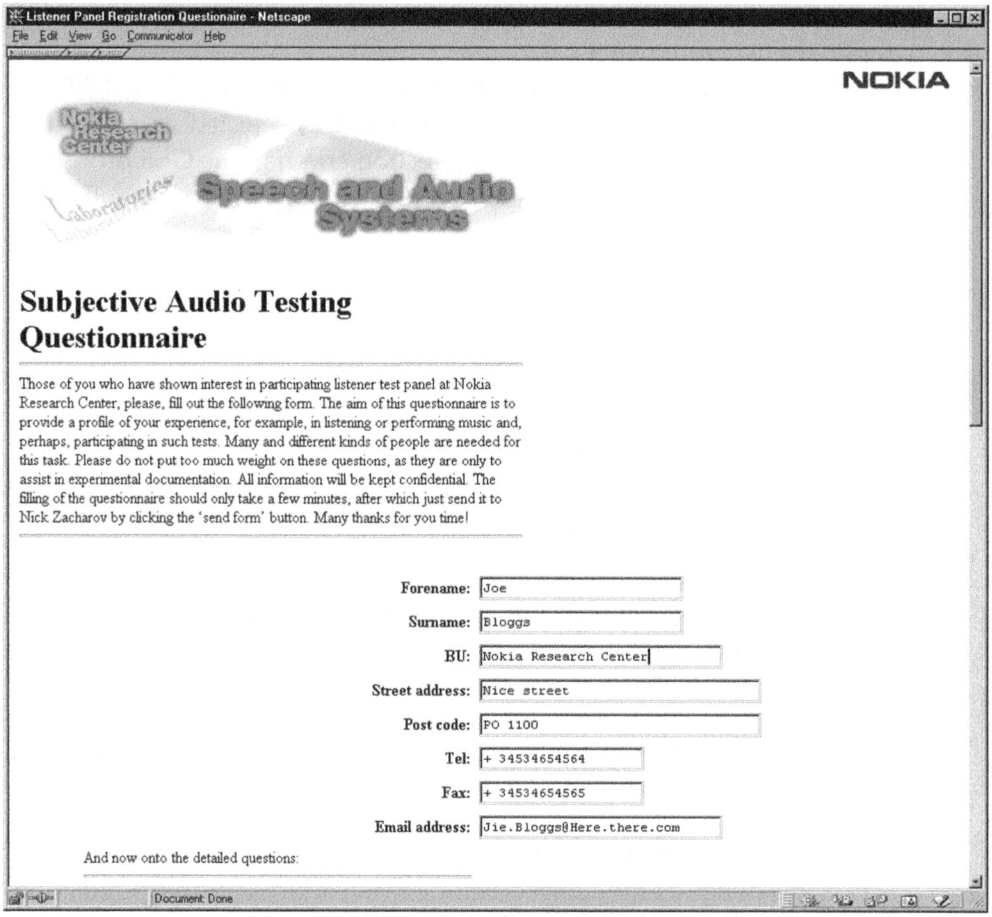

Listener Panel Registration Questionaire - Netscape

File Edit View Go Communicator Help

NOKIA

Nokia Research Center

Laboratories Speech and Audio Systems

Subjective Audio Testing Questionnaire

Those of you who have shown interest in participating listener test panel at Nokia Research Center, please, fill out the following form. The aim of this questionnaire is to provide a profile of your experience, for example, in listening or performing music and, perhaps, participating in such tests. Many and different kinds of people are needed for this task. Please do not put too much weight on these questions, as they are only to assist in experimental documentation. All information will be kept confidential. The filling of the questionnaire should only take a few minutes, after which just send it to Nick Zacharov by clicking the 'send form' button. Many thanks for you time!

Forename: Joe
Surname: Bloggs
BU: Nokia Research Center
Street address: Nice street
Post code: PO 1100
Tel: + 34534654564
Fax: + 34534654565
Email address: Jie.Bloggs@Here.there.com

And now onto the detailed questions:

Document: Done

Figure 5.3: HTML web-based questionnaire, Part 1 [290, 457].

interface presented in Figure 5.5. In each experiment, four stimuli are presented in permutation pairs with several repetitions. Ideally, six repetitions provide sufficient data for the analysis of inter- and intra-subject reliability. The four levels of stimuli are chosen such that there are quite clear upper and lower samples and the two central samples are quite closely spaced. A detailed example of the levels of stimuli employed in these experiments is presented in Table 5.3.

very simple and indirect scaling method for subjects, as shown by Lorho [275]. By using such a method, Roessler *et al.* [365] have shown that six repetitions suffice to obtain 95 % confidence intervals, while seven repetitions will yield 99 % confidence intervals.

Listener Panel Registration Questionaire - Netscape

File Edit View Go Communicator Help

Do you listen to music? Yes, daily

Do you visit music concerts, operas, ballads, plays, etc.? Yes, daily / Yes, weekly / Yes, but rarely / I don't

Do you play a musical instrument or sing?

Do you consider yourself a critical listener? Yes

How often do you go to the cinema? Yes, monthly

Do you own a Hi-fi? ☑ Yes ☐ No

If so, estimate the value of the Hi-fi (FIM): 100000

Do you own a surround sound system or TV? ☑ Yes ☐ No

Are you an audio enthusiast? Yes, I am

Have you ever used a mobile phone? Yes, daily

Have you ever used a car hands-free kit? Yes, daily

Have you ever used a desktop hands-free system? Yes, daily

Have you ever used some other kind of equipment for "phoning"? ☐ Yes ☑ No

If so, what:

Have you had a phone call with someone calling from a mobile phone? Yes, daily

Are you professionally or academically involved in audio, acoustics or Hi-fi? Acoustics

Are you professionally or academically involved in linguistics? ☑ Yes ☐ No

What is your mother tongue? English

Do you have a known history of hearing damage? ☐ Yes ☑ No

Do you have normal vision (with or without glasses)? ☑ Yes ☐ No

Have you ever previously participated in formal listening tests? ☐ Yes ☑ No

If so, how many:

Please give a brief description of these tests, if known:

Would you be willing to participate in an audiometric normal hearing test? ☑ Yes ☐ No

Would you be able to participate tests during the next summer holidays? ☐ June ☐ July ☑ August

Age: 65

What is your gender? ☐ Male ☑ Female

Send Form | Clear Form

Document: Done

Figure 5.4: HTML web-based questionnaire, Part 2 [290, 457].

Using this type of panel selection procedure, it becomes possible to screen a large number of subjects to obtain a group of 10–30 selected subjects, who can be further trained with the hope of developing their expertise.

On the basis of past experience, it is observed that, when using a student population for the recruitment of such a panel of 20–30 subjects, over 120 applicants should be evaluated. Application of the GLS procedure in this manner requires approximately one person

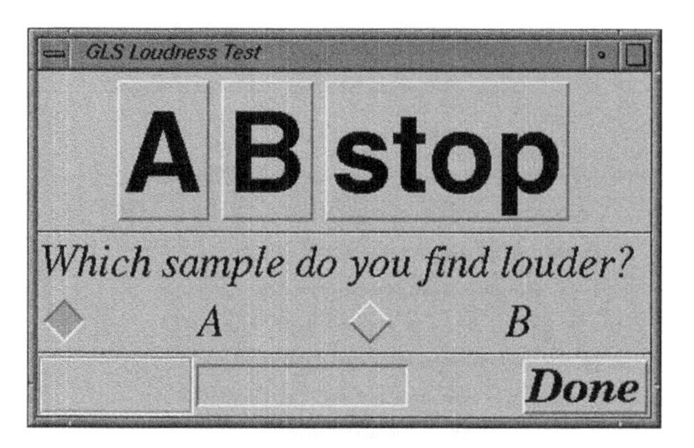

Figure 5.5: Example graphical user interface employed in the generalised listener selection (GLS) [187, 290].

month effort to select the final 20–30 subjects[22]. Experience has shown that up to 50 % of student population may not pass the pure tone audiometry requirements (see Section 5.4.3 for further details) and a further 10–25 % may not fulfil other requirements. Additional drop-outs may occur when subjects become unavailable, cannot commit to the level of effort sought, and so on.

Some analysis based upon the GLS procedure is presented in Figure 5.6 and will be discussed further in Section 5.4.3.

5.4.3 *Subject selection*

Whatever type of listening test the experimenter is planning, it is vital to consider the nature of the subjects to be employed. Even in the case of wishing to randomly sample the population at large, there will be constraints to the randomness of the sample. For example, it will be difficult to obtain a globally random sample (discussed further in Section 6.1) of the population and usually the selected subjects will fall into the subcategory, for example, of Northern Americans or Europeans. The experimenter needs to be conscious of such details and their implications. For example, an intelligibility test performed for speech coding systems in Asia may yield different results when performed in Europe.

Often, in listening tests, there is a desire to constrain or define the sample of the populations employed for a listening test in greater

[22]30 subjects is the active pool of members of the panel that can be employed for different types of experiments. Typically, 10–15 subjects are used for expert listening tests, once they are suitably trained.

Independent variable level	Exp. 1	Exp. 2	Exp. 3	Exp. 4a	Exp. 4b
1	Relative reproduction level = 8 dB	Pulse-code modulation (PCM)	PCM (44.1 kHz)	Six-channel pantophonic reproduction of decoded B-Format recording	Binaural Room Impulse Response proccessing
2	Relative reproduction level = 4 dB	Low-Delay Code Excited Linear Prediction (LD-CELP)	MPEG AAC (24 kHz)	Stereophonic reproduction of Blumlein microphone recording	Simplified HRTF model processing
3	Relative reproduction level = 3 dB	Global system for Mobile Communication Full Rate (GSM-FR)	MPEG AAC (16 kHz)	Stereophonic reproduction of omni-directional microphone recording	Stereo amplitude panning
4	Relative reproduction level = 0 dB	Global system for Mobile Communication Half Rate (GSM-HR)	MPEG AAC (8 kHz, 16 kbps)	Monophonic reproduction at +30°	Diotic presentation of speech excerpts
Sample	Pink Noise	Phonetically rich speech	Pop music	Environment recording (cafe)	Speech
Dependent variable	Loudness	Speech quality	Audio quality	Spatial quality	Spatial quality

Table 5.3: Summary of descriptions of the four levels of independent variable employed in each experiment (1-4a,b) in the generalised listener selection (GLS) procedure [187, 290]. Details of the sound sample and dependent variable are also provided for each experiment.

detail. This maybe required owing to the nature of the listening test or owing to the recommendations of a standard. Either way, a number of means are available to select subjects with differing degrees of resolution, as discussed in the following sections.

Pre-selection

A possible pre-selection process has been illustrated in Figure 5.2. The first step is to decide whether the results should apply to the population in general or to a specific group of subjects.

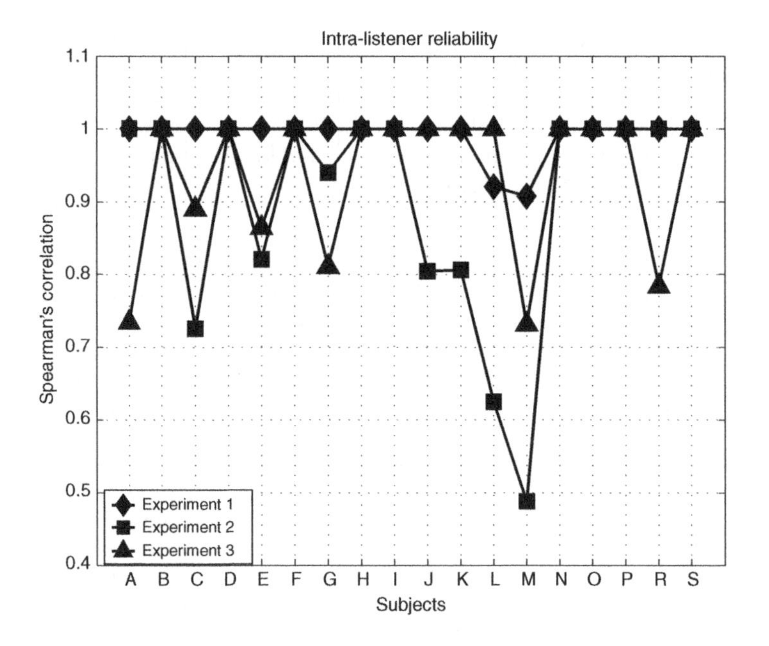

(a) Intra-subject reliability

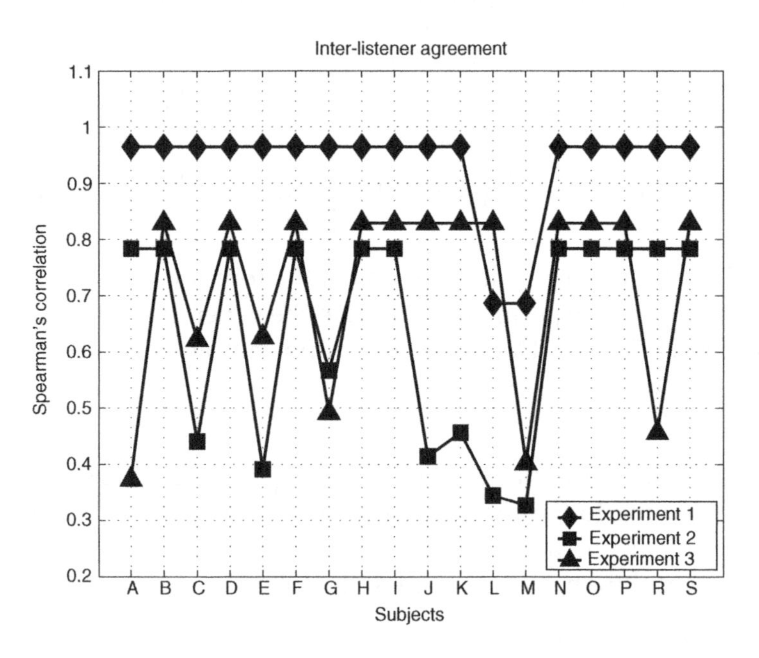

(b) Inter-subject reliability

Figure 5.6: Intra-, inter- and averaged-subject reliability results established employing the generalised listener selection (GLS) procedure [290].

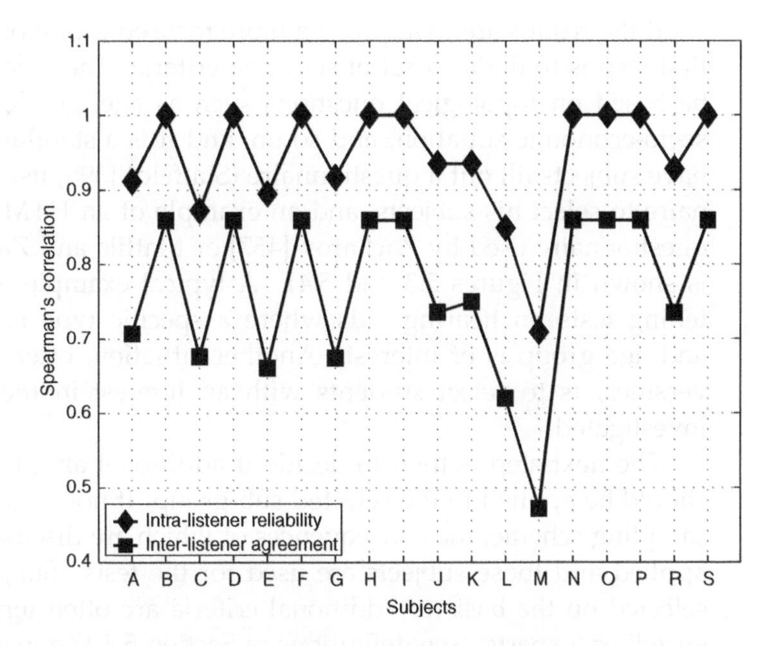

(c) Average Inter- and intra-subject reliability

Figure 5.6: Continued.

If the results are to apply to the population in general, a random sampling process is used. Statistical procedures have been developed to ensure that a 'random sampling' is actually random and describes the population in a statistically correct manner. The main argument for using a random selection of subjects from the general population is that the results must be applicable to the public in general. A typical example of this would be listening tests on standard telephone systems (see, for example, the ITU-T recommendations in the P.800 series). However, there are some major experimental implications of using a random sampling scheme. The most important is that a large number of subjects are needed, typically in the range of 100–200, to obtain statistically valid results. The reason is that subjects differ in hearing abilities, interests, ability to work as test subjects, and so on. Subjects selected on the basis of a random selection procedure have often been termed 'naïve', although a clear definition of this subject category is hard to locate in the field of audio (see a possible definition in Section 5.4.1).

If the results are to apply to a more focused group of subjects, the first step is to define a set of selection criteria. These could typically be based on topological questions such as age, gender, education, socio-economic situation, and so on, and it is a standard practice to have subjects fill out a questionnaire (Staffeldt [398] used a questionnaire to select his subjects, and an example of an HTML web-based questionnaire used by Zacharov [457] or Mattila and Zacharov [290] is shown in Figures 5.3 and 5.4). A typical example would be listening tests on hearing aids where a specific type of impairment and age group is of interest. Another situation, often used at universities, is to select students with an interest in the topic to be investigated.

The next step is then to decide if additional abilities/conditions should be applied to the selected sub-group. If not, then the random sampling scheme, the consequences of which are discussed above, is applied and those subjects are used for the tests. Subjects who are selected on the basis of additional criteria are often termed 'experienced' or 'experts' (see definitions in Section 5.4.1) depending on the additional criteria applied. The usage of the terms, however, has not been well defined and the important question is not if the person is an expert or experienced, but how the objective description or profile of the subjects is.

A number of criteria, which also have proven useful for the selection of subjects for small groups, are listed in an ASTM standard [16] and refined in ISO standard [192] aimed at the selection of subjects for sensory analysis of food quality.

- Interest and motivation–display of high interest and motivation are important for keeping the subjects in the team.
- Attitude to sound quality–it should be determined if the subjects possess strong dislikes or beliefs regarding audio quality as it is often seen that such subjects can have a strong influence on the other subjects in the group.
- Knowledge and aptitude–being a subject requires certain physical and intellectual capabilities, in particular being able to concentrate for longer periods of time. Knowledge of audio products, in general, is also a benefit.
- Ability to communicate–this ability is especially important when using descriptive analysis procedures (see Section 4.1.1).

- Availability–It is important for the planning of, especially longer, tests that subjects are fairly flexible and are not, for example, frequent travellers.
- Personality characteristics–it is important that subjects consider being a member of a listening panel or test group a serious task–meaning that they show up in time for sessions, and so on.

The experimenter should consider and tailor the selection criteria on the basis of the demands of the experiment.

Regarding the ability of subjects, experience has shown (see e.g. Bech [27, 28], Olive [329], Mattila and Zacharov [290], Isherwood *et al.* [187], Zielinski *et al.* [470], Olive [331] and Wickelmaier and Choisel [446]) that additional criteria specifically related to the subjects' listening abilities may include some or all of the following:

- Hearing acuity
- Listening skills as described by
  1. Previous listening experience
  2. Discrimination ability
  3. Intra-subject reliability
  4. Inter-subject reliability
  5. Verbalisation skills.[23]

These criteria will be discussed individually in the following section.

Hearing acuity The standard measure to consider when pre-screening subjects for listening experiments is hearing acuity, which in the present context is understood as the subject's performance in standard psycho-acoustic tests. These tests are mainly aimed at examining the functioning of the physiology of the hearing system and include threshold and discrimination tests (JND).

A large number of basic audiometric tests exist: pure tone threshold, intensity discrimination, frequency discrimination of sinusoids, temporal resolution, and so on, which are all based on simple sinusoidal signals. However, Ghani *et al.* [145] have described a series of signals that can be applied if there is a wish to test the hearing system

[23]Typically only of importance in the development of descriptive vocabularies for descriptive analysis.

for, for example, spectral shape discrimination (profile analysis [148]) and spatial acuity. Martin and Bech [282] describe the use of spatial acuity tests in the selection process for subjects for a listening team.

The basic criteria for characterising a subject's hearing ability as being normal is that it fulfils the requirements of an otologically normal person as defined in ISO [190]. This involves that the experimenter specifies the acceptable deviation of the hearing threshold at the audiometric frequencies 250–8000 Hz. It has been the rule in most listening tests reported in the literature to allow for deviations of up to 15 dB relative to the median hearing threshold level at 18 years of age if the audiometric test is conducted according to the procedure specified in ISO [189]. This means that less than 10 % of the otological normal male population at 20 years of age will be excluded on the basis of an audiometric test (see ISO [190], Annex B). The interested reader is referred to [250] for further information on audiometric testing.

Experience has, however, shown that approximately 50 % of a university's student population (male and female) will not pass an audiometric test using a 15 dB rule. This is not in agreement with the predictions of Annex B of ISO [190]; however, this phenomenon has been observed on a regular basis during the last 20 years by both the authors.

The importance of the various criteria in relation to performance in listening tests have been examined by several researchers. Toole [422] investigated the relationship between a subject's hearing threshold (no rule applied) and the standard deviation of the subject's rating in a standard listening test on loudspeakers. He found that there is a positive correlation between a subject's mean hearing thresholds (average below 1 kHz and average above 1 kHz) and the standard deviation of that subject's ratings. Bech [27] also examined this relationship for subjects fulfilling a 15 dB rule; however, he was not able to confirm Toole's findings. Thus, it is in general recommended that a 15 dB rule be applied.

Shlien and Soloudre [379] examined various psycho-acoustic measures (absolute threshold, pitch discrimination, and temporal sensitivity) and the same subject's ability to detect artifacts of low bit-rate coding systems. Their results showed no clear relationship between the performance in the two domains, but it was clear that the performance of subjects differs significantly in both areas. These results raise questions about the usefulness of the concept of an average listener or a so-called 'golden ear' representing the ultimate performance in all areas.

Shlien [378] extended the above investigation by studying a group of listeners who had extensive experience in the assessment of coding systems. He measured their hearing abilities quantified by hearing threshold, sensitivity to codec noise, bandwidth of auditory filters, frequency discrimination, and backward masking threshold. He concluded that none of the measures could be used to distinguish between 'gifted' and normal listeners.

Precoda and Meng [353] examined whether listeners assessing audio compressions schemes (low bit-rate coders) have similar strategies for their evaluations. The results support Shlien and Soloudre's questioning of the concept of an 'average' listener. Precoda and Meng found that listeners will have different strategies for evaluation of coders so care should be exercised when averaging across listener in listening tests.

Canévet [80] also examined a number of psycho-acoustic measures and their ability to describe a subject's hearing performance. However, he did not relate the performance of the subjects to their performance in other auditive tasks such as a listening test. Ghani *et al.* [145] have developed a test battery of auditory stimuli to investigate the auditory capabilities of listening panel members. Performance prediction of the panel was studied on the basis of a predictive model employing a number of psycho-acoustic measures.

The general recommendation for selection of subjects with respect to hearing acuity, despite the fact that no clear relationship between the various tests and performance in listening tests has been reported, is that subjects should pass an audiometric test that uses a 15 dB criteria (both ears).

Listening skills Previous listening experience is important, as discussed by Bech [27], Kirk [256], Gabrielsson *et al.* [133, 134, 136], and Mattila and Zacharov [290], as it will influence the subject's ability to make consistent rating and to detect small differences in sound quality.

Previous experience and interest in sound reproduction include such factors as familiarity with 'live' sound either as a listener or as an active performer, experience in critical listening to live or reproduced sound for example in the form of a Tonmeister education and finally the general aptitude for detecting sonic differences in reproduced sound.

As discussed earlier, the questionnaire (see the examples in e.g. Bech [27], Mattila [286], Mattila and Zacharov [290]) can be used as an effective means of collecting information about subjects. This method can and should be used to probe the previous experience of subjects in listening tests and questionnaires should be adapted to probe the relevant information. However, the questionnaire will not provide quantitative information regarding the subject's performance.

The discrimination abilities can be quantified by factors that measure to what extent a subject is able to discriminate between the items under test. This ability can be improved significantly by training, but subjects have been shown to differ significantly, even after extensive training. The ability is important as it has direct influence on the statistical resolution of the test.

A number of different tests exist for measuring a subject's discrimination abilities. They fall in roughly three groups:

Matching tests typically include a number of stimuli where the sensation of the attribute(s) of interest is well above the threshold. It could, for example, be a selection of sounds at different loudness levels. The subject is presented with the first stimuli set and allowed to familiarise himself with the same. Then the second set (typically including twice as many samples as in the first set) with the same stimuli (but with different identification numbers) is presented and the task is to match the stimuli in the two sets. The ability of the subject is measured by the number of correct matches–in sensory analysis, the criteria for acceptance is a minimum of 80 % correct matches.

Detection tests are based on the triangular test [408] where the subject is presented with three samples and asked to determine the two most similar samples and the one that is most different from the other two. The version used for discrimination tests has two identical samples, with the sensation level of the attribute of interest above threshold, and a third sample that is different. The ability of the subject is again the number of correct identifications of the two identical samples. In sensory analysis, the criteria is typically 100 % correct matches.

Discrimination tests are used to assess the 'resolution' of the subject's discrimination ability. Such tests are performed with an interval scale, for example, a 10-point scale. The distribution of sensory magnitudes of the selected stimuli in the test is therefore very important. The selection should include both large and small sensory differences for the attribute of interest. The large differences are not interesting per se, but should be included in order to make some of the comparisons easy for the subjects. The magnitude and the number of the smaller differences must be based on the specific purpose and the resolution that is required of the test. The discrimination ability of the subject could be measured in a number of ways, for example, by the magnitude of the F-test statistic as calculated in an analysis of variance model (see the text later in this chapter) for each of the levels of sensory differences.

A subjects consistency is often termed intra-subject reliability and it has been defined by Gabrielsson [133] as 'the consistency of ratings within each individual' and Gabrielsson further defined the reliability index as shown in Equation 5.1:

$$r_w = 1 - \frac{MS(\text{error})}{MS(\text{model})} \tag{5.1}$$

where r_w is the reliability index, MS(error) is the error variance as calculated by the analysis of variance model (ANOVA), and MS(model) is the variance related to the ANOVA model.

Another measure of the consistency of ratings often used in the literature is the variance or standard deviation of repeated ratings as it reflects the ability of the subject to repeat ratings of the same stimulus. However, Bech [27] showed that the magnitude of the standard deviation is related to the subject's use of the rating scale and suggested that the F-statistics for the loudspeaker, as calculated in an ANOVA analysis, should be considered in addition (see Chapter 6 for further details on ANOVA and F-statistics) as defined in Equation 5.2:

$$F_L = \frac{\frac{SS_L l}{df_L}}{\frac{SS_w}{df_w}} \tag{5.2}$$

where SS_L is the sum of squares for the loudspeaker factor, df_L is the degrees of freedom for the loudspeaker factor, SS_w is the with-in cell sum of squares, and df_w is the with-in cell degrees of freedom.

It is seen that the F-statistic measures the ratio of variation due to differences between the loudspeakers (plus an unknown amount of random variation) to the random variation. If the ratio is high, it indicates that the perceived differences between the loudspeakers are significantly higher than the random variation and thus reflects a true perceived difference between the loudspeakers.

The inter-subject reliability refers to the agreement between the ratings (ranking) of different subjects. The important issue is that some subjects may be very consistent in their rating, but may have a completely different view on the rank order of the stimuli. The fact that subjects tend to use different parts and ranges of the rating scale is not important in this context as, the influence of this factor can be rectified by normalising the data.

Subjects who display a deviating rank order should be interviewed to explore what the possible causes could be. If no reasonable explanation (misunderstanding of instructions etc.) can be found, it should be considered as to whether they are to be excluded from the group.

The inter-subject reliability can be measured by the correlation between the ratings of subjects for both the overall mean and the individual programme items. Gabrielsson [133] suggests various other measures, but the correlation seems to be the best measure.

It is difficult to set lower limits for acceptable values of intra- and inter-subject reliabilities; however, it is suggested that subjects be rank ordered on the basis of both the intra- and the inter-subject measures and the best subjects be selected from the top of the list. Mattila and Zacharov [290] have established a so-called GLS procedure where the subjects are ranked ordered on the basis of a weighted sum of the different measures. Both inter- and intra-subject correlation measures are employed for ranking subjects, some results of which are presented in Figure 5.6. Bivariate correlation was employed in this case. To illustrate the assessment of a subject's performance using this method, we can study a few subjects. Subject M has both poor inter- and intra-rater reliabilities over three experiments and would not be selected to the listening panel. By comparison, subject O performs well in all respects and hence he is selected to the listening panel.

Post-selection

Post-screening of subjects can be an effective, but expensive and potentially dangerous, method for 'cleaning up' the data set when applied to true experimental data. It is effective because subjects those for one reason or another do not 'behave well' can be detected and removed, thereby improving the 'signal-to-noise' ratio in the data set. It is expensive because the subjects in question have already participated in the experiment, but their results might not be useful. Finally, it is dangerous because the outlying data points could be the most interesting data for further research. However, this method is valuable when applied to the assessment of experimental data from screening experiments, as discussed in Sections 5.4.2 and 5.4.3.

Post-screening of subjects can be performed either on the basis of the data already available in the experiment or the experiment can be designed to accommodate post-screening. A good example of this is the recommended paradigm for evaluation of low bit-rate audio coding systems with small impairments as described in ITU-R Recommendation BS.1116 [207] (see Section 10.1.2 for further details).

The recommended test method presented in ITU-R recommendation BS.1116-1 [207] is the double-blind triple stimulus with a hidden reference. The test subject is presented with three stimuli (A, B, C) and the subject knows that the reference stimulus is always A. Stimulus B is either the coded version (object) or the reference, and vice versa for stimulus C. The subject's task is to compare stimuli B to A and C to A and thereafter to identify the reference as being either B or C. The reference is graded 5 on the 5-point scale and the other stimulus (B or C) is then graded on the 5-point scale.

The result of each presentation is thus given by two grades–one for the hidden reference and one for the object. If the subject cannot–on average–correctly identify the hidden reference, then the difference between the two grades (object – reference) will be distributed around zero with a mean of zero. If, however, the subject can identify the hidden reference in every trial, and thus grade the object below 5, then the average difference grade would be different from zero in a negative direction. Thus, the application of a standard one-sided t-test on the difference grades will allow for a simple check of the subject's ability to identify the reference. Assuming that there are audible differences between the codecs under test, those subjects who were merely guessing (i.e. the mean value is not significantly different from zero) could then be excluded from further analysis on the basis of this simple test. Further details on the specifics of the data analysis are provided in ITU-R Recommendation BS.1116-1 [207].

The detailed statistical issues related to the detection and exclusion of outliers are discussed more in Chapter 6.

Assessor performance evaluation for descriptive analysis

Specific to sensory profiling and descriptive analysis, techniques have been developed for the evaluation of assessor performance. Several techniques have been developed and applied, including the work of Brockhoff [69–71], Schlich [373] and also some standardised approaches presented in ISO 11035 [193] and ISO 8586-1 [192] Annex A. In this section, a review of Schlich's approach referred to as *GRAPES* will be presented, based upon single stimulus-attribute scaling experiments. While this approach has been developed for applications to non-audio applications, the authors' view that these methods can be directly applied to audio for either descriptive or discrimination tests.

The method developed by Schlich is based upon the analysis of variance of replicated sessions[24] where products have been evaluated using a number of attribute scales. The method can be viewed as a post-selection technique, that is, applied following an experiment and may also be used for monitoring and assessing panel performance. While not implicit in the paper, the method is equally applicable to single attribute experiments such as the evaluation of 'basic audio quality' or 'mean opinion score', for example. The method considers the following key questions:

- Do assessors use the scale in the same way?
- Do assessors give reproducible scores among sessions?
- Do assessors discriminate between products and do they agree on product differences?

Six performance statistics have been developed for evaluation of assessor performance and will be outlined below and described in detail in Equations 5.3–5.8.

Location The average score given by assessors

Span The average standard deviation of a score given by an assessor within a session

Unreliability Assessor unreliability based upon a measure of the session error

[24]Schlich appears to employ the term 'session' to represent a repeated presentation of all stimuli for the method to be applicable.

Source	DF	SS
PRODUCT	$I-1$	$SSP_j = K\sum_i(x_{ij.} - x_{.j.})^2$
SESSION	$K-1$	$SSS_j = I\sum_k(x_{.jk} - x_{.j.})^2$
RESIDUAL	$(I-1)(K-1)$	$SSR_j = \sum_{i,k}(x_{ijk} - x_{ij.} - x_{.jk} + x_{.j.})^2$

Table 5.4: Definition of degrees of freedom (DF) and sum of squares (SS) for ANOVA of assessor j, termed model (1) in [373].

Source	DF	SS
ASSESSOR	$J-1$	$SSA = IK\sum_j(x_{.j.} - x_{...})^2$
SESSION (ASSESSOR)	$J(K-1)$	$SSS = \sum_j SSS_j$
PRODUCT	$I-1$	$SSPP = JK\sum_i(x_{i..} - x_{...})^2$
PRODUCT * ASSESSOR	$(I-1)(J-1)$	$SSPA = K\sum_{i,j}(x_{ij.} - x_{i..} - x_{.j.} + x_{...})^2 = \sum_j SSPA_j$
RESIDUAL	$J(I-1)(K-1)$	$SSR = \sum_j SSR_j$

Table 5.5: Definition of degrees of freedom (DF) and sum of squares (SS) for ANOVA of assessor panel, termed model (2) in [373].

Drift–Mood A measure of the between–sessions error that can be associated with assessor mood, and so on.

Discrimination A measure of an assessor's discrimination skills based upon the classical F-ratio for testing the significance of the product effect for an assessor (as also presented in Section 5.4.3).

Disagreement A measure of an assessor's disagreement based upon the assessor's contribution to PRODUCT * ASSESSOR interaction F-ratio.

Two ANOVA-based models are developed for estimating the performance metric. The so-called model (1) is fitted for each assessor and is computed according to Table 5.4. Model (2) is a global model obtained by pooling the individual models (1) according to Table 5.5, where x_{ijk} is the sensory score, i is the product, j is the assessor, and k is the session.

The notation employed by Schlich differs from that employed in Chapter 6; however, to ease the comparison with the original reference, it was decided to maintain the original notation. The subscript notation employed a '.' to indicate an average over that subscript factor, for example, $x_{.j.}$ implies an average over the factors product (i) and session (k).

The statistic for location is mean score per assessor calculated by

$$\text{LOCATION}_j = x_{.j.} \tag{5.3}$$

The measure of span is the mean standard deviation per assessor within a session, given by

$$\text{SPAN}_j = \frac{1}{K}\sum_k \left[\sum_i (x_{ijk} - x_{.jk})^2/(I-1)\right]^{1/2} \tag{5.4}$$

The measure of unreliability is calculated by

$$\text{UNRELIABILITY}_j = \frac{RMSR_j}{SPAN_j} \tag{5.5}$$

where $RMSR_j$ is the root mean square associated with the sum of squares of RESIDUALS, given in Table 5.4.

The variation associated with the drift and mood per subject is given by

$$\text{DRIFT–MOOD}_j = \frac{RMSS_j}{SPAN_j} \tag{5.6}$$

where $RMSS_j$ is the root mean square associated with sum of squares of SESSION, given in Table 5.4.

The discrimination skill of each subject is calculated using

$$\text{DISCRIMINATION}_j = \frac{MSP_j}{MSR_j} \tag{5.7}$$

where MSP_j and MSR_j are the mean squares corresponding to PRODUCT and RESIDUAL, respectively, as given in Table 5.4.

The disagreement measure per assessor is estimated by

$$\textsc{Disagreement}_j = \frac{J.\,SSPA_j}{[(I-1).\,(J-1).\,MSR]} \tag{5.8}$$

where MSR is the mean squares corresponding to Product * Assessor for model 2 as given in Table 5.5 and $SSPA_j$ is the sum of squares associated with Product * Assessor as defined in Table 5.5 and calculated by

$$SSPA_j = K \sum_i (x_{ij.} - x_{i..} - x_{.j.} + x_{...})^2 \tag{5.9}$$

Two means are employed for establishing the level of performance of each assessor. The global mean of the panel is calculated; however, this does not by itself provide any absolute measure of an assessor's performance. The Fisher variance (F) ratio level is employed for both the discrimination and disagreement metrics, as given in Equations 5.7 and 5.8. Assessors whose performance exceeds the 5 % Fisher variance ratio for discrimination are considered to be discriminant. Assessors whose performance exceeds the 5 % Fisher variance ratio for disagreement are contributing to the Product * Assessor interaction, which is indicative of a level of disagreement.

In order to illustrate the value of these metrics for the assessment of an assessor's performance, an example is provided in Figure 5.7, showing an assessor's performance for an attribute 'tone colour' employed in an evaluation of headphones [274]. In this figure, the three key performance metrics out of the six performance metrics are presented, as these provide the most critical information regarding the assessor's performance, namely, unreliability, discrimination and disagreement. The heavy dotted lines in each pane are associated with the panel mean. The light dotted line in the discrimination pane illustrates the 5 % significant F-ratio for Product as given by Equation 5.7, that is, discrimination. The light dotted line in the disagreement pane illustrates the 5 % significant F-ratio for Product * Assessor as given by Equation 5.7, that is, disagreement, where the panel disagreement is defined as

$$\textsc{Disagreement} = \frac{MSPA}{MSR} \tag{5.10}$$

where $MSPA$ is the mean squares corresponding to Product * Assessor for model 2 as given in Table 5.5.

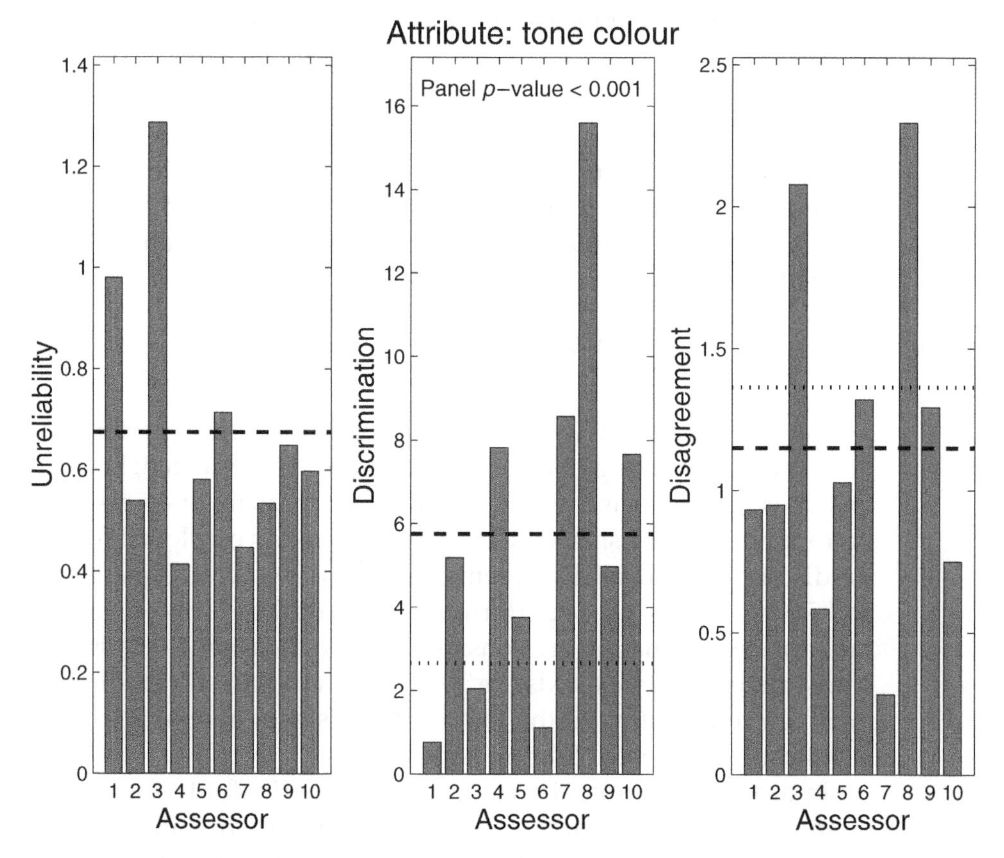

Figure 5.7: Examples of assessor performance evaluation in terms of unreliability, discrimination and disagreement. Heavy dotted line represents the panel means and the light dotted line the 5 % significant F-ratio (see text for details). Reproduced by permission of Gaetan Lorho [274].

Considering assessor 3 in Figure 5.7, it can be observed that he has the worst unreliability score of the panel and poor discrimination ability, as well as contributing significantly to the disagreement within the panel[25]. Overall, his performance would need to improve through

[25] In a sensory evaluation experiment where multiple attributes are rated by each assessor in an objective and analytical manner, it is vital to ensure that all subjects are in agreement, that is, all have a common understanding of the consensus descriptive language. In experiments where subjects are less experienced or trained, disagreement may be more commonly encountered, particularly when rating global performance metrics, such as 'basic audio quality'. As a result, the usage and interpretation of the disagreement metric may differ between descriptive analysis experiments and discrimination tests.

some training and be re-assessed before being further employed in such listening test. On the basis of current performance, this assessor would be categorised as naïve according to definitions presented in Table 5.1. Assessors 4 and 7 show very good performance according to all three performance metrics. Both the subjects have reliability scores well below the panel mean (heavy dotted line). They also have good discrimination skills, above the 5 % significant F-ratio (light dotted line) and close to the panel mean (heavy dotted line). Lastly, they do not contribute to the panel disagreement with scores below the 5 % significant F-ratio (light dotted line) and the panel mean (heavy dotted line) for the disagreement metric. Such subjects are highly skilled in this listening test application and might be considered as expert subjects.

Schlich's approach is also applicable to overall panel performance assessments and the interested reader is referred to [373] for further details on these aspects.

Verbalisation skills If the panel is planned to be employed for any form of sensory/descriptive analysis (see Section 4.1.1) or attribute development tasks, it may be of additional interest to establish how fluent and verbally skilled the subjects are. A number of different approaches to the aspect of screening have been employed. Wickelmaier and Choisel [446] applied verbal fluency tests as described in detail in [394], in the native language of the panel, to establish the degree of fluency of each subject.

Alternative methods to establish the level of fluency of subjects have been employed by Koivuniemi and Zacharov [259]. This approach was based upon the number and suitability of adjectives and synonyms elicited per subject to describe a number of different visual stimuli, that is, photographs, with different characteristics. Visual stimuli are employed here in order to avoid biasing subjects with respect to audio stimuli. A web-page approach allowed for fast and easy elicitation and processing of results.

Bang & Olufsen had applied the same idea for the selection of subjects for a viewing team. The individual subjects were presented with four music excerpts and asked to list a single word that could be used to describe their impressions. They were not allowed to list obvious words such as 'guitar' or 'female voice' as these were not considered as part of an impression, but merely as statements of facts in the reproduction. The objective measure was the number of adjectives that the subjects listed. The experimenter was present during the session and to supplement the objective measure he ranked the subjects in three categories (well qualified, less qualified and not qualified) on the basis of his impression of how they tackled the task. Another objective measure would be that described by ISO [192] where subjects are presented with odour samples with known attributes and asked to list the attributes they associate with the sample. A scoring system is recommended where the subjects gets three points for a correct identification of a known attribute, two points for identifying an attribute in general terms, one point for listing an attribute after discussion with the experimenter, and zero points for no response. Between 5–10 odour samples are presented and subjects who achieve less than 65 % of the maximum score are not considered qualified.

5.4.4 *Training and monitoring*

Training of subjects is an important issue that needs to be considered carefully by the experimenter. The training can serve several purposes:

- Familiarisation with experimental conditions
- Identification and familiarisation with new attributes
- Maintenance of training between experiments.

Prior to any listening test, it is necessary to ensure that the subjects are familiar with the experimental facilities, the stimuli and magnitude of differences that they will experience during the experiment. This can conveniently be in the form of a pre-experiment with a reduced number of stimuli, selected to illustrate the auditory attributes and the range that will be included in the experiment. Such experiments also serve the purpose of checking that the experimental set-up is working as expected and that the logistics of the subjects' participation is acceptable.

A very important part of the familiarisation process is the instructions to the subjects in their task during the experiment–any uncertainties that are left in the subjects' understanding of the task will have a direct influence on the data provided. The instruction should always be given both in written form and verbally by the experimenter. It should be written in a clear non-technical language, describing all aspects of the task and the related aspect of the set-up and user interface (see examples in Section 9.2.2). It is advisable to test the instructions on colleagues unfamiliar with the experiment–it is often experienced that what is perfectly clear to the experimenter is not at all clear when the subjects read the instructions–and the damage is done if the subjects get unclear instructions. Several examples of instructions can be found in the literature and in standards [27, 28, 207, 224, 463, 464].

Identification and familiarisation with new attributes is often needed in listening tests on low bit-rate coders and, more recently, in experiments exploring spatial sound reproduction. The standard [207] for listening tests on coders is quite explicit in describing such a training programme that typically has a duration of one or two days. If a more general training of coding artifacts is required, reference may be made to the CD [18] that the Audio Engineering Society has issued, which includes a range of artifacts representative of the current generation of coders. The 'Golden Ears' collection of compact discs developed by Moulton [310] provides a means of becoming familiar with different kinds of signal processing effects such as equalisation, reverberation, compression, delay, and so forth. Other sources of training and demonstration audio material are listed in Appendix C.

Maintenance of training between experiments or long-term training programmes aimed at developing more general listening capabilities assume that the training acquired in one experiment is transferred to the next experiment. This was investigated by Bech [28], who has a summary of the literature and presents results on a specific training programme and its influence on threshold values and associated standard deviation for timbre differences using complex signals. The transfer of training from one task to another and the long-term stability of the training are also discussed. The general conclusions are that training will lower the threshold significantly, that the threshold (after training) will be stable for a four- to six-month period and thereafter maintenance experiments are needed with a four- to six-month interval, and finally that training can be transferred from one task to another. Bech [27] refers to another experiment examining the influence of training specifically in listening tests on loudspeakers. The results suggest that a minimum of four training experiments are needed before the majority of the subjects have reached a stable level with respect to the ability to discriminate between loudspeakers and consistency of ratings. An additional three to four experiments are needed before all the subjects reach stable performance.

A number of computer-based scheme have been developed for interactive training of subjects. Many of these have focused upon timbral ear training. Brixen [67] developed a system for training sound engineers in the identification of spectral aberrations. The timbre solfege programme [269, 297] was developed for the training of Tonmeisters, and training aspects include timbre, loudness, pitch, spatial hearing, distortion and other attributes. The timbral ear training (TET) programme developed at McGill University by Quesnel and others [354–356] aims to provide subjects with interactive ear training of timbral characteristics. A development of the TET system for training of subjects in the detection and identification of aspects of artificial reverberation has been proposed by Corey [99]. Another system for training of spatial sound characteristics has been developed by Neher *et al.* [73, 319–321, 368]. Merimaa and Hess [292] provide a scheme for training of subjects for evaluation of spatial attributes of sound. Shively and House [377] report the development and application of a software-based system for the training of subjects for the evaluation of automotive sound systems. Olive [329, 330] reports the development of a software-based system for training of subjects to perform listening tests on loudspeakers. An illustration of such a training scheme is provided in Figure 5.8, which includes a means of providing subjects with feedback on their performance.

The advantages of computer-based training system is that it is possible to adapt the stimuli to meet the expertise level of the subject. Additionally, it is possible to provide subjects with vital feedback on their performance level and also to log this information for the experimenter's records. However, few of these software tools are commercially available.

If a permanent panel of subjects has been established, it is important to devise a training programme that aims at training all aspects of sound perception. Such a programme has been effective at Bang & Olufsen for many years. It includes weekly listening sessions of one or two hours where specific aspects of sound reproduction are discussed and demonstrated. This includes topics such as the influence of the position of the loudspeaker and the listener in the room, the influence of bit rate in low bit-rate coding systems, the influence of recording technique, and so on. It is also important that the training includes visits to concert, theatres, movies, and so on, to demonstrate the 'live' sound experience. The members of the listening panel should also be exposed to different qualities and brands of reproduction systems in their homes so that they are not biased towards any particular brand or type of system.

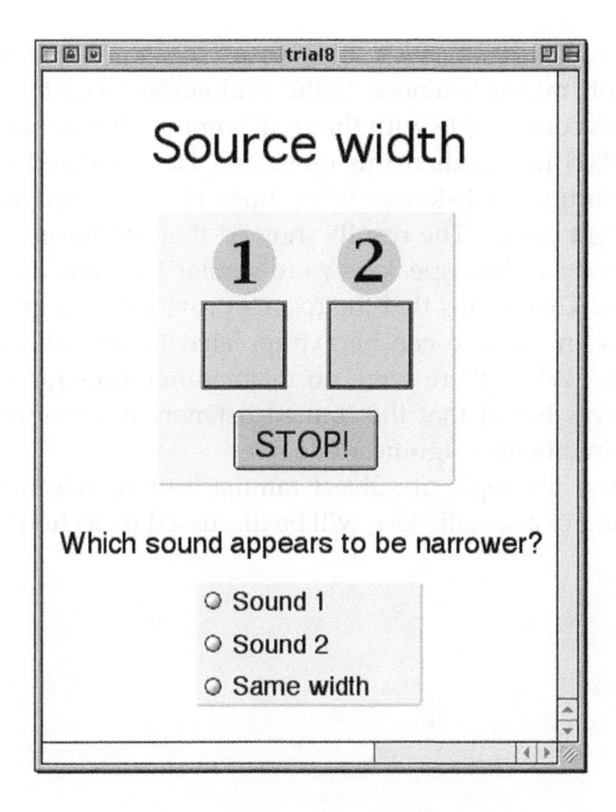

(a) Training window

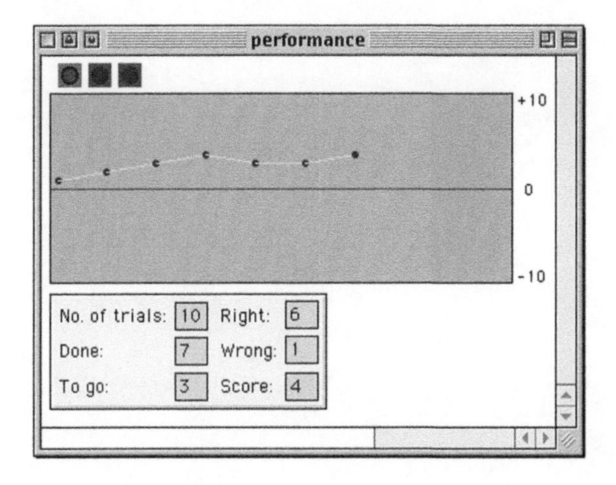

(b) Performance window

Figure 5.8: Example user interfaces of training software for spatial sound characteristics [318, 320]. Reproduced with permission of Tobias Neher.

The following question often arises in connection with permanent panels of trained listeners: is the evaluation given by trained panel members comparable with those of 'common listeners or customers? Olive [331] has studied this question. He compared the responses of 256 untrained listeners with those of 12 trained members of a permanent panel. The results showed that the mean ratings for the same group of loudspeakers were similar for trained and untrained listeners. This means that the results provided by a group of trained listeners on average can be extrapolated to an untrained group of listeners. While there were no major differences for the mean values, it was found that the trained listeners were more reliable and discriminating to a significant degree.

Lastly, the topic of subject familiarisation, which occurs at the beginning of a specific test, will be discussed in Section 9.2.2.

CHAPTER

6

Statistics

THE purpose of this chapter is to introduce the concept of statistical analysis of data from experiments on perceptual evaluation of sound. The aim is to introduce the general strategy for such analysis and the various statistical methods available and to provide guidance on method to be used in specific cases. The aim is not to give a detailed mathematical description of the procedures, but rather to present the methods such that qualified questions can be asked to a professional statistician–if needed. If you are not a professional statistician or do not have one on your team, it is recommended that arrangements be made such that one can be consulted on a regular basis. It is especially important that the statistical parts of an experimental design be discussed thoroughly before any subjects are invited to participate in the experiment.

Perceptual Audio Evaluation – Theory, Method and Application Søren Bech and Nick Zacharov

The focus of this chapter is on hypothesis testing and to a lesser degree on the estimation of the various variables. This is motivated by the fact that the purpose of most listening tests is to detect and quantify significant differences between the loudspeakers, for example, and not (yet) to quantify the relationship between the dependent and independent variables.

A multitude of books exist on the various statistical analysis procedures, to which the interested reader is referred. A selection that the authors have found useful, covering most of the issues discussed, includes [155, 163, 253, 268, 328, 396].

The sections have been organised to reflect the order of the processes that an experimenter follows: statistical design of the experiment, testing of the statistical assumption of the obtained data and, finally, the detailed statistical analysis. This organisation might not be optimal when seen from a purely statistical point of view, but as this book addresses the experimenter and not the statistician, the needs of the former have prevailed.

The importance of detailed statistical planning of the experiment and subsequent analysis of the results is often not acknowledged by audio engineers (nor practitioners from other disciplines as well). The physical variables (the test set-up, the stimuli, etc.) are usually controlled to an acceptable degree; however, the results are often only reported by simple mean values without confidence intervals, for example. Such a lack of statistical information may lead to incorrect scientific or application decisions.

The basic purpose of a listening test is to objectively quantify the variations in auditory impression caused by the presented stimuli. The variation in the stimulus set is created by the controlled combinations of the experimental independent variables. The research question asked by the experimenter is whether the variation in the subjective impression is a result of this intended variation or is it more likely to be a random variation. The purpose of the statistical planning of the experiment and analysis of the subjective responses is to provide a statistical validation of the answer to this question.

The statistical analysis is based on the assumption that the t, ith score, $Y_{t,i}$ provided by a subject consists of two parts: one part μ_t representing the reaction to the ith repetition of the stimuli t defined by the controlled experimental variables (the treatments) and the other, an error part $\epsilon_{t,i}$, representing a random variation caused by a range of uncontrolled variables:

$$Y_{t,i} = \mu_t + \epsilon_{t,i} \tag{6.1}$$

The random variation caused by uncontrolled variables is related to the subject themselves and/or the experimental set-up and environment. The variables related to the subject are physiological variables such as variations in sensitivity, background level of intrinsic neural activity, and so on, and cognitive variables such as mood, expectation, interpretation and bias effects (see Section 4.2.4). The variables related to the experimental set-up and the environment in which the set-up is established are, for example, changes in the temperature of the room, in the lighting conditions in the room, background noise, and so on. Toole [421–423] and Bech [26] have more detailed discussions of these variables and their influence in relation to listening tests (see also chapter 5). Keppel and Wickens [253] (Section 2.1) provide a more general discussion of the variables that cause the random fluctuations in the subjects' responses.

The variables contributing to the score of a subject can, in general, be classified as follows:

- **Permanent or stable abilities.** This includes, for example, the subject's hearing threshold for the attribute under study or the general impression of a loudspeaker that remains constant during the experiment.
- **The treatment effect.** This is the influence of the variables controlled by the experimenter.
- **Internal variability.** This includes the variables related to temporary changes (during the experiment) in the subject's physiology and psychology.
- **External variability.** This includes changes in the external environment (the experimental set-up).

The model describing a subject's score can be further elaborated as the mean μ_t can be divided into an overall mean μ representing the mean of all treatments in the experiment plus a treatment effect corresponding to a specific treatment t:

$$\mu_t = \mu + \alpha_t \tag{6.2}$$

By combining the above equations, the following statistical model for the observed t, ith score from a subject is obtained:

$$Y_{t,i} = \mu + \alpha_t + \epsilon_{t,i} \tag{6.3}$$

where $Y_{t,i}$ is the ith score of treatment t, μ is the mean of all treatments, α_t is the treatment effect for treatment t, and $\epsilon_{t,i}$ is the experimental error.

The experimental error is often assumed to be normally distributed $N(0, \sigma^2)$ [143, 253]. Thus, to establish a statistically reliable estimate of the mean and component, it is necessary to obtain a number of repeated responses either from a group of subjects in the form of one response from each subject or the same subject in the form of several responses–or both.

The question, as noted above, to be answered through the statistical analysis is: Is the variation in the subjective impression introduced by the controlled variables, sufficiently large compared to the random variation (error) to be able to conclude that the controlled variables have a real influence on the impression?

Another purpose of the statistical analysis is to provide an estimate of how generalisable the conclusions are. This is due to the need to establish a generalised statement regarding how the subject's responses are influenced by the controlled variables.

The following sections discuss the statistical design of listening tests and the data analysis.

6.1 Statistical experimental design

The previous section introduced the issue of uncontrolled variables in listening tests and the need for either 'control' or awareness of these variables in the planning of an experiment. This section introduces the statistical design of experiments, the purpose of which is to ensure that the conclusions of the experiment is actually related to the variables under investigation and not to some uncontrolled variables confounded with the primary variables.

It is noted that the discussion assumes that it is possible to control the various variables and, for example, randomise the presentations. This is often (always) the situation in the laboratory environment but can be quite difficult (impossible) in field experiments where other techniques must be used. The statistical design of experiments is a common area for almost all research disciplines, so a large number of books and papers have been published on this topic. The authors have found Box *et al.* [63], Hicks [162] and Montgomery [300] quite useful, and each contains numerous references for the interested reader. Many statistical analysis programs, like Statgraphics, have experimental design modules that will help design the experiment and apply the correct analysis of variance (ANOVA) model after the experimentation, for example. The discussion in this section is not detailed as this is a broad field, and only the main arguments and lines of reasoning are presented.

Kish [257] (Section 1.2) divides the variables in experimental research into four classes:

Explanatory variables These are the factors under investigation, and they are divided into two classes:

- **Independent variables.** These are variables controlled by the experimenter. The independent variables can be divided into two types according to how they are varied by the experimenter. The first type is termed *qualitative* or *categorial variables*. These variables represent variation in kind or type; examples would be 'Loudspeaker' with five levels (i.e. five different loudspeakers) or 'Programme' with four levels (i.e. four different programme sections). The second type is the quantitative variables, which can be continuously varied in magnitude by the experimenter; examples would be reproduction level in sound pressure level (SPL) (dB) or base angle (degrees) in a stereophonic set-up. See Chapter 5 for a more detailed discussion of the independent variables
- **Dependent variable.** This is the response(s) provided by the subject (i.e. 'basic audio quality' or 'perceived horizontal direction of the sound source'). See Chapter 4 for a more detailed discussion of the dependent variables.

The explanatory variables are defined by the formulation of the research question(s), which again are based on previous scientific experiments, theories, literature or simply a good idea.

Controlled variables These are the experimental variables known to the experimenter but not included in the explanatory variables. These variables could be blind versus sighted,[26] the

[26]In this context blind and sighted refer to whether or not the experimenter and/or subject are aware of the nature of the variable in question.

subject's fatigue, the position of the loudspeaker in the room, equalisation of loudness, and so on. Some of these are typically held at a constant level (e.g. the experiment is conducted 'blind', loudness differences are equalised and the room temperature is held constant) and are therefore not a specific part of the statistical planning. Some are controlled by including them in the experimental design and some others are controlled by randomisation.

Disturbing variables These are uncontrolled (unknown) variables that can or will be confounded with the explanatory variables. It is one of the major tasks of the experimental design to either move these variables into being controllable or apply a randomisation scheme that will break any systematic relationship between them and the independent variables. These variables are also termed *nuisance variables* in the literature (see for example Keppel and Wickens [253] and Toole [421]).

Randomised variables The variance caused by these variables will add to the error or residual variance if the experimental design employs a proper randomisation scheme.

The purpose of the experimental design is, as noted, to control the above variables and Hunter [173] classifies the design of the experiment into two categories:

- Treatment design
- Allocation of stimuli design.

The treatment design is the basic specification of the treatment structure, that is, which stimuli to use and how to administer them in the experiment, independently of the subject. The allocation of stimuli design then specifies how the stimuli are administered to the individual subject.

There are two basic types of treatment designs:

- Full factorial
- Fractional factorial.

The full factorial design includes all possible combinations of the independent variables. The statistical literature also refers to this as a factorial design, in which the variables are completely crossed. The number of stimuli in a full factorial design increases quite rapidly as it is determined by the following formula:

$$N = (\text{number of levels per variable})^{(\text{number of variables})} \tag{6.4}$$

The full factorial is the optimal design, when seen from a statistical point of view, as all stimuli can be randomly accessed and distributed to the measuring instrument (subject) as required and all variables can be statistically tested. However, given the (often) large number of stimuli, it is not always practically possible to use such designs, and the experimenter thus needs a design with fewer stimuli–the so-called fractional designs.

Let us consider an example in which the experimenter is interested in quantifying the differences between four loudspeakers[27] (A, B, C and D). The experimenter knows either by experience or from the literature that the characteristics of the listening room will have an influence on the assessment of the loudspeakers and further that the position of the loudspeaker within the listening room will also influence the assessment. The experimenter thus decides to conduct the listening test at four different test sites, each with a particular type of listening room, and in four positions in each listening room. By doing this, he or she hopes to obtain a more generalisable mean for each loudspeaker and further, if possible, to be able to quantify the impact of the influence of position and room. To simplify the example, the experimenter decides to use only one programme item. The experiment now includes one independent categorical variable: loudspeaker with four levels (the treatment), and two controlled variables: test site with four levels and position with four levels. The test site and position are typically referred to as blocking variables, which will be discussed in greater detail later in this section.

A full factorial design would thus include 4 (loudspeakers) $\times$ 4 (test sites) $\times$ 4 (positions) $=$ 64 stimuli to be assessed by each subject. However, two other constraints apply. Firstly, there is only time for assessing one loudspeaker in each of the four positions (so no time for moving the loudspeakers between positions). Secondly, the duration of the test is one week and the four test sites are only available for the same week, so the tests will have to run in parallel.

The question then is whether it is possible to conduct the test under these constraints and still produce meaningful results? The

[27]The reader should note that the use of this type of experimental design is not limited to loudspeakers. For example, loudspeakers could be replaced by low bit-rate codec type, position by programme and test site replaced by presentation order.

Position	Test site			
	1	2	3	4
1	A	B	C	D
2	A	B	C	D
3	A	B	C	D
4	A	B	C	D

Table 6.1: The first proposal for distribution of loudspeakers to test sites and positions. The row number indicates the position of the loudspeaker at each test site and the entries are the four loudspeaker types.

main problem is to control the influence of the test site (listening room) and the position effects and at the same time reduce the number of experiments to 16 stimuli (one loudspeaker in each of the four positions in four test sites).

First, it is necessary to have four identical versions of each loudspeaker, one for each test site, so it is assumed that these can be provided, and second, a fractional design must be employed.

A first proposal could be to ship all four versions of a loudspeaker to the same test site, as shown in Table 6.1. This design allows a quantification of the position effect by comparing the row means (averaging across test sites and loudspeakers); however, this assumes that the differences between positions are independent of both the loudspeaker and the test site as all loudspeakers and test sites are included in all row means. It is known, however, that different loudspeakers react differently in different positions in different rooms (see Olive *et al.* [333] and Bech [29]), so this design will not include all the known position effects. Another obvious problem with this design is that the average for the test site (column sums) is also the average for the loudspeaker type, so the differences between test sites cannot be separated from the differences between loudspeakers. The statistical term for this is '*confounding*', and in this case the test site variable is confounded with the loudspeaker variable.

A second proposal would be to use a so-called randomised design in which the loudspeakers are randomly shipped to and randomly positioned at each site. A possible outcome is shown in Table 6.2. The idea behind this design is that the random distribution across test sites and positions should tend to average out the influence of test site and

Position	Test site 1	2	3	4
1	C	A	D	A
2	A	A	C	D
3	D	B	B	B
4	D	C	B	C

Table 6.2: The second proposal for distribution of loudspeakers to test sites and positions. The row number indicates the position of the loudspeaker at each test site and the entries are the four loudspeaker types.

Position	Test site 1	2	3	4
1	B	D	A	C
2	C	C	B	D
3	A	A	D	B
4	D	B	C	A

Table 6.3: The third proposal for distribution of loudspeakers to test sites and positions. The row number indicates the position of the loudspeaker at each test site and the entries are the four loudspeaker types.

position on the individual loudspeaker means. However, it is seen that, for example, loudspeaker A is tested only in three of the four sites and in two of the four positions, which means that there will still be a confounding between the loudspeaker means and test site and position.

This suggests a third distribution of loudspeakers, as shown in Table 6.3. This design is called a *randomised block design* because the distribution of loudspeakers is now controlled for test site (all four types of loudspeaker occur once in all four test sites), and the randomisation only takes place within the blocks (the position of the loudspeaker in the room). The loudspeaker averages in this design will include all four test sites but not all positions (e.g. loudspeaker A is not tested in position 2).

So the best solution would seem to be a design in which both the position and the test site variables are controlled. Such a design is

	Test site			
Position	1	2	3	4
1	B	D	A	C
2	C	A	B	D
3	A	C	D	B
4	D	B	C	A

Table 6.4: A Latin square design for distribution of loudspeakers to test sites positions within test site. The row number indicates the position of the loudspeaker at each test site and the entries are the four loudspeaker types.

called a *Latin square design* and is characterised by having each loudspeaker type appear once and only once in each position and each site. The term *square* stems from the fact that these designs can only be used when the design is 'square', that is, the number of levels for each restriction (position and site) equals the number of treatments (loudspeakers).

A Latin square design is shown for the loudspeaker test in Table 6.4. The use of this design ensures that the loudspeaker means are not confounded with either positions or sites. The influence of test sites can be evaluated by comparing column means, and the influence of position can be evaluated by comparing row means. Again, the assumption is that the differences between the loudspeakers are independent of the position and listening rooms.

The statistical model for a subject's response can thus be extended:

$$Y_{t,k,m,i} = \mu + \eta_t + \alpha_k + \gamma_m + \epsilon_{t,k,m,i} \tag{6.5}$$

where $Y_{t,k,m,i}$ is the i'th score of treatment 't, k, m', μ is the overall mean of all treatment combinations, η_t is the loudspeaker effect for loudspeaker 't', α_k is the position effect for position 'k', γ_m is the test site effect for test site 'm', and $\epsilon_{t,k,m,i}$ is experimental error.

This model is termed an *additive model* and assumes that the influence of the variables can be added, implying that the difference between two loudspeakers is the same independent of which positions and rooms they have been evaluated in. This assumption is not valid, as shown by Olive [333] and Bech [29]. However, the Latin square is the best design under the given constraints, but it illustrates that a reduction in the number of experiments reduces the statistical information that can be tested. The reduction in the number of experiments, and thus statistical information, also has an impact on the statistical power of the experiment. This issue of statistical power is discussed later in this section.

The above situation had the same number of levels for each of the variables, which allow for application of a Latin square design. However, it is also possible to use other square designs even if the number of levels is not the same for all variables. In these cases, the so-called Youden squares are employed. For a more detailed discussion of Latin and Youden squares, see for example, Box *et al.* [63].

The use of a Latin square design has the consequence, as discussed above, that the statistical model is additive, so it is not possible to test any interaction between the independent variables (the concept of interaction is discussed in more detail in section 'Analysis of variance model for more than one variable'). The consequence of blocking an experiment is thus that some of the effects in the statistical model cannot be tested. The non-testable effects, however, can be predetermined for the different designs, so the basic design rule when designing blocked experiments is to 'sacrifice' interactions for the ability to block the experiment.

The primary goal of most experiments is to be able to statistically test the independent variables, and secondly, to test any interactions. The main issue for the experimenter, in case of blocking an experiment, is thus to identify which of the variables and/or interactions that could be used for blocking. To aid this task, it may be very useful to use a graphical representation of the experiment, the so-called design plots. These plots employ the idea that the levels of the independent variables can be viewed as coordinates in a space with dimensions corresponding to the number of variables. A two-variable experiment thus corresponds to the standard two-dimensional ($x - y$) space, and with two levels of each variable, four points in space are relevant ((0,0), (0,1), (1,0), (1,1)). Using this system, it is very easy to get an overview of the consequences of not conducting certain combinations of the variables. Normally it would become somewhat difficult to use this graphical system for more than three variables; however, Barton [24] gives a good introduction to the technique and further demonstrates how the system can be extended to more than three main variables. For a basic statistical discussion of fractional factorials, both Hicks [162] and Box *et al.* [63] provide detailed discussions of these types of experiments, so it will not be pursued further here.

The discussion above has been concerned with the treatment design. However, another important issue to consider when designing an experiment is the design for allocation of stimuli to the subjects.

There are two basic design types: between-subjects design and within-subjects design. In the between-subjects design, each subject receives only one of the stimuli specified in the treatment design, whereas in the within-subjects design all subjects receive all stimuli specified. A combinations of the two designs is also possible where, for example, the same within-subjects design is conducted at several laboratories using local subject groups. The laboratory variable thus becomes a between-subject variable and such designs are termed *mixed designs*. The statistical analysis of the two design types also differ and this is discussed in the Section 'Analysis of variance model for more than one variable'.

Keppel and Wickens [253] (Section 17) provide a good discussion of the advantages and limitations of the two design types, and they note that there are three main advantages of the within-subjects designs compared to the between-subject designs:

- More efficient use of the subject's time. Each subject will in principle have to be in the laboratory only once, so a large number of responses can be collected during a single visit. In contrast, the between-subjects design collects only one observation per subject.
- The differences between subjects will be more 'controllable' as seen from a statistical point of view. Each subject assesses all treatment conditions, implying that any differences in the subjects will apply to the entire stimulus set.
- The random error term will be smaller for the within-subject design compared to the between-subjects design. The reason is that while the error term for the between-subject design includes both the variation due to difference between the subjects plus the variation of the disturbing variables, the within-subject design only includes the variance from the disturbing variables.

These advantages explain why the basic within-subjects design is the most often applied design in perceptual evaluations. So the following discussion will focus on some of the problems such designs will present to the experimenter.

A basic within-subjects design is the completely randomised full factorial approach in which the subject(s) will be presented with, and assess, all possible combinations of the independent variables in a completely randomised order in the same session. This design ensures (or should ensure) that all disturbing variables become randomised and thus are part of the error variance term in the ANOVA model.

However, it is well known that subjects will tire during a session, that context effects will appear, and that other bias effects (discussed in Section 4.2.4) will influence the assessment of the stimuli as a function of presentation order. It is therefore often necessary to consider the presentation order within a session more carefully, and it must often be considered a restriction on randomisation within a session.

If the order of presentation was a restriction, it would be of interest to ensure that loudspeaker A was not always evaluated after loudspeaker B, loudspeaker C after D, and so on. This added restriction can be handled by using the so-called balanced Latin squares (or Williams squares according to Hunter [173]) in which the presentation order and the subject are the two main restrictions (instead of test site and position in Table 6.4) and, in addition, each treatment appears once and only once after each of the other treatments. This type of square can be constructed using the following rules:

- If n is even, construct the first row according to the following order: $1, 2, n, 3, n-1, 4, n-2, \ldots$
- The second row is constructed by adding 1 to the first row and taking modulo n.
- If n is uneven, two balanced squares are used. Each is constructed according to the rules for an even n, but the second square has the order in each row inverted.

Applying this scheme to a listening session in which four loudspeaker pairs (A–D) are to be evaluated in succession, a presentation order as shown in Table 6.5 is obtained.

The full factorial design discussed so far often presents a logistic problem caused by the rapid increase in the number of stimuli as a function of the number of variables and their levels. This means that it will quite often be necessary to divide the complete experiment into several blocks to keep the maximum session duration below 30–40 min. This raises the issue of how to distribute the stimuli across sessions such that the blocking will not be confounded with the independent variables. This problem is similar to the one discussed

Subject number	Presentation number			
	1	2	3	4
1	A	B	D	C
2	B	C	A	D
3	C	D	B	A
4	D	A	C	B

Table 6.5: A balanced Latin square design for the distribution of loudspeakers pairs (A–D) to presentation order for four subjects.

earlier with the distribution of loudspeakers to test sites and positions as restrictions (blocking) of the design. So a balanced Latin square design with session number and subjects as restrictions could be used to solve the problem. Another possibility is to use the so-called incomplete block designs (see for example Hicks [162]).

It is very important to carefully consider the 'treatment' and 'allocation of stimuli' designs before the experiment is conducted. There is, however, an additional third design parameter that must also be considered before conducting the experiment: that of the number of observations for each stimuli. The number of explanatory variables and the treatment design determine the number of stimuli. However, the number of observations per stimuli, which is equal to the number of subjects times the number of repetitions, can be determined when three questions have been answered:

- What is the smallest subjective difference that needs to be resolved?
- What is the variance of the ratings?
- What are the probability levels of Type I (α) and Type II (β) errors?

The answers to the first two questions can be based either on experience or a small-scale pilot test. The author's experience is that well-trained subjects typically produce scores with an associated variance of approximately 0.4 when using a 0–10-point scale with one decimal resolution (see for example Bech [27]). However, the third question requires some serious considerations by the experimenter. The following section introduces the concept of Type I and II errors and associated probabilities.

The basic purpose of scientific experimentation is to reject the null hypothesis (H_0), thereby showing that the alternative hypothesis (H_a) is likely (see Chapter 2 for an introduction to formulating a

True state	Outcome of evaluation: H_0 rejected	Outcome of evaluation: H_0 accepted
H_0 false	Correct decision	**Type II error** (probability $= \beta$)
H_0 true	**Type I error** (probability $= \alpha$)	Correct decision

Table 6.6: Definition of Type I and Type II errors in general statistical decision making.

hypothesis). In statistical terminology, the two hypothesis could be defined as shown below:

$$H_0 : \mu_1 = 5 \tag{6.6}$$

$$H_a : \mu_1 > 5, \tag{6.7}$$

where μ_1 is the mean for loudspeaker 1. The null hypothesis states that the mean value of the ratings of loudspeaker 1 is equal to 5 scale units on a scale of 0–10, and the alternative hypothesis is that it is greater than 5.

The purpose of the experiment is thus to obtain observations that support a conclusion to either accept or reject H_0. However, because of the variability in the experimental data, such a conclusion can only be made with a certain probability that it is wrong. Two types of errors can be made:

- Type I error with associated probability α. This is the error made when H_0 is rejected but is actually true.
- Type II error with associated probability β. This is the error made when H_0 is false and should be rejected but is accepted.

The various possibilities in statistical decision making are illustrated in the Table 6.6 (see for example Keppel and Wickens [253] (Section 3.3) or Lawless and Heymann [268] (Appendix 1)).

The probability of making Type I error is normally the main concern of the experimenter and is often the only level reported. The interest in Type I errors has its roots (some of them) in the legal arena, where the null hypothesis is that the person is innocent. Thus, most legal systems want to minimise the risk of sentencing an innocent person, corresponding to rejecting H_0, when it is actually true (Type I error). Certain legal systems also prefer to use a principle in which 'one must prove the guilt of the accused beyond any reasonable doubt' and are therefore willing to accept a higher risk of reporting 'not guilty', corresponding to accepting H_0 when it is in fact false (Type II error).

True state	Outcome of evaluation: Audible difference reported	Outcome of evaluation: No audible difference reported
Audible difference	Correct decision	**Type II error** (probability = β)
No audible difference	**Type I error** (probability = α)	Correct decision

Table 6.7: Role of Type I and Type II errors in statistical analysis of listening tests.

This discussion can also be transposed to audio, where a long debate once focused on the fact that optimising Type I error favoured the 'engineering approach' versus the 'musicians approach'. The issue at hand was the audibility of the differences between cable, amplifiers and other types of electronic equipment, where musicians argued that they could hear differences, and engineers, using their (perhaps limited) physical background, argued that there should (could) be no difference (see Leventhal [271] and comments [375]). The audio discussion corresponds to a table of decisions as shown in Table 6.7 for a null hypothesis of 'no audible difference'.

The audio discussion focusses on a very relevant issue, that is, whether one, the other or both of the error types should be favoured?

It is correct that the implications of the two error types can favour certain interests over another. For instance, in the example given above on the audibility of differences between cables, the Type I error represents the engineer's risk and the Type II error represents the musician's risk. Thus, it is important to consider which of the two error types is of primary importance for the given experiment.

In order to optimise or balance the risk correctly between the two types of errors, it is necessary to be familiar with the definitions and parameters that influence the α and β levels. The following discussion will hopefully help clarify these issues.

To explain the statistical definitions of Type I and II error concepts, the common steps in hypothesis testing should be considered first (Hicks [162] (Section 2.3)). However, before going into details, it is worth considering the basic idea behind statistical hypothesis testing.

One of the main purposes of statistical analysis is to estimate how general the results of the experiment are. Ideally, the conclusions of an experiment based on a limited sample of the population (loudspeakers, programmes, subjects, etc.) should be valid for the population in general. The statistical hypothesis is thus always formulated for the population in general, and the population statistics (mean, standard deviation, etc.) are denoted in Greek letters. The test of the hypothesis is, as noted, based on the statistics of a limited sample of the population. These statistics are termed *test* or *sample statistics*, and they are denoted by uppercase Latin letters. The general idea is thus to test whether the test statistics of the limited sample can be concluded to be representative of the population statistics defined by the hypothesis.

The standard steps in hypothesis testing are thus:

- Establish the hypothesis. This was done previously, and according to it, the null hypothesis (H_0) is that the population mean rating of loudspeaker 1 is equal to 5 and the alternative (H_a) is that it is greater than 5.
- Set the significance level of the test α (equal to the probability of a Type I error). A typical level is 0.05 (see Cowles and Davis [100] for a discussion of why 0.05 is often used).
- Choose a test statistic. The obvious choice would be the mean of the ratings of loudspeaker 1 ($\overline{Y}_1$). However, the test statistic will be compared to a normalised distribution, so a normalised version is a better choice. If the hypothesis is true, then $\overline{Y}$ is normally distributed with a mean $\mu_1 = 5$ and standard deviation $\frac{\sigma}{\sqrt{n}}$, where σ is the standard deviation of the population. If σ is unknown, then the sample standard deviation S_1 is used instead and the test statistic then follows a Student's t-distribution. The test statistics is thus given by:

$$t_{\text{obs}} = \frac{\overline{Y}_1 - \mu_1}{\sqrt{S_1^2 \times \frac{1}{n}}}, \tag{6.8}$$

  where n is the number of observations in the sample. The variance of a series of observations is calculated as shown in Equation 6.14.
- Determine the sampling distribution of the test statistic when H_0 is true. The t-test statistics follows a Student's t-distribution with $n - 1$ degrees of freedom. The degrees of freedom are equal to the number of observations that are free to vary independently. So in this case there are n observations, but one of these has been used to estimate the mean value, as a result, only $n - 1$ are free to vary.
- Set up a critical region where H_0 will be rejected in ($100 \times \alpha$ %) of the samples when H_0 is true. The cut-off value is determined via a Student's t-distribution using $n - 1$ degrees of freedom and the α level.
- Choose a random sample of n observations, compute the test statistic, and if it is higher than the cut-off value, reject H_0, otherwise accept H_0.

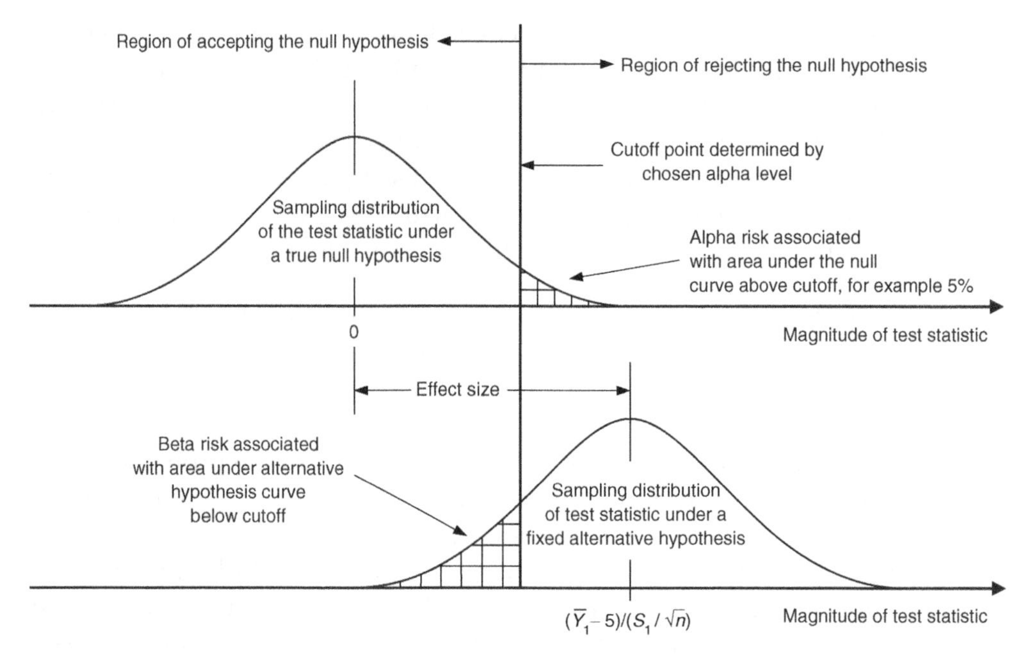

Figure 6.1: Definition of Type I (α) and Type II (β) errors in statistical testing. See text for a further discussion.

The procedure is illustrated in Figure 6.1. The top part shows the sampling distribution assuming H_0 is true. The critical region is defined by the cut-off value on the basis of the chosen α level such that the area under the curve above the cut-off value represents $100 \times \alpha$ percent of the distribution. The test statistic is then calculated and if the value is higher (in a so-called one-tailed test) than the predetermined cut-off level, H_0 is rejected with a $100 \times \alpha$ percent probability of being wrong.

Now assume that the null hypothesis is rejected, which means that a true sampling distribution exists with a higher mean value, as shown in the lower part of Figure 6.1. It is seen that if the null hypothesis is accepted on the basis of the previously determined cut-off point (i.e. the value of the test statistic is smaller than the cut-off value), there is a certain probability β, corresponding to the area under the alternative distribution, to the left of the cut-off point, that this value of the test statistic would be observed under the alternative distribution. Many statistical text books use the term power in connection with the Type II error level, and power is defined by:

$$\text{Power} = 1 - \beta \tag{6.9}$$

Power is thus the probability of rejecting the null hypothesis H_0 when it is false (see Table 6.6) or, in the audio relation discussed previously, the probability of rejecting the hypothesis 'no audible difference' when in fact there is an audible difference (see Table 6.7).

The common steps in the hypothesis testing just discussed only considers the α value. If the β value is of importance (and it should be) and the experimenter concludes that it is too large, there are three variables that can be used to control the weighting of probabilities for Type I and Type II errors:

Probability of Type I error, α. It can be seen in Figure 6.1 that changing the α level, for example, to a higher value (move the cut-off point to the left) will decrease the β level.

Number of observations, n. The Student's t-distribution will be more narrow for an increase in the number of observations, which means that the overlap between the two distributions will decrease and thus decrease the β level.

Effect size of interest. It can also be seen that if the effect size is increased for a given α level, it follows that the β level will decrease.

The α level is usually set at either 5 % or 1 % on the basis of the traditions within the field (see Cowles and Davis [100] for a discussion of choice of α level), and pilot experiments can often provide initial estimates of the effect size and variance of the population. This leaves the number of observations as a 'free' variable for controlling the β level. As an example, assume that the experimenter wants to detect an effect size of 0.5 scale units on the rating scale and that α is 0.05, β is 0.10, and an estimate of variance of ratings is 0.4; the following procedure can be used for calculating the necessary number of observations (Hicks [162] (Section 2.5)):

- The variance of the distribution is now known, so the Student's *t*-distribution used above is replaced by the standard normal distribution $N(0, 1)$ such that the value of the test statistic under H_0 can be calculated by

$$N(0,1)_{1-0.05} = \frac{\overline{Y}_1 - 5}{\sqrt{0.4 \times \frac{1}{n}}} = 1.645 \qquad (6.10)$$

- The same applies to the test statistic under H_a:

$$N(0,1)_{0.10} = \frac{\overline{Y}_1 - 5.5}{\sqrt{0.4 \times \frac{1}{n}}} = -1.28 \tag{6.11}$$

- Solving the two equations for n provides the number of observations needed to obtain the requested α and β levels ($n \sim 14$).

The effect size was specified directly in scale units in the example above, however, it is often more useful to have the effect size specified in a measure that is independent of the experimental details such as scale units and number of observations. An often used measure is the standardised difference between means (see for example Keppel and Wickens [253] (Section 8.1)):

$$d_{12} = \frac{\overline{Y}_1 - \overline{Y}_2}{S_{12}}, \tag{6.12}$$

where $\overline{Y}_1$ and $\overline{Y}_2$ are the mean values of groups one and two and S_{12} is the pooled standard deviation of the scores in the two groups:

$$S_{12} = \sqrt{\frac{SS_1 + SS_2}{df_1 + df_2}}, \tag{6.13}$$

where SS_1 and SS_2 are the sum of squares (see Equation 6.55 for a definition of sum of squares) for groups one and two with corresponding degrees of freedom df_1 and df_2.

The above example for calculation of sample size assumes that the approximate effect size was known, for example, from a pilot experiment. However, if this is not the case, it would be useful to have the power calculations as a function of effect size, as well as number of observations. Keppel and Wickens [253] (Section 8.3) have a discussion of this and present a useful set of formulas for performing these calculations.

Calculation of power for a given design as a function of the number of observations and/or effect size is highly recommended as it provides a good overview of the possibilities for optimising the design. Staffeldt [398, 399] presents good examples of this strategy for listening tests on loudspeakers. Calculations of power for various designs can be performed by a number of software packages (see for example `http://www.ncss.com/` for a commercially available program and http://members.aol.com/johnp71/javastat.html for freely available programs). Thomas and Krebs [417] have an excellent review of software (1997) for power analysis[28].

Once the experimenter has decided on a treatment design, determined the number of observations per stimuli, and designed the allocation of stimuli to the subjects, the experiment is, statistically speaking, ready for execution (the other details, as discussed elsewhere in the book, should of course also be taken care of). After collecting the observations, the statistical analysis is performed, on the basis of which the conclusion to either accept or reject H_0 will be made. The next section discusses the statistical analysis of observations.

6.2 Statistical analysis

6.2.1 Classification of data type

Before conducting the statistical analysis, the type or properties of the collected data should be considered. Data fall into two categories, namely, *quantitative* and *categorical* data. Quantitative data reflects a continuous quantity or the degree of a given property, and an example would be ratings reported on interval scales (or scales that are assumed to have interval properties). Such data are usually normally distributed and analysed using the standard t-test, ANOVA, and so on, and this class of methods are often referred to as *parametric* statistics. Categorical data are based on categorical scales such as 'prefer' versus 'not prefer' or ranks like the 9-point hedonic scale, and they are often binomially or chi-squared distributed. The statistical techniques used to analyse such data are often referred to as *non-parametric* statistics.

However, when seen from a purely statistical point of view, such a division of the statistical methods does not make sense as many of the so-called parametric methods can also be used to analyse categorial data and non-parametric methods can be used to analyse quantitative data as well. These multi-purpose methodologies, however, are outside the scope of this book, so for the basic analysis of data it is still of importance to know whether the data must be considered as quantitative or categorical.

[28]The programmes discussed have of course been updated, but the principles used for the discussion are still relevant and thus might be useful for analysis of present-day programmes.

So the question is when can data from an assumed interval scale be considered continuous? This is discussed in Section 4.2.1, where the conclusion was that responses based on interval scales including 11 or more levels could be analysed by parametric procedures provided the statistical assumptions are not violated severely. This means that results based on the 10-point fidelity scale and 5-point impairment scale, both of which have a one decimal place resolution leading to 100- and 50-point scales, respectively, could be analysed using parametric statistics.

The 10- and 5-point rating scales are those that are most frequently applied in listening tests, so the following chapters will be more focused on parametric methods than categorical methods. Examples of such scales are presented in Table 4.3 and Figure 4.7.

Another important aspect that needs to be considered before the analysis is whether the data can be divided into dependent and independent variables. A typical situation is that the experimenter wishes to examine the influence of a set of independent variables on the dependent variable (e.g. the subject's impression of sound quality). In this case, the so-called dependence techniques must be applied. Another situation is when the experimenter asks the subject a series of questions (e.g. in the form of a questionnaire) and subsequently wants to examine the relationship between the questions. In this case, all the dependent variables need to be analysed simultaneously, and the interdependence techniques must thus be applied.

The main purpose of most listening tests is to explore the relationship between a number of independent variables and a single dependent variable. The following text will thus focus on the dependence techniques; however, readers interested in interdependence techniques such as principal components analysis (PCA), multidimensional scaling (MDS) and factor analysis (FA) should consult, for example, Hair *et al.* [155].

Before applying a dependence technique, the number of dependent variables that an analysis should include must be considered. The typical situation is that the relationship between the dependent and independent variables is analysed for each dependent variable separately. This calls for the so-called univariate analysis methods such as standard ANOVA and regression analysis. However, if, for example, the dependent variables belong to the same group of variables (e.g. spatial impression) and thus, logically, could interact, they must be analysed together and the so-called multivariate techniques need to be applied. Most listening tests will only ask the subject a single question, so the univariate analysis technique is most frequently applied, and the following text will thus be primarily focused on this technique.

6.2.2 *Levels of analysis*

A typical statistical analysis is divided into three levels:

Descriptive level This includes checking statistical assumptions, identifying outliers and summarising the observations in the form of means and associated variances.

Inferential level This includes estimating whether the independent variables have a real influence on the dependent variable(s), that is, rejecting or accepting the experimental hypothesis.

Measurement level This includes estimating the relationship between the independent and dependent variables.

A complete overview of the possibilities is presented in Figure 6.2. The descriptive and inferential levels are the most relevant for analysis of results from listening tests. The measurement level is used only at a later stage of the analysis when, for example, the relationship between the subjective impression and physical measures is investigated. The following sections will thus discuss only the dependence techniques and the corresponding descriptive and inferential levels.

The reader should note that although the descriptive and inferential levels are listed and described as separate and sequential steps, real data often requires an iterative process including both levels a number of times. The typical sequence is that the descriptive analysis is conducted, a model is established and fitted to the data in the inferential process and then a descriptive analysis is performed on the residual, a new model is established and the new set of residuals is analysed, and so on.

6.2.3 *Descriptive level*

The purpose of the analysis at this level is to obtain an overview of the collected data. This includes checking the statistical assumptions, identifying outliers and summarising the observations in the form of mean values and associated variances.

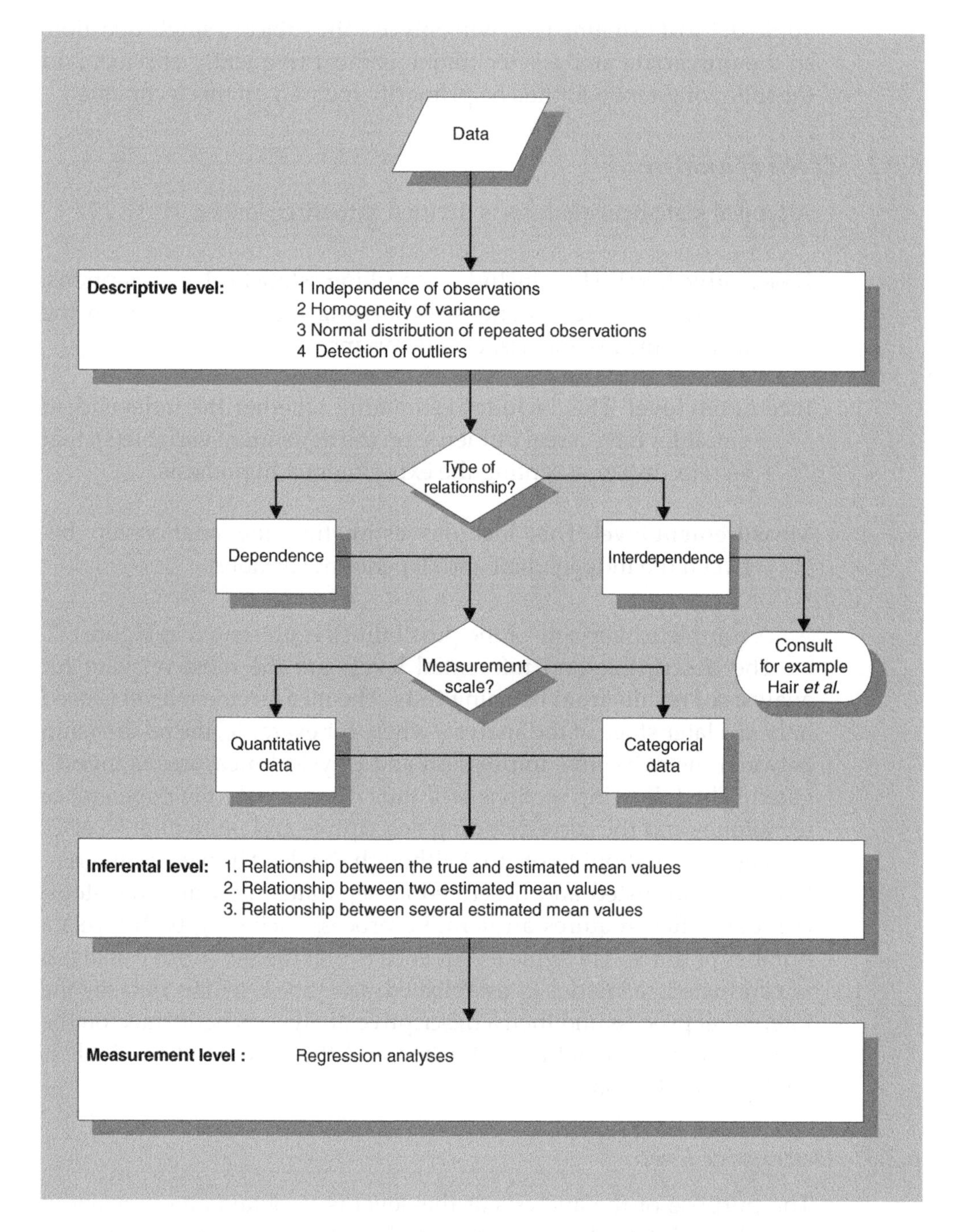

Figure 6.2: Flow chart for typical statistical analysis.

In the first section of this chapter, it was assumed that a score from a subject included a part determined by the controlled variables plus an error term and further that the error term was normally distributed. If the assumption is correct, it follows that repeated scores can be summarised by a mean value and variance or standard deviation for each experimental treatment. It is also assumed, although implicit, that the individual responses are independent. Furthermore, if there are several independent variables, it is assumed that the variances of the means are equal for the tested combinations of the independent variables. Finally, the issue of 'outlying' observations must be discussed. Outliers are observations that for one reason or another deviate significantly from other observations obtained under similar conditions.

Thus, there are four major statistical issues that need to be addressed before any analysis is conducted:

- Independence of observations
- Homogeneity of variance across tested combinations of independent variables
- Normal distribution of repeated observations
- Detection of outliers.

These four issues are discussed below in the order of importance according to Belle [434] (Section 1.4). See also Hair *et al.* [155] (Chapter 2) and Keppel and Wickens [253] (Section 7) for further discussions of the four issues.

Independence of observations Independent observations require that all the bias effects that were discussed in Section 4.2.4 are controlled by means of the statistical experimental design, represented by dummy variables in the statistical model and tested for influence. A good example of a dummy variable is the order of stimulus presentation. This variable can be used to examine the observations for influence of time, and if found to be a significant variable, it can be included as a so-called covariate (see Section 6.2.4).

In general, it is very difficult to test for the independence of the observations after they have been collected, so it is very important that all possible bias and other variables are considered before the experiment is started. The consequence of having correlated observations is discussed, for example, by Belle [434] (Section 3) and it can be shown that the true Type I error (α) increases with increasing degree of correlation. The reported significance level will thus be too optimistic if not corrected for any correlation. The general rule is to include all known variables in the experimental design and then use a randomised presentation order to reduce the influence of the unknown variables.

Homogeneity of variance across tested combinations of independent variables

This is one of the basic assumptions for a classical analysis of variance. It originates from the fact that the statistical theory and computations become much simpler if this is fulfilled. There are solutions if the assumption must be rejected, but it will then require more elaborate statistical procedures.

In the statistics literature, the term *cell* is often used. It refers to the group of observations belonging to a specific combination of independent variables. The 'coordinates' of the cell are defined by the levels of the independent variables, and the observations obtained for this combination of levels are 'in' the cell. The assumption of homogeneity of variance across tested combinations of independent variables thus is that the observations in each cell should originate from distributions with equal variance.

The degree of heterogeneity of variance can be tested by a number of procedures. One was formulated by Levene [270], and it is standard in most statistical programmes (see for example [396]). Keppel and Wickens [253], in Section 7.4 of their book, discuss a number of different tests and possible transformations of the data if the assumption is not met. Another procedure for comparing the within-variance between cells is to use a so-called spread-versus-level plot, where the spread per cell is plotted against the tested combinations (levels) of the independent variables (see for example Figure 6.15). The plots can also be used to estimate possible transformations of the data if homogeneity is not present (see for example Hair *et al.* [155] (Chapter 2) or Belle [434] (Section 1.12)).

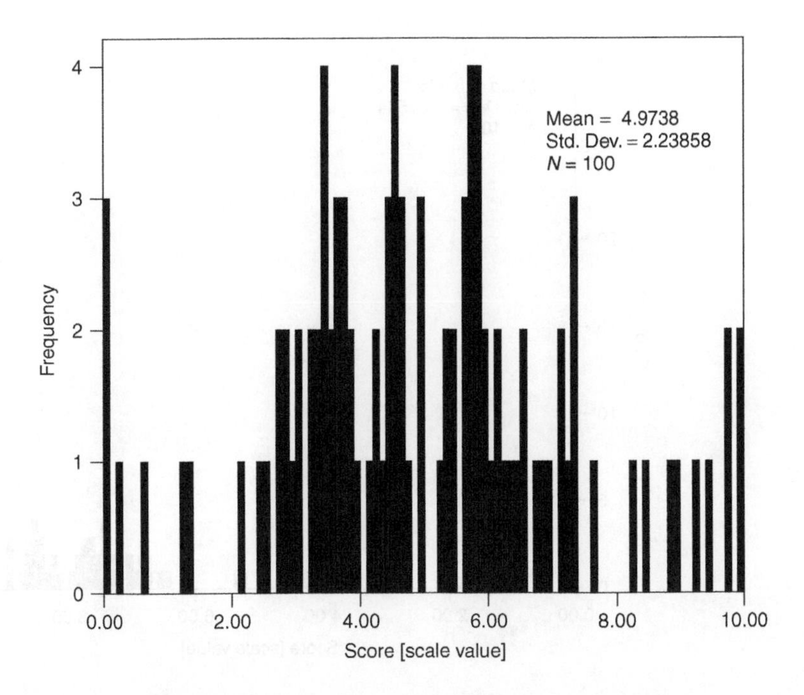

Figure 6.3: Histogram of 100 repeated assessments by subject 1 providing normal distributed scores. The subject assessed the timbral quality of the reproduced sound from one loudspeaker and one programme on a scale of 0 to 10, including one decimal place. The histogram is based on an interval width of 0.1 scale units.

Normal distribution of repeated observations It is very important to note that the normality assumption is related to repeated observations for combinations of the independent variables, not across the independent variables themselves. This corresponds to an analysis of the residuals or error term of the analysis of variance model (ANOVA) (see Equation 6.63). The distribution of the original data, for say a group of four loudspeakers, will be determined by the four mean values and can or should not be normally distributed, but the repeated ratings for each loudspeaker should be.

The examination of the statistical distribution of the observations should preferably be performed both graphically and

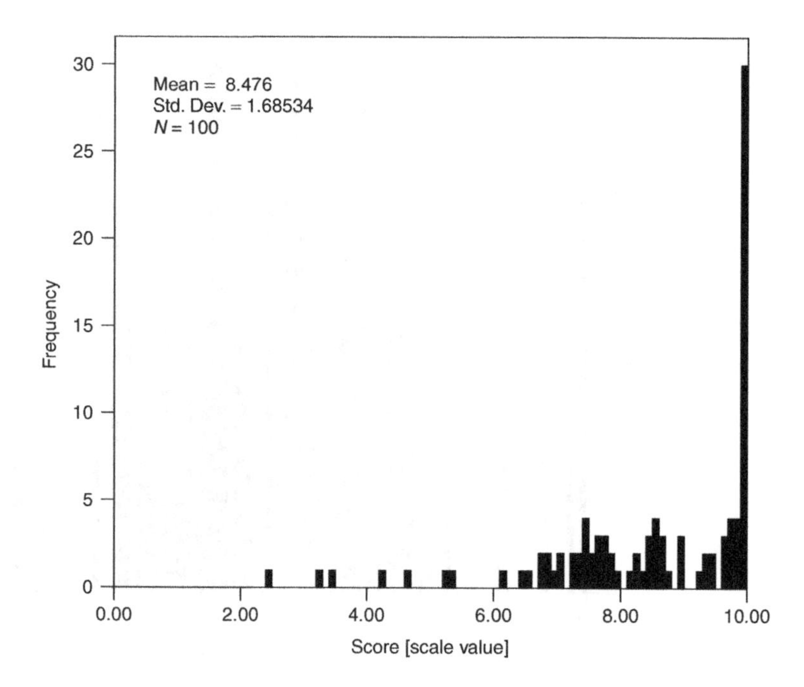

Figure 6.4: Histogram of 100 repeated assessments by subject 2 providing non-normal distributed scores. The subject assessed the timbral quality of the reproduced sound from one loudspeaker and one programme on a scale of 0 to 10, including one decimal place. The histogram is based on an interval width of 0.1 scale units.

statistically. The steps and different techniques are illustrated by two examples in the following text. Figures 6.3 and 6.4 show frequency distributions (histograms) of the scores for two hypothetical subjects. The subjects were both asked to rate the timbral quality of reproduction of a loudspeaker for one programme on an interval scale of 0 to 10 with one decimal place. The test was repeated 100 times. Note that an assessment is not likely to be repeated 100 times in a real experiment, but here it serves as an illustration.

The histogram for subject 1, shown in Figure 6.3, seems to be following a normal distribution, and this is confirmed by the test results in Table 6.8. A number of statistical tests exist for testing the deviations from normality. The two most common

	Kolmogorov–Smirnov			Shapiro–Wilk		
	Statistic	*df*	Sig.	Statistic	*df*	Sig.
Subject 1	0.06	100	0.200	0.979	100	0.111
Subject 2	0.183	100	0.000	0.838	100	0.00

Table 6.8: Test of normality for hypothetical subjects.

(i.e. they are included in most statistical programs) are the Kolmogorov–Smirnov and Shapiro–Wilk. The hypothesis is that the distribution is normal, and this can only be rejected for subject 1 at fairly high probability levels ($p = 0.20$ and $p = 0.111$). So it can be concluded that the distribution is normal, and the results can be represented by the estimated arithmetic mean value and standard deviation. These are found in Table 6.9 together with a number of other useful statistics:

95 % confidence interval. The confidence interval is further discussed in Section 6.2.4.

5 % trimmed mean. The trimmed mean is computed by ordering the values from smallest to largest and deleting the 5 % top and bottom observations, and then recalculating the mean value. The trimming ensures that unusually small or large values will not affect the mean value. The distribution in Figure 6.3 indicates that there could be too many observation of the value 0; however, the trimmed mean is similar to the arithmetic mean value, so the higher number of '0's does not influence the mean calculation. If the trimming procedure has a noticeable (significant) effect on the mean value, the data should be examined more carefully.

Median. The median is calculated by ordering the samples from the largest to the smallest and the middle value is then the median. This value is used when the distribution is not normal. Both the arithmetic mean and the median are measures of central tendency, and if the distribution is perfectly symmetric they are identical. If the median and mean are different, the symmetry of the distribution should

	Descriptor		Statistic	Std. error
Subject 1	Mean		4.97	0.22
	95 % Confidence interval	Lower bound	4.53	
	for Mean	Upper bound	5.42	
	5 % Trimmed mean		4.97	
	Median		4.70	
	Standard deviation		2.24	
	Interquartile range		2.77	
	Skewness		0.14	0.24
	Kurtosis		0.10	0.48
Subject 2	Mean		8.48	0.17
	95 % Confidence interval	Lower bound	8.14	
	for mean	Upper bound	8.81	
	5 % Trimmed mean		8.66	
	Median		8.70	
	Standard deviation		1.69	
	Interquartile range		2.51	
	Skewness		−1.35	0.24
	Kurtosis		1.94	0.24

Table 6.9: Statistical descriptors of the results for subjects 1 and 2.

be checked. This is done by examining the skewness and kurtosis values.

Standard deviation. The standard deviation is the square root of the sample variance calculated by

$$S = \sqrt{\frac{\sum_i (Y_i - \overline{Y})^2}{n-1}} \tag{6.14}$$

where Y_i is the ith observation, $\overline{Y} = \frac{1}{n} \times \sum_i^n Y_i$ and n is the number of observations in the sample.

Percentiles and interquartile range. The median value is the value at which 50 % of the observations are higher and 50 % are lower. The median is also termed the 50th *percentile or the second quartile.* Percentiles divide the distribution into groups of specific sizes. The 25th percentile is thus the value at which 25 % of the observation falls below and 75 % falls above. The interquartile range is the difference between the 25th and 75th percentile.

Skewness and kurtosis. Skewness measures whether a distribution has longer or shorter tails than those of a normal distribution and kurtosis measures whether the peak is higher or lower than that of a normal distribution. The values will fluctuate around zero even for a normal distribution, and critical ranges can be calculated as shown below (see for example Hair *et al.* [155] (Chapter 2)):

$$\text{skewness} \leq z_{\frac{\alpha}{2}} \times \sqrt{\frac{6}{n}} \vee \text{skewness} > z_{1-\frac{\alpha}{2}} \times \sqrt{\frac{6}{n}} \qquad (6.15)$$

$$\text{kurtosis} \leq z_{\frac{\alpha}{2}} \times \sqrt{\frac{24}{n}} \vee \text{kurtosis} > z_{1-\frac{\alpha}{2}} \times \sqrt{\frac{24}{n}} \qquad (6.16)$$

where z follows a unit normal distribution ($N(0,1)$) and n is the number of observations in the sample. A positive skewness value larger than the indicated value is significant and indicates too long a right tail and a significant negative value indicates too long a left tail. A significant positive kurtosis value indicates that the tails are too long and a significant negative value that they are too short. In the example shown in Table 6.9, for subject 1, both measures are within the critical range.

A number of visualisation techniques exist for further examination of the distribution of data. The histogram has already been mentioned and a supplement would be a so-called stem-and-leaf plot, as shown in Figure 6.5 for the observations in Figure 6.3. A stem-and-leaf plot is an extended version of the histogram shown in Figure 6.3. The observations are split into two columns; a stem and a leaf and extreme (outliers) values (none present in Figure 6.5) will be put into separate stems at the beginning and end of the plot. It contains a 'frequency' column that is the number of observations for the stem and leaves in the same row. This plot can be used to examine whether the subject is using the same leaf value constantly–which could indicate something about the (mis)usage of the scale.

```
Subject 1 Stem-and-leaf plot

 Frequency    Stem and leaf

     4.00        0 .  0002
     1.00        0 .  6
     2.00        1 .  23
      .00        1 .
     2.00        2 .  14
     6.00        2 .  577889
    10.00        3 .  0022334444
    11.00        3 .  55666777889
     7.00        4 .  1223444
    11.00        4 .  55556667999
     5.00        5 .  23344
    13.00        5 .  6667777888899
     6.00        6 .  011234
     5.00        6 .  55789
     6.00        7 .  112333
     1.00        7 .  6
     2.00        8 .  24
     2.00        8 .  89
     2.00        9 .  24
     3.00        9 .  779
     1.00       10 .  0

 Stem width:       1.00
 Each leaf:        1 case(s)
```

Figure 6.5: Stem-and-leaf plot of data from subject 1 providing normal distributed ratings.

Another useful plot is the box plot, shown in Figure 6.6 for the observations in Figure 6.3. The different parts of a box plot are explained in Figure 6.7. The box plot (also called a *box-and-whisker plot*) provides a quick overview of the distribution of the data including outlying and extreme observations. The box represents the range occupied by 50 % of the observations plus an indication of the median. If the median is in the middle of the box, it indicates a symmetrical distribution and the standard mean value can be used instead. Observations more

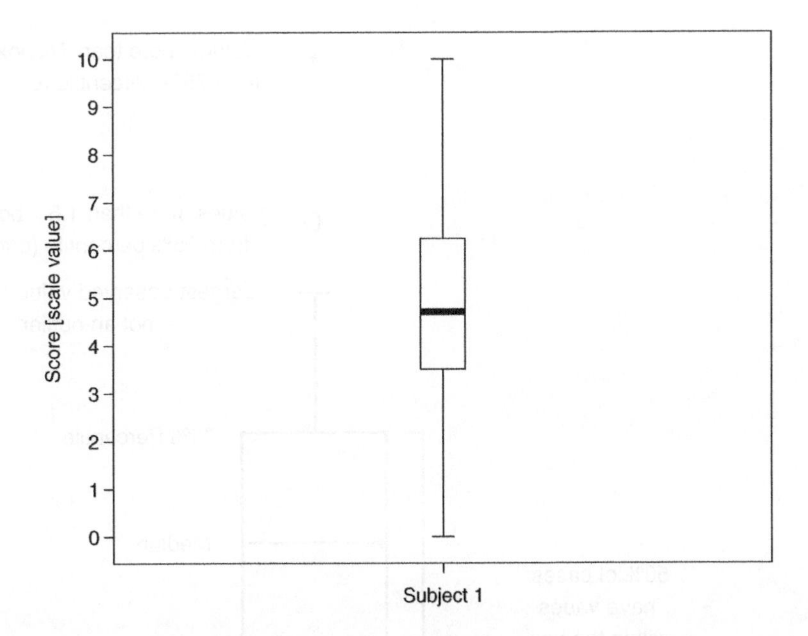

Figure 6.6: Box plot of data from subject 1 providing normal distributed ratings.

than 1.5 times the box length from the 25 % or 75 % percentile, termed *outliers*, are marked and hence are extreme observations corresponding to more than three times the box length from the 25 % and 75 % percentiles.

The box plot in Figure 6.6 shows, as expected, that the distribution is symmetrical and that there are no outliers.

The distribution of the ratings of the second hypothetical subject is shown in Figure 6.4. The first impression is that the distribution is not normal, and this is confirmed by the test of normality shown in Table 6.9. The histogram suggests that the subject used the score 10 for a large number of stimuli and rarely found a score below 6 appropriate. The lack of normality means that it is more appropriate to represent the central tendency by the 5 % trimmed mean or median value. The skewness and kurtosis are both outside the critical range, which confirms the lack of symmetry and further that it is the left-hand tail that is too long. The stem-and-leaf plot in

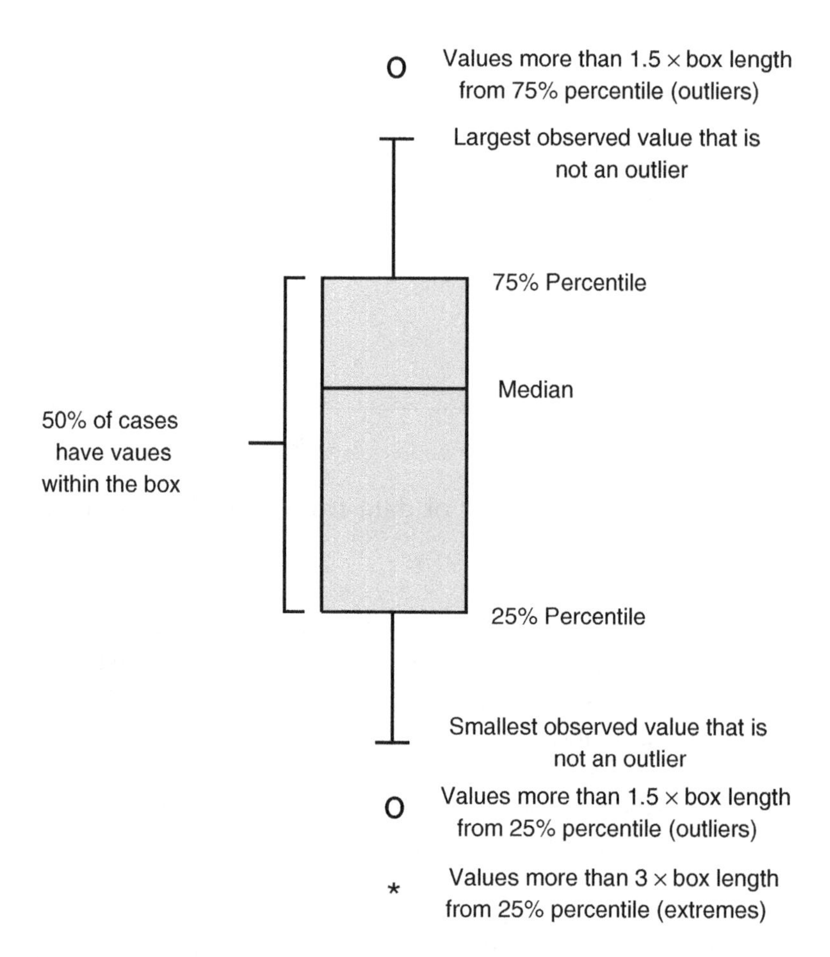

Figure 6.7: Description of the different parts of a box plot.

Figure 6.8 shows that there are three outliers and confirms that a large number (28) of 10s have been awarded. The box plot in Figure 6.9 again confirms the skewness of the distribution and identifies the three outliers so they can be further investigated.

```
Subject 2 Stem-and-leaf plot

 Frequency    Stem and leaf

     3.00 Extremes      (≤3,4)
     1.00          4 .  2
     1.00          4 .  6
     2.00          5 .  23
      .00          5 .
     2.00          6 .  14
     6.00          6 .  577889
    10.00          7 .  0022334444
    11.00          7 .  55666777889
     7.00          8 .  1223444
    11.00          8 .  55556667999
     5.00          9 .  23344
    13.00          9 .  6667777888899
    28.00         10 .  0000000000000000000000000000

 Stem width:      1.00
 Each leaf:       1 case(s)
```

Figure 6.8: Stem-and-leaf plot of data from subject 2 providing non-normal distributed ratings.

Detection of outliers. Detection, identification of the cause and handling of outliers represent interesting problems that must be dealt with before the main analysis as they can have strong influence on the outcome of the analysis.

The tradition in statistics is to label an observation 'outlier' if its value is higher than 2.5 times the standard deviation. This assumes a symmetrical distribution of observations, so it is recommended to also use the box plot as a further aid. The box plot shown in Figure 6.9, however, suggests that values more than 1.5 times the box length from the 25 % and 75 % percentiles are labelled *outliers* and values more than three times the box lengths are termed *extremes*. The actual label assigned to deviating observations is of course not important; however, the mathematical definition of 'deviating' is and should be carefully reported.

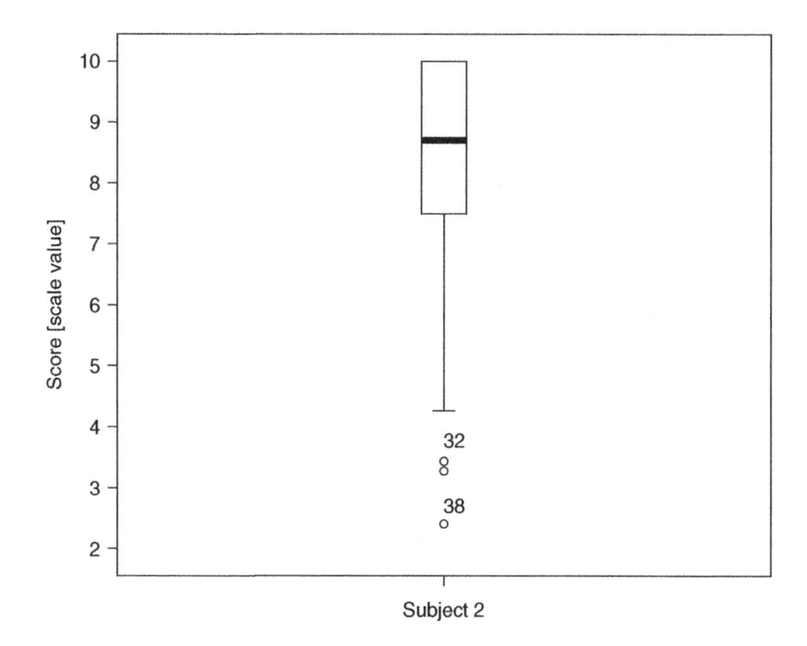

Figure 6.9: Box plot of data from subject 2 providing non-normal distributed ratings.

Once observations have been identified as outliers, it is important to classify them according the assumed cause. In general, outliers are caused by the following four factors (see Hair *et al.* [155]):

Procedural error. For example, the observation has been coded wrongly by the experimenter, the subject pressed a wrong button, and so on. Such data should be coded as missing values in the final data file.

Extraordinary event that can be explained. For example, if the subject, when interviewed, explains that he really disliked the specific type of music and thus rated all examples of the programme very low. The type and magnitude of the problem should determine whether the observations should be retained or not. If, as in the example, it is a subject who has a 'deviating' taste in music, it is recommended to exclude the subject or at least perform two sets of analyses: one with and one without the subject.

Extraordinary event that cannot be explained. For example, if the subject is not able to explain why a specific observation of a series received 'outlying' values. This can be a very difficult situation: On one hand the observations are recorded as outliers; however, if deleted, it restricts the generality of the population (e.g. subjects, music, loudspeakers, etc.) in question. Again, it is advisable to conduct two sets of analyses to quantify the magnitude of the influence of the outlying observations.

For specific combinations of variables. Observations that are within the normal range for the individual independent variables; however, for a certain combination of variables they become outliers. This is also a very difficult situation, similar to the previous one. In general, if there is no 'extraordinary' event, the observations should be retained.

Thus, handling outliers is not always straightforward and could require a substantial amount of work. However, it is very important that the outlier problem is handled before the main analysis is conducted.

The discussion above has been concerned with the various assumptions that must be met for the statistical analysis to be valid. It cannot be emphasised enough how important this part of the analysis is, and it should be mandatory for all experimenters to examine the data set for conformance with the statistical assumption before any statistical analysis is conducted.

So the obvious questions are: Do the results of standard listening test normally meet the assumptions and what should be done if they do not?

The authors have conducted and analysed a substantial number of listening tests, and in most cases the statistical assumptions have been fulfilled. However, in some cases (see for example [471]) they are not fulfilled; it is then crucial to be able to test the usefulness of alternative routes for the analysis. Many papers have been published on specific details of a certain deviation, but it is often difficult to relate such results to more general cases. There are, however, a few exceptions such as Cochran [93], Box *et al.* [63], Stonehouse and Forrester [409] and Miller [296] that include specific guidance.

Cochran discusses the effects of non-normality, outliers, homogeneity of the variance and correlation between observations. He concludes that the general consequence of deviations from the assumptions is that the reported significance levels (p values) are too optimistic. Cochran concludes that the most important deviations are extreme skewness, presence of outliers, relationship between the variance and, for example, the mean and, of course, errors or anomalous behaviour in the experiment or parts of it. Box [63] discusses random sampling and related independence of the observations and the effects of deviations from normality. Stonehouse and Forrester specifically examine the t-test and its categorical version–the U-test (see Equation 6.48 for a definition)–for robustness under combinations of different types of violations. They conclude that the t-test is very robust to most violations except when the distributions have different numbers of observations, different variances and severe skewness. In general, the t-test is very robust to departure from normality of the distributions as long as they are not skewed but are equal in both variance and form. The categorical alternative is found to be surprisingly sensitive to departures from normality and equal variances, although this happens for situations other than where the t-test is sensitive. The conclusion is that the U-test is not the categorical alternative to the t-test as often claimed in the literature. Miller [296] has a very instructive and comprehensive discussion of the consequences of departure from the normality assumption. Another route is to use newer and more general methods such as the general linear model (GLM) or mixed models, which deal with data that do not follow the standard assumptions. These models are, however, outside the scope of this text and readers are referred to Keppel and Wickens [253] (Section 14) for GLM and Brown and Prescott [74] for mixed models.

It is not within the scope of the present text to give specific guidance as to what alternative routes should be taken as the degree and causes of violation are specific to each individual case. Further action is thus very much based on a specific evaluation and test of the consequences that other possibilities include. Interested readers are referred to the references cited above.

However, if the examination of the assumptions leads to the conclusion that the violations are unacceptable, there are at least two alternative routes. The first is to use a method of analysis that is independent of the underlying distribution, as discussed in Section 6.2.4.

The other possibility is to transform the data such that the resulting distribution fulfils the statistical assumptions. The topic of transformation of data is large, complicated and somewhat out of the scope of this book. However, if a transformation is requested, a number of good references exist that the reader should consult [155, 163]. A few procedures are known in which a *priori* transformation of the responses is needed. This includes the magnitude estimation procedure (see Section 4.2.1 for details of the procedure), which produces responses that need to be logarithmically transformed to obtain a normal distribution. Other typical transformations are shown in Table 6.19.

6.2.4 *Inferential level*

The purpose of the analysis at this level is to be able to make conclusions regarding the relationship between the true population mean and the estimated sample mean and to compare two or more estimated sample means of the experiment. This subsequently translates into concluding whether the experimental variables have a significant influence on the perception reported by the subjects. The analysis can be based on either parametric or categorical procedures, depending on the conclusions of the analysis at the descriptive level. Note that the term *mean* represents the central tendency of the distribution of observations and the statistical definition thus depends on the type of distribution.

This section is divided into three parts:

- Relationship between the true and estimated mean values
- Relationship between two estimated mean values
- Relationship between several estimated mean values.

Relationship between the true and estimated mean values

The basic question: What is the relationship between the estimated mean values based on the selected population and the true mean values of the population of interest? In the listening test context, this corresponds to estimating the relationship between a mean rating of a loudspeaker based on the experimental selection of programmes and subjects compared to a mean rating based on all available programmes and subjects in the population of interest for the manufacturer or experimenter.

Parametric tests In order to apply parametric tests, it must be verified that the ratings are provided on a continuous scale and further that the distributions are normal or can be transformed into normal distributions. It is further assumed that the tested programmes and subjects represent a random but representative sample of programmes and subjects of the population specified.

The relationship between the true mean and estimated mean is given by the confidence interval of the mean. The confidence interval has the following statistical meaning: If a series of identical listening tests were conducted on a single loudspeaker using different samples from the same populations of programmes and subjects and for each sample a mean and a 95 % confidence interval were calculated, then in the long run, 95 % of these confidence intervals would include the true mean value.

A $100 \times (1 - \alpha)$ % confidence interval is constructed on the basis of the formula below.

$$\overline{Y} - t_{(1-\alpha/2)} \times \left(\frac{S}{\sqrt{n}}\right) \text{ to } \overline{Y} + t_{(1-\alpha/2)} \times \left(\frac{S}{\sqrt{n}}\right) \tag{6.17}$$

where $\overline{Y}$ is the arithmetic mean of the observations, t is the t-value corresponding to the t-distribution with $n - 1$ degrees of freedom, α is the significance level equal to (1 − (degree of confidence)), S is the estimated standard deviation of the sample of observations and n is the number of observations in the sample. Note that the quantity $\frac{S}{\sqrt{n}}$ is called *the standard error of the mean*.

Typically, the degree of confidence is set to 95 %, but other values can be used depending on the researcher's need for certainty. The t-distribution is used for smaller samples ($n < 50$) and the normal distribution for larger samples, in which case a value of 1.96 is used for the 95 % interval.

Categorical-based tests The confidence interval is still the measure of the relationship between the true population mean and the estimated mean. For data that cannot be described by a normal distribution, the median or the most frequent value is used to describe the central tendency. The confidence interval can then be given [268] for the median of n scores as the interval between the rank-ordered observations numbered $(n+1)/2 \pm 0.98 \times \sqrt{n}$. If the data are continuous and close to normally distributed, the 95 % confidence interval for the median is given by:

$$Y_{\text{med}} - 1.253 \times t_{1-\frac{\alpha}{2}} \times \frac{S}{\sqrt{n}} \text{ to } Y_{\text{med}} + 1.253 \times t_{1-\frac{\alpha}{2}} \times \frac{S}{\sqrt{n}}, \tag{6.18}$$

where Y_{med} is the estimated median and t is the t-value for $n-1$ degrees of freedom.

Another alternative is the semi-interquartile range, which is equal to half the difference between the 75th and 25th quartiles.

The reader should note that Gardner and Altman [140] have an excellent discussion of confidence intervals for both quantitative and categorical data.

Relationship between two estimated mean values

In listening tests, the purpose is often to compare different versions of the same loudspeaker or to benchmark against competitors. Assume that a series of listening tests have been conducted, in which ratings have been obtained for two loudspeakers for the same selection of programmes. The question is: Are the two loudspeakers significantly different or is the observed difference just due to chance?

Parametric tests A distribution of ratings (collected through repetition of the experiment), similar to that shown in Figure 6.3, exists for each loudspeaker. After having examined the distributions, as described above, the means and variances can be estimated and the observed difference in means can be statistically tested to determine whether the two estimated mean values are different.

Remembering that the statistical null hypothesis should be formulated with the purpose of rejecting it (see Section 6.1), the null hypothesis H_0 and the alternative H_a are as follows:

$$H_0 : \mu_1 = \mu_2 \tag{6.19}$$

$$H_a : \mu_1 \neq \mu_2, \tag{6.20}$$

where μ_1 is the mean for loudspeaker 1 and μ_2 is the mean for loudspeaker 2.

The consequence of accepting H_0 is that the difference between the means will be zero and that the distribution of differences between mean values follows a so-called t-distribution.

The calculation of the degrees of freedom for the t-test depends on the conditions under which the ratings have been obtained. In listening tests, two situations are commonly encountered as follows:

- Each subject has evaluated both loudspeakers. In this case the test is the 'paired comparisons' t-test, and the test value is based on the difference scores and calculated as follows:

$$t_{\text{obs}} = \frac{(\text{Mean of difference scores})}{\sqrt{S_{\text{diff}}^2 \times \frac{1}{n}}}, \tag{6.21}$$

  where S_{diff}^2 is the variance of the difference scores and n is the number of differences. The degrees of freedom for the t-test is $n - 1$.
- Loudspeaker 1 has been evaluated by one group of subjects and loudspeaker 2 by another group. In this case, the test is the 'independent groups' t-test, and the t-value is calculated as follows:

$$t_{\text{obs}} = \frac{\overline{Y}_1 - \overline{Y}_2}{\sqrt{S_p^2 \times (\frac{1}{n_1} + \frac{1}{n_2})}}, \tag{6.22}$$

  where S_p^2 is the pooled variance for loudspeakers 1 and 2 and calculated by the following formula:

$$S_p^2 = \frac{(n_1 - 1) \times S_1^2 + (n_2 - 1) \times S_2^2}{n_1 + n_2 - 2} \tag{6.23}$$

  where S_1^2 and S_2^2 are the estimated variances and n_1 and n_2 are number of observations for loudspeakers 1 and 2, respectively. The degrees of freedom for the t-test is $df = (n_1 + n_2 - 2)$.

On the basis of the observed mean values, standard deviations and number of observations for the two loudspeakers, a t-value can be calculated and the statistical question to be answered is: If H_0 is accepted, what is then the probability of observing the calculated t-value?

A large positive or negative value is not very probable if the two means are equal. Thus, if a large positive or negative value is obtained, it suggests that the two tested mean values are not equal, and H_0 must be rejected at a certain probability level. See also the discussion of α and β in Section 6.1.

This leads to the definition of the so-called critical range for the observed t-value defined by:

$$t_{\text{obs}} < t(df)_{\frac{\alpha}{2}} \vee t_{\text{obs}} > t(df)_{1-\frac{\alpha}{2}} \tag{6.24}$$

The critical range is calculated using a table for the t- or Student distribution. An often used significance level is $\alpha = 5\,\%$ so if the observed t-value is outside the critical range it follows that the null hypothesis H_0 can be safely rejected, and the conclusion is that the two mean values are significantly different at the defined significance level α (see Section 6.1 for a further discussion of the meaning of α and β values.).

The null hypothesis tested above (Equation 6.20) is a so-called two-tailed test. It is two-tailed because the test is not concerned with the direction of the difference between the mean values, and therefore, the critical range includes two areas of the probability distribution, each representing one-half of the significance level. This is the standard test in research, but in more applied settings the main question could be to know whether loudspeaker 1 was better than loudspeaker 2, and thus, a one-tailed test is used where the whole significance level is applied in one direction. This leads to the null hypothesis and its alternative:

$$H_0 : \overline{Y}_1 \leq \overline{Y}_2 \tag{6.25}$$

$$H_a : \overline{Y}_1 > \overline{Y}_2 \tag{6.26}$$

with a critical range:

$$t_{\text{obs}} > t(df)_{1-\alpha} \tag{6.27}$$

and similarly for the other direction:

$$H_0 : \overline{Y}_1 \geq \overline{Y}_2 \tag{6.28}$$

$$H_a : \overline{Y}_1 < \overline{Y}_2 \tag{6.29}$$

and critical range:

$$t_{\text{obs}} < t(df)_{\alpha} \tag{6.30}$$

A note of caution is relevant at this point. It is standard procedure when reporting results to include a table or figure of mean values including the 95 % confidence intervals. However, it has been observed that such a figure is used to determine the mean values that are significantly different on the basis of the (wrong) assumption that if the confidence intervals of two means are overlapping it follows that the means are not significantly different. This is, however, not correct as Barr [22] and, more recently, Nelson [322] have shown. Barr illustrated that the length of the confidence intervals must be corrected before the criterion of overlapping intervals can be used to conclude on the degree of significance. The correction to be applied is given by:

$$z' = \frac{\sqrt{n_1 + n_2}}{\sqrt{n_1} + \sqrt{n_2}} \tag{6.31}$$

The corrected confidence intervals for the two means are thus given for the first mean:

$$\begin{aligned} &\overline{Y} - \left(\frac{\sqrt{n_1 + n_2}}{\sqrt{n_1} + \sqrt{n_2}}\right) \times t_{\alpha/2} \times \left(\frac{S}{\sqrt{n_1}}\right) \text{ to} \\ &\overline{Y} + \left(\frac{\sqrt{n_1 + n_2}}{\sqrt{n_1} + \sqrt{n_2}}\right) \times t_{(1-\alpha/2)} \times \left(\frac{S}{\sqrt{n_1}}\right) \end{aligned} \tag{6.32}$$

and for the second mean:

$$\begin{aligned} &\overline{Y} - \left(\frac{\sqrt{n_1 + n_2}}{\sqrt{n_1} + \sqrt{n_2}}\right) \times t_{\alpha/2} \times \left(\frac{S}{\sqrt{n_2}}\right) \text{ to} \\ &\overline{Y} + \left(\frac{\sqrt{n_1 + n_2}}{\sqrt{n_1} + \sqrt{n_2}}\right) \times t_{(1-\alpha/2)} \times \left(\frac{S}{\sqrt{n_2}}\right) \end{aligned} \tag{6.33}$$

A more simple approach would of course be to report the outcome of the relevant t-tests.

Categorical-based tests A number of tests are available for the analysis of categorical data, of which the two best known are probably the binomial and chi-square tests. There is a growing acknowledgement of the need for categorical tests, especially in areas like sensory evaluation of food, market surveys, economics, and so on. A number of recommendable books on this topic have been published in the recent years [9, 165, 352, 395], and the interested reader is advised to consult them.

In order to select the appropriate test, it is necessary to know the number of categories the data have been coded in (e.g. Yes/No or Acceptable/Slightly Annoying/Very Annoying/Not acceptable) and whether they are based on dependent or independent samples. The

Number of categories in rating scale	One sample	Two samples: Independent observations	Two samples: Dependent observations
Two	Binomial	Hypergeometric	See one sample
More than two	Chi-squared	Mann–Whitney U	Chi-squared

Table 6.10: Distributions for analysis of categorical data.

usual situation in listening tests is that all subjects evaluate all combinations of the independent variables, and if the test is performed using, for example, a paired comparisons technique where both stimuli are rated simultaneously, the ratings will be dependent. Alternatively, if the test is performed using a single stimulus procedure, the samples may be assumed to be independent[29]. Further, it is important for the correct choice of procedure to specify whether it is a one-sample situation or a two-sample situation. A one-sample situation examines whether a set of observations obtained for a single combination of the independent variables conforms to a given expected distribution. The two-sample situation examines whether sets of observations obtained for two combinations of the independent variables are significantly different.

Some of the underlying distributions to be used for the different situations are illustrated in Table 6.10.

One sample–two categories The one sample and two categories situation could be a listening test employing one loudspeaker, where the subjects have been asked to state whether they like it or not, for a selection of programmes. The question of interest to the experimenter is whether the obtained distribution of ratings is significantly different from a distribution based on a chance probability of preference for the loudspeaker - that is, the subjects are split 50/50 in terms of choice.

The null hypothesis H_0 and the alternative H_a are as follows:

$$H_0 : p = 0.5 \tag{6.34}$$

$$H_a : p \neq 0.5 \tag{6.35}$$

[29]It should be noted that this assumption should always be verified. A certain degree of dependency is likely if, for example, training or types of time-related effects occur.

The test measure n_{like} is the number of observations in the 'like' category and is binomially distributed ($B(n, 0.5)$) if H_0 is true. The two-tailed critical range for the test measure is defined as

$$n_{\text{like}} \leq B(n, 0.5)_{\frac{\alpha}{2}} \vee n_{\text{like}} > B(n, 0.5)_{1-\frac{\alpha}{2}} \tag{6.36}$$

The critical values are calculated using a table for the binomial distribution $B(n, 0.5)$ and the chosen significance level.

The binomial distribution can be approximated by the normal distribution if the number of observations is sufficiently high. A rule of thumb is that both np and $n(1-p)$ should be higher than 15. The parameters to be used in the normal distribution are mean value $= n \times p$ and variance $= n \times p \times (1-p)$.

This leads to the test measure z:

$$z = \frac{P_{\text{obs}} - p - (\frac{0.5}{n})}{\sqrt{\frac{p \times (1-p)}{n}}}, \tag{6.37}$$

where $P_{\text{obs}} = \frac{n_{\text{like}}}{n}$ and $\frac{0.5}{n}$ is the so-called continuity correction accounting for the fact that the binomial distribution is not continuous. The test measure follows a unit normal distribution, and the two-tailed critical range is thus defined as:

$$z \leq N(0,1)_{\frac{\alpha}{2}} \vee z > N(0,1)_{1-\frac{\alpha}{2}} \tag{6.38}$$

One sample–more than two categories Imagine that the subjects in the example above instead of being asked to state whether they liked the loudspeaker were asked to classify it into one of several categories: 'like very much', 'like', 'acceptable', 'dislike' and 'dislike very much'. The question of interest to the experimenter is still whether listeners in general opt for a specific opinion about the loudspeaker or, alternatively, if there is an almost equal spread of opinions across the categories.

The null hypothesis H_0 and the alternative H_a are as follows:

$$H_0 : p_j = p_0, \ j = 1, \ldots, k \tag{6.39}$$

$$H_a : p_j \neq p_0, \ j = 1, \ldots, k, \tag{6.40}$$

where $p_0 = \frac{1}{k}$ is the expected, evenly distributed probability for an observation to fall in any category, and k is the number of categories.

The test measure is based on the differences between the number of observed (n_j) and expected ($\frac{n}{k}$) observations in each category and is defined as:

$$z = \sum_{j=1}^{k} \frac{(n_j - \frac{n}{k})^2}{\frac{n}{k}} \tag{6.41}$$

The test measure is chi-squared (χ^2) distributed with $n - 1$ degrees of freedom. The two-tailed critical interval is thus

$$z \leq \chi^2(n-1)_{\frac{\alpha}{2}} \ \vee \ z > \chi^2(n-1)_{1-\frac{\alpha}{2}} \tag{6.42}$$

Two dependent samples–two categories A common situation is that the subject is presented with two loudspeakers and asked to indicate one of the two that he or she prefers. This could be repeated for a number of subjects and programmes. Each loudspeaker thus receives a number of 'preferred' and a number of 'not preferred' ratings. As the test is paired, the sum of the ratings for the two loudspeakers is equal to the total number of ratings. The problem is thus reduced to a one-sample problem testing one of the loudspeakers using a null hypothesis $H_0 : p = p_0 = 0.5$ corresponding to the equal probability of choosing one or the other loudspeaker in the pair.

Two dependent samples–more than two categories Again, two loudspeakers are presented simultaneously to the subject, but now the subject is asked to assign a category to each of the loudspeakers for each of the programmes tested. The categories assigned to the loudspeakers are thus dependent as the loudspeakers were present at the same time. The question of interest to the experimenter is whether the loudspeakers can be considered to be significantly different. The hypothesis is thus that both loudspeakers have the same distribution of 'ratings' across the categories, that is, they have been assigned to all categories an equal number of times.

If the categories are assigned numbers and the loudspeakers are ranked according to their categories, the hypothesis translates into all loudspeakers having the same mean ranking. This is the idea behind Friedman's test, which has a test measure defined by

$$z = \frac{12}{N \times j \times (j+1)} \times \left(\sum_{i=1}^{j} T_i^2 \right) - 3 \times N \times (j+1), \tag{6.43}$$

where N is the number of matched observations (number of times the group of loudspeakers have been rated), j is the number of loudspeakers (which in the present case is two) and T_i is the sum of ranks for each loudspeaker.

The test measure is chi-squared distributed with $j-1$ degrees of freedom, giving the following two-tailed critical range for the test measure:

$$z \leq \chi^2(j-1)_{\frac{\alpha}{2}} \vee z > \chi^2(j-1)_{1-\frac{\alpha}{2}} \tag{6.44}$$

Two independent samples–two categories In this situation, there are two groups of subjects that each have performed a one sample, two categories 'prefer it or not' test, and in each test it has been concluded that the loudspeaker is 'preferred' to a significant degree. The experimenter then wants to test whether the probability of preference is equal for the two loudspeakers. This leads to the null and alternative hypotheses

$$H_0 : p_1 = p_2 \tag{6.45}$$

$$H_a : p_1 \neq p_2, \tag{6.46}$$

where p_1 p_2 are the probabilities of 'prefer' for loudspeaker 1 and 2, respectively.

The test measure is $t = n_{\text{prefer1}} + n_{\text{prefer2}}$ equal to the sum of 'prefer' observations for both loudspeakers. The test measure is distributed according to a hypergeometric distribution $H(t, n_1, n_1 + n_2)$, where n_1 and n_2 are the total number of observations for loudspeakers 1 and 2, respectively.

The two-tailed critical range for the test measure is:

$$t \leq H(t, n_1, n_1 + n_2)_{\frac{\alpha}{2}} \vee t > H(t, n_1, n_1 + n_2)_{1-\frac{\alpha}{2}} \tag{6.47}$$

Two independent samples–more than two categories This situation is equivalent to the independent group t-test discussed above where loudspeaker 1 has been evaluated by one group of subjects and loudspeaker 2 by another group. Both groups have used the same rating procedure and a rating scale with more than two categories. The experimenter would like to know whether the two loudspeakers can be considered significantly different, and thus the hypothesis is that the two sets of ratings are equal.

The Mann–Whitney U-test and the Wilcoxon test are both a possibility for testing this hypothesis for categorical data. Both procedures are based on ranks and are independent of the underlying distributions, but it is assumed that both distributions have the same shape (see discussion of the robustness of the U-test to violations of the assumptions in Section 6.2.3). It has been shown that the U- and Wilcoxon tests are equivalent [396], so only the Mann–Whitney U-test will be examined here.

The first step is to rank order the loudspeakers within each data set and then to rank order the combined data set. The sum of the rankings of the smallest data set (in terms of number of observations) based on the combined rank order is then calculated; the test statistic U is calculated as

$$U = n_1 \times n_2 + \left(\frac{n_1 \times (n_1 + 1)}{2}\right) - R_1, \tag{6.48}$$

where n_1 is the number of samples in the smallest group and n_2 the number in the largest group. R_1 is the sum of the ranks assigned to the smaller group. The distribution of U must have the correct form, and this is tested by comparing the calculated U value with $\frac{n_1 \times n_2}{2}$. If the calculated U is larger, a transformation value must be used:

$$U = n_1 \times n_2 - U_1, \tag{6.49}$$

where U_1 is the value that was found to be too high.

The critical range for U can be found in tables (see for example [268]) and is defined as $U \leq U(n_1, n_2)_\alpha$ for a one-tailed test and as $U \leq U(n_1, n_2)_{\frac{\alpha}{2}}$ for a two-tailed test.

The above formula applies for small groups of observations as the largest group of observations should be smaller than 20. If the largest group is larger than 20, then U can be converted into a score following a unit normal distribution:

$$z = \frac{U - \frac{n_1 \times n_2}{2}}{\sqrt{\frac{n_1 \times n_2 \times (n_1 + n_2 + 1)}{12}}} \tag{6.50}$$

and the critical range is

$$z < N(0,1)_{\frac{\alpha}{2}} \ \vee \ z > N(0,1)_{1-\frac{\alpha}{2}} \tag{6.51}$$

Relationship between several estimated mean values

Most listening tests include several loudspeakers and programmes. This means that the number of mean values that should potentially be compared increases quite quickly. The relationship between the number of pairs and the number of means is given by

$$N = \frac{n \times (n-1)}{2} \tag{6.52}$$

where N is number of pairs and n is number of mean values. A high number of pairs means not only that it can be quite a task to conduct all the necessary t-tests but also that special compensation techniques must be applied to maintain the chosen test level (α) in all the tests (this issue will be discussed in detail later in this section). Thus, it would be advantageous if a preliminary analysis could be conducted with the aim of identifying whether any significant differences exist within the group of means. Such a test is termed an analysis of variance (ANOVA), and this section introduces this type of analysis and associated test paradigms. It is not the intention of this section to provide a detailed discussion of the ANOVA but only to introduce the technique so that the interested reader can seek further insight. A large number of references can be found in this area; however, as an introduction to the area of a standard ANOVA, the authors recommend Box *et al.* [63], Hoaglin *et al.* [163] and Kepper and Wickens [253], and for the multivariate versions of ANOVA Bernstein [54] and Hair *et al.* [155]. The reader should be aware that the statistical terminology used in the references depends to a certain degree on the area of application (e.g. psychology versus engineering), so it is advisable to check the definitions in each area. One example would be the use of either 'factor' or 'variable' as a label for the treatments effects in an ANOVA. The authors have no specific preference, so the term 'variable' is used.

The chapter is focussed on the analysis of a so-called randomised block design with a complete factorial experiment randomised within each block. In the listening tests, such a design would have subjects as blocks and each subject would be presented with all combinations of the variables in a random order in a single session. The first example will only include one subject, so this reduces the design to a complete randomised factorial design.

The chapter does not discuss the categorical-based methods as ANOVA is the *de facto* standard for the analysis of results from listening tests, and it was thus considered important to introduce the reader to ANOVA as the general technique. This is not to suggest that categorical-based tests are not relevant, and the readers are encouraged to familiarise themselves with the topic. An excellent introduction to the area is Hollander and Wolfe [165].

Basic one variable ANOVA model Let us assume that a pilot-test including five loudspeakers and one programme item was conducted to obtain an impression of the differences in sound quality between the loudspeakers. One subject[30] participated in the test and rated each loudspeaker 10 times in randomised order using a single presentation paradigm. A rating scale of 0 to 10 with one decimal place was used. The observations provided by the subject are assumed to be independent[31]. A simulated data set for this experiment has been produced and will be referred to in the following text. The data set includes ratings for four programmes, but ratings for only one programme will be considered to begin with. The null hypothesis H_0 and the alternative H_a are defined as follows:

$$H_0 : \mu_1 = \mu_2 = \mu_3 = \mu_4 = \mu_5 \tag{6.53}$$

$$H_a : \text{not all } \mu_i\text{'s are equal} \tag{6.54}$$

where $\mu_t; t = 1,\ldots,5$ is the mean for loudspeaker t

The discussion in Section 6.2.4 would suggest that to fully examine H_0, 10 t-tests should be conducted according to the stated formula for the 'paired comparisons'. However, before doing so, it would be preferable to run a test to determine whether H_0 must be rejected for at least one pair as this would prevent one from conducting a series of unnecessary t-tests.

The paired t-test applied in the previous sections is based on the ratio of the average of difference scores and the variation of the difference scores. The variation of the difference scores represents a so-called error variance (see discussion of a statistical

---

[30]The usage of a single subject is for example only and not recommended in practise.

[31]The assumption of independent observations will later be shown to not hold, but for the present set of examples, it is useful to accept the assumption. The reader should also note that the described scenario is not very likely to occur in reality, but again, as an example it serves a purpose.

model for subject scores in Section 6.2) caused by, for example, bias phenomena related to providing ratings on a scale. If the ratio is large enough, it can be concluded that the observed differences are not the results of random variations and H_0 is rejected.

The rationale behind the ANOVA is the same as that for the t-test: Compare the deviation between the sample means of the treatments to the random deviation within the samples and test whether the magnitude of the ratio is higher than a certain critical value.

The deviation between sample means of the treatments is, in the example above, the deviation related to differences between the mean values of the tested loudspeakers. The random deviation is the deviation caused by the repeated ratings for each loudspeaker. The deviation between the loudspeaker mean values is often termed *between-groups* deviation and the deviation related to the repeated ratings for each loudspeaker is often termed *within-groups* or *error deviation*.

The deviations are calculated as so-called sum of squares (SS) and the between-groups SS is given by

$$SS_G = \sum_{t=1}^{k} n_t \times (\overline{Y}_t - \overline{Y})^2 \tag{6.55}$$

The subscript G is used to indicate 'groups' (loudspeakers), $\overline{Y}_t$ is the mean for group (loudspeaker) t, $\overline{Y}$ is the overall mean of all ratings, n_t is the number of observations in group t and 'k' is the number of groups[32].

The within-group $SS_{R,k}$ for group k is given by

$$SS_{R,k} = \sum_{i=1}^{n_k} (Y_{k,i} - \overline{Y}_k)^2 \tag{6.56}$$

where n_k is the number of observations in group k, Y_i is the ith observation in group k and $\overline{Y}_k$ is the mean for group k (loudspeaker k). The subscript R is used to indicate 'residual', which is the often used term for error variation.

[32]The reader should note that the multiplication by n_t is due to the fact that the between-group deviation is the same for all observations within a group.

The total within-group SS_R is the found by summation of the SSs for the k groups:

$$SS_R = \sum_{i=1}^{k} SS_{R,i} \tag{6.57}$$

The total within-group SS_R could also have been found using the relation:

$$SS_{\text{total}} = SS_G + SS_R, \tag{6.58}$$

where SS_{total} is the total SS based on all observations and the overall mean:

$$SS_{\text{total}} = \sum_{t}^{k} \sum_{i}^{n_k} (Y_{t,i} - \overline{Y})^2 \tag{6.59}$$

The division of the total squared deviations into components related to the treatments and the residual is fundamental to ANOVA as the SS's are converted to estimates of variance by dividing by the corresponding degrees of freedom. These estimates are termed *mean squares* (MS), so the between- and within-group MS are given by:

$$MS_G = \frac{SS_G}{k-1} \tag{6.60}$$

$$MS_R = \frac{SS_R}{N-k}, \tag{6.61}$$

where N is the total number of observations.

The ratio of the estimated 'between-groups' and the 'within-groups' MS can now be established and tested. If H_0 is true and the error variance is normally distributed, it follows that the ratio of two variances is $F(df_{\text{num}}, df_{\text{denom}})$ distributed with df_{num} and df_{denom} degrees of freedom, where df_{num} is defined by the numerator as the between-groups degrees of freedom $(k-1)$ and df_{denom} by the denominator as within-groups degrees of freedom $(N-k)$.

The rationale for testing is the same as that for the t-test: The probability of obtaining F values higher than a specified value can be determined by the F-distribution for a given set of degrees of freedom. The experimenter thus specifies a given test level α

(e.g. 5 %) and the obtained F_α value specifies the critical range. Note that the ratio of variances only produces positive values, so the F-test is a non-directional test. The decision rule is thus:

$$F_{\text{observed}} \geq F_\alpha(df_{\text{num}}, df_{\text{denom}}) => \text{reject } H_0$$

or

$$F_{\text{observed}} \langle F_\alpha(df_{\text{num}}, df_{\text{denom}}) => \text{accept } H_0 \qquad (6.62)$$

Keppel and Wickens [253] (Section 3.2) discusses the evaluation of the F value further.

For the example above, the critical range is defined by a $F(4, 45)$ distribution and is equal to 2.579 at the 5 % level. This means that H_0 can be rejected at a 5 % level if the ratio of variances is higher than 2.579.

The general principle in the process described above is to divide the total variation in the data set into a number of variance components, which in this case is the 'between-groups' and the 'within-groups' variance (see Equation 6.58). The variance caused by the treatment (between-groups) is then compared to the error variance (within-groups), and the comparison will show whether the treatment(s) cause(s) a statistically significant change in the response compared to random variation.

The formal version of this principle can be described by an ANOVA model, which specifies the components responsible for the variation in the data. Note that this is the same model (see Equation 6.3) used in the introduction to describe the basic model for the t,i'th score from a subject:

$$Y_{t,i} = \mu + \alpha_t + \epsilon_{t,i}, \quad \epsilon_{t,i} \sim N(0, \sigma^2), \qquad (6.63)$$

where μ is the overall mean, α_t is the between-groups treatment effect and $\epsilon_{t,i}$ represents the within-groups or error effect.

The formulation of the ANOVA model in Equation 6.63 leads to an alternative formulation of the null hypothesis that is often used in statistical texts:

The hypothesis formulated in the beginning of the Section (see Equation 6.54) was based on the individual means $(\mu_1, \mu_2, \mu_3, \mu_4, \mu_5)$, however, with the introduction of an overall

Dependent variable: rating	Sum of squares	*df*	Mean square	*F*	Sig.
Between-Groups	219.628	4	54.907	65.553	.000
Within-Groups	37.692	45	.838		
Total	257.321	49			

Table 6.11: ANOVA table of simulated data based on five loudspeakers and one programme assessed by 10 subjects.

mean and the individual treatment effects, the null hypothesis can be formulated in terms of the treatment effects α_t:

$$H_0 : \alpha_1 = 0, \alpha_2 = 0, \alpha_3 = 0, \alpha_4 = 0, \alpha_5 = 0 \tag{6.64}$$

which is sometimes given in the form of squared treatment effects:

$$H_0 : \sum_t \alpha_t^2 = 0 \tag{6.65}$$

The ANOVA is often conducted using one of the many software programs available (SPSS, SAS, etc.), and a typical output from a statistical analysis program (SPSS) is shown in Table 6.11 for the pilot experiment (discussed above) including five loudspeakers, one programme, one subject and 10 repetitions.

The table includes an identification of the dependent variable (rating) and then six columns. The first column identifies the source of the variation (the α_t and $\epsilon_{t,i}$ effects in the ANOVA model), followed by the sum of squares (SS) corresponding to each of the effects, the degrees of freedom for each element, the 'mean square' equal to SS/df, the magnitude of the ratio of the variances or the F-test value and, finally, the significance level corresponding to the F-test value and associated degrees of freedom.

Before the results of the ANOVA are studied in detail, the assumption of variance homogeneity across the five loudspeakers must be checked. This is done via Levene's test (see Section 6.2.3), and the results show that the Levene statistic [Levene(4,45) $= 2.046$; $p = 0.104$] is not significant and the assumption of variance homogeneity is accepted.

The second row includes the results for the 'between-groups' variable, and it shows that the p ('Sig.') value corresponding to the F-test value that is smaller than 0.001, corresponding to a highly significant 'between-groups variable'. This means that H_0 must be rejected, and the conclusion should be that of all the tested loudspeakers, at least two are significantly different.

It is noted that the total number of degrees of freedom for 'Total' is 49 and not 50, corresponding to one subject rating five loudspeakers 10 times. The reason is that one degree of freedom has been used to estimate the overall mean in the ANOVA model in Equation 6.63.

The next step is to identify the pairs with a significant difference in mean values. This could be performed by testing each pair, one by one at a specified significance level, using the standard t-test described in Section 6.2.4. The problem, however, is that the specified significance level is only correct for one t-test. If more than one test is conducted, the test level across the entire set of comparisons increases. For example, assume that 100 in significantly different (determined previously) pairs are available and are tested at a 5 % significance level. The applied 5 % level means that at least five of the tested pairs will erroneously be concluded to be significantly different. The relationship between the number of comparisons and the corresponding overall Type I error level is formally given by (see for example Keppel and Wickens [253] (Section 6.1)):

$$\alpha_{FW} = 1 - (1 - \alpha)^c, \tag{6.66}$$

where α_{FW} is the family-wise Type I error equal to the probability of making at least one Type I error in the family of tests when the null hypothesis is true, α is the Type I error test level for the individual tests and c is the number of tests.

Consider the example above of conducting 10 t-tests at a 5 % level. This would correspond to a true test level of $\alpha_{FW} = 1 - (1 - 0.05)^{10} = 0.40$.

The relationship between the number of comparisons and change in test level is further discussed by, for example, Ryan [369], Wilson [453], Petrinovich and Hardyck [343] and Keppel and Wickens [253]; the reader is advised to consult these references as this issue is often overlooked or not understood by researchers.

Dependent variable: rating
Bonferroni

(I) Loudspeaker	(J) Loudspeaker	Mean difference (I-J)	Std. error	Sig.	95% Confidence interval	
					Lower bound	Upper bound
1.00	2.00	-1.88252[a]	.40929	.000	-3.0908	-.6743
	3.00	2.15280[a]	.40929	.000	.9445	3.3611
	4.00	-3.77312[a]	.40929	.000	-4.9814	-2.5649
	5.00	-2.68383[a]	.40929	.000	-3.8921	-1.4756
2.00	1.00	1.88252[a]	.40929	.000	.6743	3.0908
	3.00	4.03532[a]	.40929	.000	2.8271	5.2436
	4.00	-1.89060[a]	.40929	.000	-3.0989	-.6823
	5.00	-.80131	.40929	.565	-2.0096	.4070
3.00	1.00	-2.15280[a]	.40929	.000	-3.3611	-.9445
	2.00	-4.03532[a]	.40929	.000	-5.2436	-2.8271
	4.00	-5.92592[a]	.40929	.000	-7.1342	-4.7177
	5.00	-4.83663[a]	.40929	.000	-6.0449	-3.6284
4.00	1.00	3.77312[a]	.40929	.000	2.5649	4.9814
	2.00	1.89060[a]	.40929	.000	.6823	3.0989
	3.00	5.92592[a]	.40929	.000	4.7177	7.1342
	5.00	1.08929	.40929	.107	-.1190	2.2976
5.00	1.00	2.68383[a]	.40929	.000	1.4756	3.8921
	2.00	.80131	.40929	.565	-.4070	2.0096
	3.00	4.83663[a]	.40929	.000	3.6284	6.0449
	4.00	-1.08929	.40929	.107	-2.2976	.1190

[a] The mean difference is significant at the .05 level.

Table 6.12: Multiple comparisons of means using a Bonferroni correction. The ANOVA of the data is shown in Table 6.11.

A number of modified t-tests have been suggested to account for the problem (see for example [63,162,253,268,343,369,453]), and the reader is advised to seek additional information when required. A simple and often applied correction is the Bonferroni correction for small groups of paired tests, and it simply states that the test level for multiple tests should be equal to $\frac{p_0}{n}$, where p_0 is the level applied if only one t-test is conducted and n is the number of t-tests to be conducted. The results of a Bonferroni corrected multiple comparisons are shown in Table 6.12. The second row shows the comparison between loudspeakers 1 and 2. The difference between the mean values is -1.88, with a 95 % confidence interval of (see the last two columns) $[-3.09; -0.67]$. This is well below zero and significant with a probability level, as shown in the fourth column ('Sig.'), below 0.001. The following rows show the comparison between loudspeakers 1 and 3, 1 and 4 and 1 and 5, and then follows comparisons for the other loudspeakers. The results show that only loudspeaker pairs 2 and 5 and 4 and 5 are not significantly different.

For larger groups of paired comparisons, the Bonferroni corrected test level becomes very small, so it is better (see for example the discussion in Keppel and Wickens [253] (Section 6.4)) to use Tukey's honestly significant difference (HSD) procedure. The Tukey procedure is based on a critical difference D_{Tukey}, which is compared to the observed difference between a pair of means. If the observed difference is larger than D_{Tukey}, it follows that the difference is significant at the applied α level. The critical difference is given by:

$$D_{\text{Tukey}} = q_a \times \sqrt{\frac{MS_R}{n}} \qquad (6.67)$$

where q_a is a quantity that can be found in a table of the Studentised Range statistics, with the entry parameters a equal to the number of pairs to be compared, degrees of freedom for MS_R and test level α. MS_R is the error variance and n is the number of observations for each mean. The Studentised Range statistics is defined and discussed in, for example, Hoaglin *et al.* [163].

ANOVA model for more than one variable The example examined in the previous section included five loudspeakers, evaluated using one piece of programme material. The generality of the results thus depends on how representative the chosen programme material is or whether it can be assumed that the perceived sound quality of a loudspeaker is independent of programme material. However, it has been shown (see for example Toole [421], Olive [333], Bech [29], Zacharov [459]) that programme material, in loudspeaker listening tests, is one of the most influential variables[33]. Most listening tests are thus

[33]Note that although all the references refer to listening tests on loudspeakers, programme dependency has also been observed in, for example, tests of sound quality of low bit-rate codec. It is recommended to include the programme variable in general, but of course situations can be and have been observed where the programme variable is less influential.

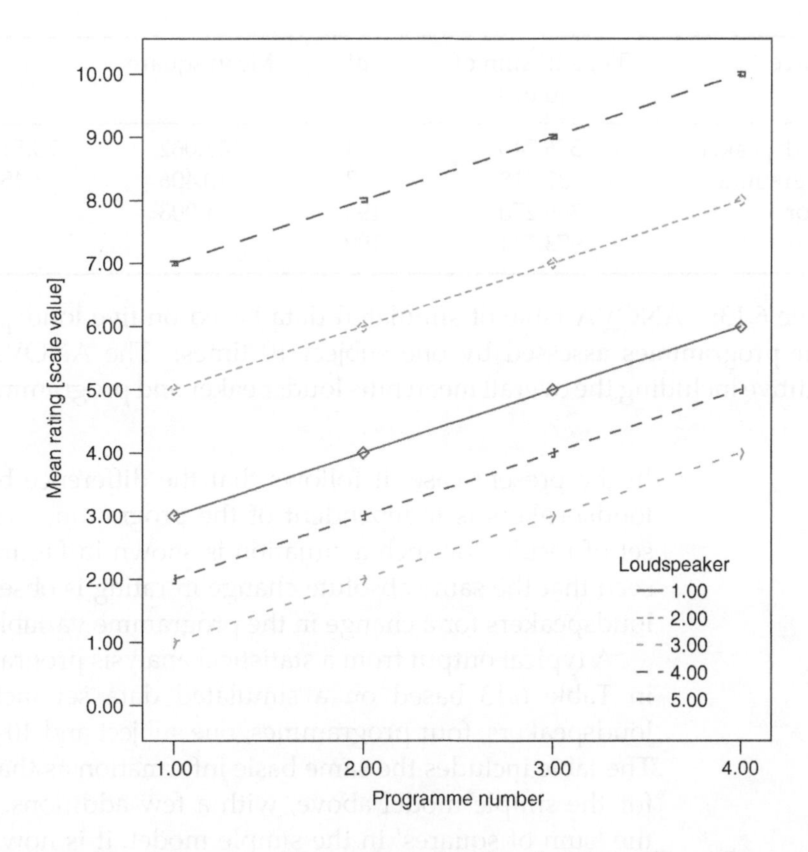

Figure 6.10: Mean quality ratings for five loudspeakers, four programmes, one subject and 10 repetitions. The results are based on simulated data and can be described by an additive ANOVA model including the overall mean plus loudspeaker and programme as variables.

conducted using several programmes, and this introduces an additional treatment variable in the ANOVA model, accounting for the variation caused by the different programmes:

$$Y_{t,j,i} = \mu + \eta_t + \beta_j + \epsilon_{t,j,i}, \quad \epsilon_{t,j,i} \sim N(0, \sigma^2), \tag{6.68}$$

where μ is the overall mean, η_t is the loudspeaker effect, β_j is the programme effect and $\epsilon_{t,j,i}$ represents the error.

This model is termed an *additive model* because it assumes that a change in one of the variables is reflected in the response, independent of the level of the other variables in the model.

Source	Type III sum of squares	*df*	Mean square	*F*	Sig.
Loudspeaker	576.246	4	144.062	75.518	0.000
Programme	31.218	3	10.406	5.455	0.001
Error	366.270	192	1.908		
Total	973.734	199			

Table 6.13: ANOVA table of simulated data based on five loudspeakers and four programmes assessed by one subject 10 times. The ANOVA model is additive including the overall mean plus loudspeaker and programme variables.

In the present case, it follows that the difference between the loudspeakers is independent of the programme. A simulated set of results for such a situation is shown in Figure 6.10. It is seen that the same absolute change in rating is observed for all loudspeakers for a change in the programme variable.

A typical output from a statistical analysis program is shown in Table 6.13 based on a simulated data set including five loudspeakers, four programmes, one subject and 10 repetitions. The table includes the same basic information as that discussed for the simple model above, with a few additions. Instead of the 'sum of squares' in the simple model, it is now a 'Type III sum of squares'. A number of different calculations of the sum of squares have been formulated, each for a specific purpose and context. The most common and default method in most statistical programs is the type III as it handles both balanced (equal number of observations for all variable combinations) and unbalanced designs[34].

The assumed normal distribution of the residuals should be examined, and Kolmogorov–Smirnov (KS) and Shapiro–Wilk (SW) tests should be applied. These tests were conducted, as discussed in Section 6.2.3, and both tests confirm ($p_{KS} = 0.20$ and $p_{SW} = 0.111$) that the residuals are normally distributed. It is noted that the central limit theorem, with a sample size of 193 observations, will ensure that the assumption is fulfilled.

[34]In certain 'nice' cases, such as the present with an equal number of observations in all cells, there is no difference between the different types of SS (see for example the SPSS manual [397] or Keppel and Wickens [253] (Sections 14.2 and 14.3)).

The results show that it was a good idea to include the programme variable as it is highly significant [$F(3, 192) = 5.455; p = 0.001$].

However, it is known by experienced recording engineers, for example, that the rank order of loudspeakers with respect to sound quality can change significantly for different programme types. This means that a change in the programme variable will influence the ratings of the loudspeaker, depending on the loudspeaker. An example of this is shown in Figure 6.11, where loudspeakers 1, 3 and 5 span a similar range of ratings for all programmes, but the ratings for loudspeaker 2 are independent of programme and the ratings for loudspeaker 4 have the same absolute, but negative, change in ratings as that seen for loudspeakers 1, 3 and 5. So the rank order of the loudspeakers changes as a function of programme. The additive model is thus not sufficient to describe the results of the experiment, and a so-called interaction variable needs to be included in the model. Keppel and Wickens [253] (Section 10.3) define 'interaction effect' for two independent variables as:

'An interaction is present when the effects of one independent variable on behaviour change at the different levels of the second independent variable'.

An ANOVA[35] model including the interaction is shown below for the *t,j,i*'th score:

$$Y_{t,j,i} = \mu + \eta_t + \beta_j + \gamma_{t,j} + \epsilon_{t,j,i}, \quad \epsilon_{t,j,i} \sim N(0, \sigma^2) \tag{6.70}$$

where μ is the overall mean, η_t is the loudspeaker effect, β_j is the programme effect, $\gamma_{t,j}$ is the effect accounted for by the interaction between loudspeakers and programmes and $\epsilon_{t,j,i}$ represents the error.

The results of an ANOVA analysis is shown in Table 6.14.

[35]It is noted that the notations used in Equation 6.70 are often used in mathematically oriented papers; however, another form of notation is also used, as shown below:

$$Y_{t,j,i} = \mu + \mathrm{LOUD}_t + \mathrm{PROG}_j + \mathrm{LOUD} \times \mathrm{PROG}_{t,j} + \epsilon_{t,j,i}, \quad \epsilon_{t,j,i} \sim N(0, \sigma^2) \tag{6.69}$$

where LOUD_t is the loudspeaker effect, PROG_j is the programme effect, $\mathrm{LOUD} \times \mathrm{PROG}_{t,j}$ is the interaction between loudspeaker and programme and $\epsilon_{t,j,i}$ is the error.

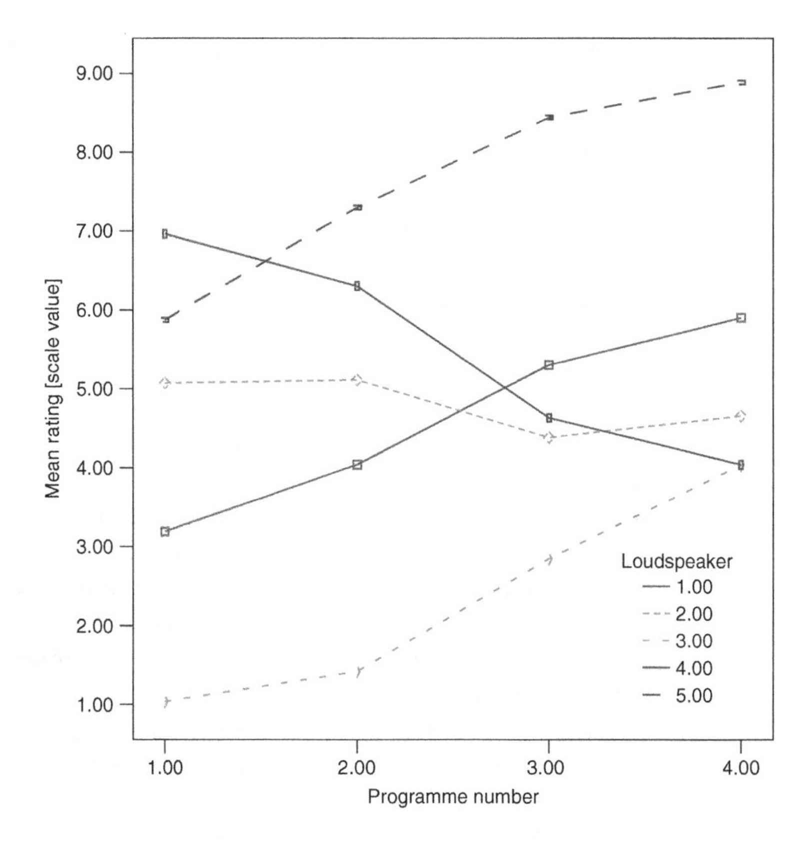

Figure 6.11: Mean quality ratings for five loudspeakers and four programmes. The results are based on simulated data and can be described by an ANOVA model including the overall mean, loudspeaker, programme, and their interaction as variables.

The only new item in the table is the interaction variable, and it is seen that it is highly significant [$F(12, 180) = 15.357; p < 0.001$], so the assumed interaction between loudspeaker and programme has been shown to be significant.

It is noted that the SS_{error} has been more than halved because of the addition of the interaction term. The reason is that the variance due to the interaction was pooled with the error variance in the simple model without the interaction term. This has the consequence that the magnitude of the other F-tests has increased. The inclusion of the interaction

Source	Type III sum of squares	*df*	Mean square	*F*	Sig.
Loudspeaker	576.246	4	144.062	143.279	0.000
Programme	31.218	3	10.406	10.349	0.000
Loudspeaker × programme	185.287	12	15.441	15.357	0.000
Error	180.983	180	1.005		
Total	973.734	199			

Table 6.14: ANOVA table of simulated data based on five loudspeakers and four programmes assessed by one subject 10 times. The ANOVA model includes the overall mean plus variables–loudspeaker, programme and their interaction.

term has thus increased the resolution of the ANOVA quite considerably. The distribution of the residuals is still normal, with Kolmogorov–Smirnov and Shapiro–Wilk tests confirming ($p_{KS} = 0.20$ and $p_{SW} = 0.153$) this.

Status of the 'subject' variable in an ANOVA model The last addition to the ANOVA model that will be discussed in this section is the 'subject' variable. One of the main assumptions of the ANOVA model is that of independent observations. This means that all 200 observations in the example analysed above have been assumed to be independent. This was justified on the basis of the randomised presentation order. However, it raises the question whether ratings (both repeated and for different stimuli) from the same subject can be assumed to be independent even for randomised presentations? Nunnally and Bernstein [328] discuss this problem, and the general conclusion is that scores from the same subject will be correlated to a certain degree. This introduces the problem of how to handle dependent observations in the ANOVA model.

Now assume that instead of 1 subject providing all 200 observations, it was 10 subjects providing one rating of each of the 20 loudspeaker and programme combinations. This introduces the additional problem of modelling differences between subjects as both Poulton [351], for the general case, and Bech [27], for the specific case of listening tests, have shown that subjects use the rating scale differently. Subjects will be different in terms of overall mean, overall range employed and range of repeated ratings. Brockhoff and Skovgaard [72] and Brockhoff [69] have shown that this also applies to sensory testing of food quality.

These differences between subjects will inflate the error variance in the ANOVA model, and the resolution of the test is lower than it would be if this variance could be removed by adding a subject variable. The ANOVA model should thus include a 'subject' variable that is able to take the differences between subjects into account, as well as the effects of having correlated scores from the same subject.

The previous discussion of including an extra variable in to the ANOVA model raised the issue of including interactions between the variables, and it was shown that the sensitivity of the F-tests (the error variance decreased) was improved by including interactions. The addition of a subject variable thus adds the possibility of including the interaction between the subjects and the other variables, thereby potentially further improving the sensitivity of the tests. However, it should be remembered that the 'cost' of including the additional interactions is a loss of degrees of freedom for the error variance, which could mean a decrease in the resolution. Thus, if the interaction variables are found not to be significant, it is recommended that the ANOVA model is reformulated only to include significant variables to obtain the maximum sensitivity of the F-test.

The inclusion of a subject variable in the ANOVA model further adds to the possibility of including the so-called three-way interaction term between loudspeaker–programme–subject. However, these and higher-order interactions are difficult to interpret in relation to the experiment and very rarely represent significant effects, so it is standard practise to not include them. Further, if the experiment does not include replications, it is normal to use these higher-order interactions as the best estimate of the error variance.

The simplest approach is thus to add 'subject' as a main variable plus its interactions in the ANOVA models examined so far. This will extend the model to include differences in mean value between subjects and is the approach recommended by Lawless and Heymann [268], and the revised model is shown

Source	Type III sum of squares	*df*	Mean square	*F*	Sig.
Loudspeaker	576.246	4	144.062	140.715	0.000
Programme	31.218	3	10.406	10.164	0.000
Subject	10.469	9	1.163	1.136	0.344
Loudspeaker × programme	185.287	12	15.441	15.082	0.000
Loudspeaker × subject	36.579	36	1.016	0.992	0.493
Programme × subject	23.366	27	0.865	0.845	0.684
Error	110.568	108	1.024		
Total	973.734	199			

Table 6.15: ANOVA table of simulated data based on five loudspeakers and four programmes assessed by 10 subjects. The ANOVA model includes the overall mean plus variables–loudspeaker, programme and subject plus second-order interactions.

below for the t, j, s, i'th score.

$$Y_{t,j,s,i} = \mu + \eta_t + \beta_j + \gamma_{t,j} + \alpha_s + \delta_{t,s} + \eta_{j,s} + \epsilon_{t,j,s,i}, \quad \epsilon_{t,j,s,i} \sim N(0, \sigma^2) \tag{6.71}$$

where μ is the overall mean, η_t is the loudspeaker effect, β_j is the programme effect, $\gamma_{t,j}$ is the effect accounted for by the interaction between loudspeakers and programmes, α_s represents the effect of subjects, $\delta_{t,s}$ represents the interaction between loudspeaker and subjects, $\eta_{j,s}$ represents the interaction between programme and subjects and $\epsilon_{t,j,s,i}$ represents the error.

The results of an analysis using this model is shown in Table 6.15. The results show that none of the subject-related variables are significant, and a comparison with the results in Table 6.14 shows that the variance removed from the error variance was not enough to compensate for the loss in degrees of freedom. The MS_{error} thus increases from 1.005 to 1.024 when the subject variables are included. It is therefore recommended to use the simpler model in this situation. However, it has been found, in most experiments analysed by the authors, that the subject variable and its interactions are often significant and thus lead to an increase in sensitivity and more accurate analysis of the results.

The next issue to be considered in connection with adding the subject variable is the potential for correlation between scores from the same subject. This introduces a more general topic that in textbooks is often discussed under the heading of 'fixed versus random-effect models'. The term *fixed* variable describes variables, like the loudspeaker and programme variables discussed above, that have specific fixed levels, represented by the selected loudspeakers and programmes, which eventually can be used in replicated versions of (or new) experiments. An ANOVA model including only fixed variables is termed a *fixed model*. The other possibility is a 'random' variable, which has levels randomly chosen from the relevant population. If the ANOVA model only includes random variables, it is a *random model*; however, most models include both fixed and random variables, and these models are termed *mixed models*[36].

Including a random variable in an ANOVA model has two main implications:

- The tested levels of the variable (e.g. different subjects) represent a true random sample from the population of interest.
- Correlation between scores from the same subject is included in the ANOVA model.

The first implication is often used as the main argument for modelling a variable as 'random' because it follows that the statistical inferences regarding this variable can be extrapolated to the entire population from which the sample has been drawn. The statistical inferences based on a fixed variable, however, only apply to the tested levels of the variable.

The conclusions based on an analysis with, for example, the loudspeaker as a fixed variable are thus only valid for the loudspeakers in the group and cannot be generalised to the entire population of loudspeakers. The idea of having a random sample of loudspeakers is appealing as it would increase the benefit of the results of the experiment. However, in most listening tests, it is impossible to have a truly random sample of loudspeakers and programmes, so these variables are often (always) modelled as fixed variables.

Specifying a variable (e.g. the loudspeaker variable) as fixed in an ANOVA model implies that the treatment levels are

[36]Note that the term *'mixed'* was also used for the combination of a between-subject and a within-subject experimental design.

assumed to be fixed constants and that the treatment effects add up to zero, as shown in the formula:

$$\sum_{t=1}^{5} \eta_t = \sum_{t=1}^{5} (\eta_t - \mu) = 0 \qquad (6.72)$$

where η_t is the loudspeaker variable (it is assumed that there are five loudspeakers) and μ is the overall mean.

The hypothesis being tested for fixed variables is (see also Equation 6.64)

$$H_0 : \eta_t = 0 \text{ (for all } t) \qquad (6.73)$$

The subject variable, however, could potentially be a random variable as one could assume that a random sample from a nation's population would be representative (at least for auditory capabilities) for the population in general–or at least a large part of it. Further, as discussed above, it is generally assumed that scores from the same subject are not independent.

Assuming that the subject variable is a random variable and that it is $N(0, \sigma_S^2)$ distributed, with σ_S^2 being the variance of the population of interest, the hypothesis being tested is:

$$H_0 : \sigma_S^2 = 0 \qquad (6.74)$$

The issue of whether to use the subject variable as fixed or random has been the topic of much debate[37] The basic arguments have often been focussing on the problems in obtaining a random sample of subjects from a given population. The argument for this is that no matter what size of population the sample is based on, it will always be a random sample from that population and thus will be of interest to be able to extrapolate it. The basic argument against this is that in many situations a given

[37] Most psychologists would argue that subjects must be treated as a random factor, and in analysis of medical experiments, this is routine procedure (see for example Brown and Prescott [74]). However, it was not until recently that, for example, the sensory scientists in the food industry started including subjects as a random variable in the model. The work of Brockhoff and Skovgaard [72] was one of the first to establish the formal statistical basis for including subjects as a variable in the analysis of sensory data from food evaluation. In [72] the subjects are modelled as a fixed variable, but in Brockhoff [69] they are modelled as a random variable.

subject group will participate in many experiments and thus acquire training, which will disqualify them as representative (a random sample) of the original group from which they were sampled. The issue of generalising from a random sample is a complicated issue and somewhat outside the scope of the book. Keppel and Wickens [253] (Section 1.3) discuss the issue and distinguish between statistical generalisation, which is related to the true randomness of the sample, its size, and so on, and non-statistical generalisation, which is related to the knowledge of the research area and the variables being examined.

It is difficult to give general guidance, but it is the author's view that in most listening tests it is justified to include subjects as a random variable. The exception could be if the subjects have been highly trained for a particular type of experiment.

The issue of obtaining a random sample of subject has, as noted often, been the focus of the discussion, but it is important to realise that the choice of a random versus fixed subject variable also has implications for the tests and assumptions in the ANOVA model. The statistical considerations are rarely discussed in the auditory literature, but they are, in the author's view, more important than the sampling issue.

The ANOVA model that will be used for this discussion is a simplified version of the model used when introducing the subject variable (see Equation 6.71) as the interactions have been left out. This will not limit the principal aspects of the discussion as an interaction that is a combination of a fixed and random variable is a random variable.

$$Y_{t,j,s} = \mu + \eta_t + \beta_j + \alpha_s + \epsilon_{t,j,s}, \quad \epsilon_{t,j,s} \sim N(0, \sigma^2) \tag{6.75}$$

where μ is the overall mean, η_t is the fixed loudspeaker effect, β_j is the fixed programme effect, α_s is either fixed or random subject effect–if random, it is assumed to be independent and normally distributed $\alpha_s \sim N(0, \sigma_S^2)$–and $\epsilon_{t,j,s}$ represents the error.

The statistical differences between fixed and mixed ANOVA models can be clarified by studying[38]:

[38]The discussion of differences between fixed and mixed ANOVA models is strongly inspired by the excellent lecture notes prepared by the statistics groups at the Royal Veterinary and Agricultural University in Denmark. These notes are, unfortunately, only available for students at the University. Interested readers are referred to Brown and Prescott [74] for a further discussion.

Quantity	Fixed model	Mixed model
$E(Y_{t,j,s})$	$\mu + \eta_t + \beta_j + \alpha_s$	$\mu + \eta_t + \beta_j$
$\text{var}(Y_{t,j,s})$	σ^2	$\sigma^2 + \sigma_S^2$
$\text{cov}(Y_{t,j,s}, Y_{t',j',s'})$	0	$\sigma_S^2 (\textit{if } s = s')$
$(t \neq t', j \neq j'))$		$0\ (\textit{if } s \neq s')$

Table 6.16: Comparisons of fixed and mixed ANOVA models.

- the expected value $E(Y_{t,j,s})$ of the t,j,sth observation $Y_{t,j,s}$.
- the variance $\text{var}(Y_{t,j,s})$ of the t,j,sth observation.
- the relation between two different observations (the correlation/covariance).

The comparisons are summarised in Table 6.16.

The first row shows that the average value of $Y_{t,j,s}$ is the sum of the expected values for the fixed variables, but is zero for the random variable, as defined by Equation 6.74. The variance of $Y_{t,j,s}$ for the fixed model is given by the error variance (σ^2); however, for the mixed model it is increased by the variance of the subject variable σ_S^2. This result is derived, using standard rules, below:

$$\begin{aligned} \text{var}(Y_{t,j,s}) &= \text{var}(\mu + \eta_t + \beta_j + \alpha_s + \epsilon_{t,j,s}) \\ &= \text{var}(\alpha_s + \epsilon_{t,j,s}) \\ &= \text{var}(\alpha_s) + \text{var}(\epsilon_{t,j,s}) \\ &= \sigma_S^2 + \sigma^2 \end{aligned} \tag{6.76}$$

The variance estimate of the error variance in the mixed model thus includes the variance relevant for the fixed variables (σ^2) plus the variance related to the random variable (σ_S^2). The statistical inferences (e.g. based on the confidence interval) thus includes the 'properties' of the subject population, and the conclusions of the experiment are thus assumed to be applicable to the subject group. The error variance in the fixed model only includes the variance (σ^2) relevant for the particular set of tested levels of the fixed variables, and conclusions are thus only relevant for these levels.

One of the main assumptions for the fixed model is that all observations are independent, so the covariance is zero between observations. This can be illustrated by the covariance matrix

$$\begin{pmatrix} \sigma^2 & 0 & . & . & . & 0 \\ 0 & \sigma^2 & . & . & . & 0 \\ . & . & . & . & . & . \\ . & . & . & . & . & . \\ . & . & . & . & . & . \\ 0 & 0 & . & . & . & \sigma^2 \end{pmatrix}$$

Figure 6.12: Covariance matrix for N independent observations for a single subject. The matrix includes N rows and N columns and the diagonal entry is the variance of the underlying distribution of each observation.

for a single subject, as shown in Figure 6.12. It includes N rows and N columns corresponding to all stimuli combinations that the subject has evaluated. The entry is the covariance between observations, and it is seen that it includes only variance elements in the diagonal as the observations are assumed to be independent. It is also noted that the variance component is the same for all diagonal elements, in accordance with the assumption of variance homogeneity. This covariance structure is termed *diagonal* in statistics programs.

The addition of a random variable introduces another covariance structure for the mixed model as the covariance between observations by different subjects ($s \neq s'$) is zero; however, observations by the same subject ($s = s'$) are now correlated and have a covariance σ_S^2. This is shown as follows:

$$\begin{aligned}
&\text{cov}(Y_{t,j,s},\ Y_{t',j',s'}) \\
&\quad = \text{cov}(\mu + \eta_t + \beta_j + \alpha_s + \epsilon_{t,j,s},\ \mu + \eta_{t'} + \beta_{j'} + \alpha_{s'} + \epsilon_{t',j',s'}) \\
&\quad = \text{cov}(\alpha_s + \epsilon_{t,j,s},\ \alpha_{s'} + \epsilon_{t',j',s'}) \\
&\text{(only random variables are correlated)} \\
&\quad = \text{cov}(\alpha_s,\ \alpha_{s'}) + \text{cov}(\alpha_s,\ \epsilon_{t',j',s'}) \\
&\qquad + \text{cov}(\epsilon_{t,j,s},\ \alpha_{s'}) + \text{cov}(\epsilon_{t,j,s},\ \epsilon_{t',j',s'})
\end{aligned} \tag{6.77}$$

Considering observations from different loudspeakers ($t \neq t'$), different programmes ($j \neq j'$) and different subjects ($s \neq s'$), it follows from the independence assumptions of the mixed model

$$\begin{pmatrix} \sigma^2 + \sigma_S^2 & \sigma_S^2 & \cdot & \cdot & \cdot & \sigma_S^2 \\ \sigma_S^2 & \sigma^2 + \sigma_S^2 & \cdot & \cdot & \cdot & \sigma_S^2 \\ \cdot & \cdot & \cdot & \cdot & \cdot & \cdot \\ \cdot & \cdot & \cdot & \cdot & \cdot & \cdot \\ \cdot & \cdot & \cdot & \cdot & \cdot & \cdot \\ \sigma_S^2 & \sigma_S^2 & \cdot & \cdot & \cdot & \sigma^2 + \sigma_S^2 \end{pmatrix}$$

Figure 6.13: Covariance matrix for N observations for a single subject under the assumption of a random subject variable. The matrix includes N rows and N columns and the diagonal entry is the variance of the underlying distribution of each observation plus the variance (σ_S^2).

that these covariances will be zero. However, if observations from the same subject ($s = s'$) are considered, it follows that the covariance will be given by the first term in Equation 6.77:

$$\mathrm{cov}(Y_{t,j,s}, Y_{t',j',s'}) = \mathrm{cov}(\alpha_s, \alpha_{s'}) = \mathrm{var}(\alpha_s) = \sigma_S^2 \quad (6.78)$$

The covariance matrix for a single subject will thus appear as shown in Figure 6.13, where it is seen that all off-diagonal elements are equal to σ_S^2. This form of the covariance matrix is termed a *compound symmetry*.

A small note on the covariance matrix and its interpretation is relevant at this point. The covariance matrix in Figure 6.13 is based on the assumption of homogeneity of variance and homogeneity of correlation. This is likely to be fulfilled if the experiment uses a completely randomised design and a reasonable number of stimuli ensuring that order effects, and other effects, will not be important[39]. The assumption is called *sphericity* and can be tested by Mauchly's test. It should be noted that Mauchly's test is not very robust for small sample sizes. If the assumptions are not met, corrections have been formulated, called *Greenhouse-Geisser* [149] and *Heynh–Feldt* [174] corrections, that may be

[39] It is difficult to give exact guidance on a sufficient number of stimuli, however, the general rule (see for example Miller [295]) that subjects can maximally distinguish between seven stimuli when presented simultaneously can be used as guidance for the minimum number of stimuli in a test.

applied. A more detailed discussion of the test for sphericity and the corrections is outside the scope of this book. Both the Mauchly's test and the corrections are included in SPSS's standard analysis package, and the interested reader is advised to consult the original references or Keppel and Wickens [253] (Section 17) for a discussion of applicability. Another option, if the assumptions are not met, is to use multivariate analysis of variance (MANOVA) models. This type of model includes several dependent variables that can be correlated, and there are no assumptions regarding the structure of the covariance matrix. The use of the MANOVA model to analyse mixed models even if there is only one dependent response variable is based on the observation that the repeated scores from the same subject can be viewed as a vector of individual scores, where each score is from an individual dependent variable. The reader is recommended to consult, for example, Hair *et al.* [155] and Keppel and Wickens [253] (Section 17) for a further discussion on MANOVA models.

While the request for a completely randomised presentation order is manageable for small sample sizes, it can quite quickly become a logistic problem if the maximum session duration must be limited to approximately 30–40 min to avoid fatigue in the subjects. In this case, observations could be separated in time, and it is thus quite likely that observations close in time will be more correlated than the observations more separated in time. Such a situation will violate the sphericity assumption; however, it is possible to change the structure of the covariance matrix to accommodate such a situation, and a number of different standard versions of the covariance matrix are options in most analysis software. The analysis of mixed models including a choice of difference structures for the covariance matrix can be performed in SPSS using either the 'linear mixed models' or the 'GLM repeated measures'.

The choice of using random, fixed or a combination of variables in an ANOVA model also has consequences for the subsequent testing of the variables. The issue is to use the correct denominator for the F-tests.

If all variables are fixed, the denominator in all tests is the estimated error variance. However, if, for example, the subject variable is modelled as random, the correct denominator is now the interaction between the variable being tested and the subject variable.

The choice of the correct denominator can be mathematically formalised as it can be shown that the expected mean square (EMS) (EMS $= SS/df$) for each variable is a weighted sum of the EMS related to the variable itself plus EMSs for other variables in the ANOVA model. The rule is that the EMS of the numerator must only have one additional element compared to the denominator. The composition of the EMS terms can thus be calculated for each variable in the model, and on the basis of this EMS scheme the correct denominator can be determined for each variable. For a more detailed discussion of this, see Hicks [162] (Section 10). Most statistical analysis programs have an automatic feature that will do this.

The simulated data set used previously has been analysed using a model shown in Equation 6.71, but with the subject variable as a random variable. The results are shown in Table 6.17, and it is seen that a second line, termed *error* has been added under each variable. This is the denominator used for the F-test, and it is seen that it is different for the different variables and that the composition of each error variable is identified below the table. The conclusion regarding the significance of the subject variable has not changed by changing the status from fixed to random, but the F value and significance level have changed compared to the fixed model analysis shown in Table 6.15. This also applies to the loudspeaker and programme variables. The composition of the EMS for the denominator in the F-tests is calculated by the statistics program and listed, as shown in Table 6.17. However, if more detailed knowledge of the EMS compositions is required, it is possible (in SPSS) to provide the calculations, as shown in Table 6.18.

On the basis of the table it is now possible to compose the EMS for each variable, and an example is given below for the programme variable ($\mathrm{EMS_{prog}}$) and its 'error' ($\mathrm{EMS_{prog\times sub}}$):

$$\mathrm{EMS_{prog}} = 5 \times \sigma^2_{\mathrm{prog\times sub}} + \sigma^2_{\mathrm{error}} + Q(\mathrm{prog, loud} \times \mathrm{prog}) \quad (6.79)$$

$$\mathrm{EMS_{prog\times sub}} = 5 \times \sigma^2_{\mathrm{prog\times sub}} + \sigma^2_{\mathrm{error}} \quad (6.80)$$

Source		Type III sum of squares	*df*	Mean square	*F*	Sig.
Loudspeaker	Hypothesis	576.246	4	144.062	141.780	0.000
	Error	36.579	36	1.016[a]		
Programme	Hypothesis	31.218	3	10.406	12.024	0.000
	Error	23.366	27	0.865[b]		
Subject	Hypothesis	10.469	9	1.163	1.356	0.311
	Error	9.543	11.126	0.858[c]		
Loudspeaker × programme	Hypothesis	185.287	12	15.441	15.082	0.000
	Error	110.568	108	1.024[d]		
Loudspeaker × subject	Hypothesis	36.579	36	1.016	0.992	0.493
	Error	110.568	108	1.024[d]		
Programme × subject	Hypothesis	23.366	27	0.865	0.845	0.684
	Error	110.568	108	1.024[d]		

[a] MS(loudspeaker × subject)
[b] MS(programme × subject)
[c] MS(loudspeaker × subject) + MS(programme × subject) − MS(error)
[d] MS(error)

Table 6.17: ANOVA table of simulated data based on five loudspeakers and four programmes assessed by 10 subjects. The ANOVA model includes the overall mean plus variables loudspeaker, programme and subject plus second-order interactions. The subject variable is modelled as a random variable.

Expected Mean Squares[a,b]

Source	Variance component				
	Var(subject)	Var(loudspeaker × subject)	Var(programme × subject)	Var(Error)	Quadratic term
Intercept	20.000	4.000	5.000	1.000	Intercept, loudspeaker, programme, loudspeaker × programme
Loudspeaker	.000	4.000	.000	1.000	loudspeaker, loudspeaker × programme
Programme	.000	.000	5.000	1.000	programme, loudspeaker × programme
Subject	20.000	4.000	5.000	1.000	
Loudspeaker × Programme	.000	.000	.000	1.000	loudspeaker × programme
Loudspeaker × subject	.000	4.000	.000	1.000	
Programme × subject	.000	.000	5.000	1.000	
Error	.000	.000	.000	1.000	

[a]For each source, the expected mean square equals the sum of the coefficients in the cells times the variance components, plus a quadratic term involving effects in the quadratic term cell.

[b]Expected mean squares are based on the Type III sums of squares.

Table 6.18: Composition of expected mean squares (EMS) for the ANOVA shown in Table 6.17.

So the F-test of the programme variable is the ratio of the two EMS values, and it is seen that it is the magnitude of the quadratic term, involving the programme and the interaction between programme and loudspeaker, that is tested.

Reducing the number of variables in the ANOVA model The above discussion on various forms of the ANOVA model has only included the full model with all variables and interactions. The results of the ANOVA will then show which of the included variables have a significant influence on the dependent variable. The next step in the analysis is to conduct the necessary comparisons of the levels of the significant variables to conclude which of these are responsible for the significant effect. This is not a problem if the variables are clearly significant, but in some cases the significance level might be just on the borderline. In general, but especially in borderline cases, it is a good idea to re-run the ANOVA without the non-significant variables included. The reason is that the degrees of freedom will increase for the error variance, but the added variance will often not compensate for this, and the result is that the MS_{error} decreases in magnitude and thus increases the resolution of the tests.

Furthermore, it is good practice to report an ANOVA model only including significant variables.

Test of ANOVA model assumptions In Section 6.2.3, it was discussed how to check the statistical assumptions behind various statistical tests. The issue of outliers was also discussed. This section discusses the same issues, with special emphasis on the ANOVA model, and provides some examples on the basis of the example discussed above.

The discussion in Section 6.2.3 was focussed on the distribution of repeated observations, homogeneity of the variance across stimuli, and so on. For the ANOVA model, however, it is the residuals that are of interest. The assumptions, in order of importance, that need to be checked are thus:

- Independence of residuals
- Homogeneity of the variance of the residuals
- Normal distribution of residuals
- Loss of subjects.

The residuals to be investigated are based on the difference between the observed data and those predicted by the ANOVA model. The type and 'size' of the ANOVA model must thus be considered before analysing the residuals. The following checks of the assumptions are only valid for fixed models as the theory for mixed models is still under development. Any random variables could (should) be checked with simple measures in a separate analysis. The ANOVA model should also be the final (reduced) version including only significant variables. Most statistical analysis packages include an option under the ANOVA analysis for saving the residuals for further analysis. There are often several options for the type of residuals that are to be saved:

- **Unstandardised.** These are the residuals in raw form.
- **Standardised.** These are the residuals normalised with the estimated standard deviation, so they have a mean of 0 and a standard deviation of 1.
- **Studentised.** This is the residual divided by an estimate of its standard deviation that varies from case to case, depending on the distance of each case's values of the independent variables from the means of the independent variables.

All three versions can be used, but normally, preference is given to the standardised version because if the assumptions are fulfilled, the errors will be normally distributed, and it is easy to check (visually) whether about 95 % of the residuals fall between −2 and +2 and to spot whether there are any outliers. The theoretically most correct version is the studentised version as the variance of the unstandardised residuals is not homogeneous and the studentised version has been corrected for this. The following examples will be based on the studentised residuals:

The independence of the residuals is normally quite difficult to examine, but in case of repeated measurements it can be performed by the test for sphericity, as mentioned earlier. In other cases, a good starting point is visual inspection of the relationship between the residuals and any sequencing variable such as time.

The homogeneity of the variance of the residuals can be checked by visual inspection of plots of observed versus expected residuals and observed residuals versus variable levels. The observed versus expected residual values have been plotted in Figure 6.14 on the basis of a model including all fixed variables (loudspeaker, programme, subject and their two-way interaction). It is seen that there is no relationship between the magnitude of the residual and the size of the observation.

The relationship between the residuals and the various variables are shown in Figure 6.15.

The results show that the distribution of the residuals is similar for all variable levels. The results thus indicate that the variance of the residuals is homogeneous across variable levels.

The distribution of the residuals are shown in Figure 6.16, where it is seen, as expected base on the previous tests, that they are normally distributed. This is confirmed by the so-called normal probability plot of the residuals shown in Figure 6.17. The probability plot compares the studentised residuals with the normal distribution, and if all residuals are on the diagonal, it follows that the distribution is normal. A closer inspection of the deviations from normality can be seen in the so-called detrended probability plot shown in Figure 6.18, where the deviation from the diagonal is shown.

If the analysis of the residuals shows that they do not fulfil the assumptions, the possibility of transforming the data should be

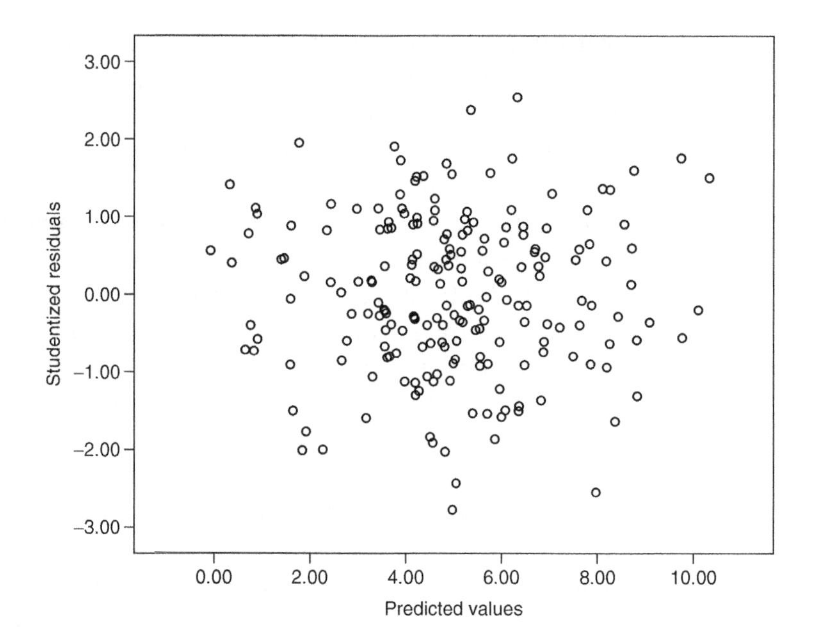

Figure 6.14: Studentised residuals versus predicted values.

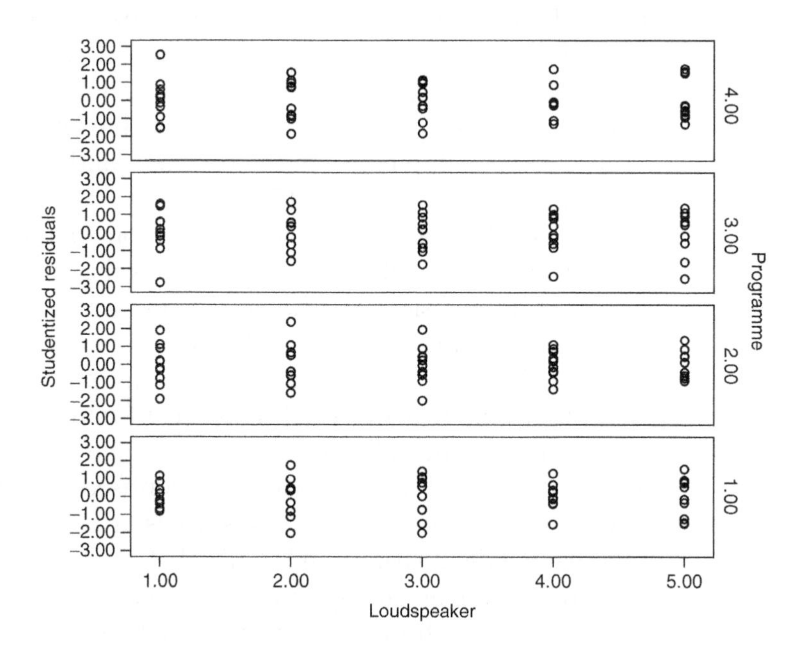

Figure 6.15: Studentised residuals versus variable levels.

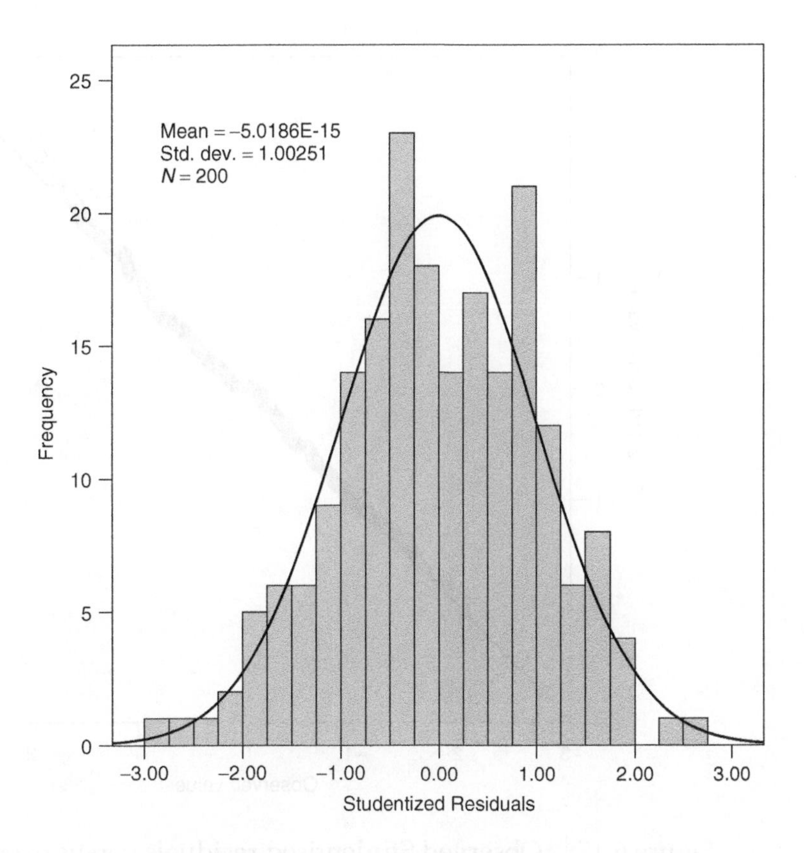

Figure 6.16: Distribution of Studentised residuals. The expected normal distribution $N(0,1)$ is indicated by the solid line.

considered. In general it is difficult to give specific recommendations, but if it can be determined that there is a relationship between the size of the observations and their variability, its often possible to make simple transformations (see also Keppel and Wickens [253] (Section 7.4)). Table 6.19 lists a number of distributions that have a relationship between the mean and the variance, and corresponding transformations that will transform these distributions closer to normality are also shown.

Keppel and Wickens [253] (Section 7.4) list a number of examples of experimental paradigms where these distributions are likely to be encountered. Two of these are relevant for listening test situations: The distribution of the time interval between presentation of the stimulus and recording of the response will often have a standard deviation that is proportional to the

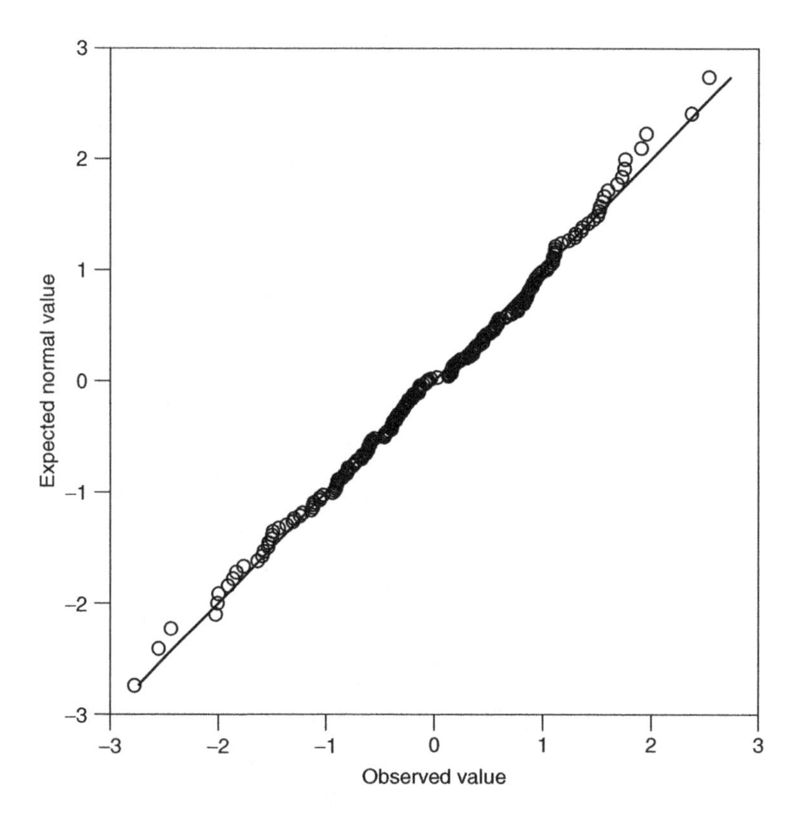

Figure 6.17: Observed Studentised residuals versus expected values based on the normal distribution.

mean. Such data can be transformed by applying a logarithmic transformation $Y_{i,j} = \log(Y_{i,j} + 1)$ instead of the square root indicated in the table. Another situation of interest is whether the score is a proportion (e.g. the number of times a loudspeaker is preferred) as the variance of such data depends on the true value of the proportion. The distribution of the scores is binomial and the correction to apply is the arcsine of the square root of the original observations, as shown in the first line in the table.

One of the experimenter's major concerns is to lose subjects either before or in the middle of the experiment. Although this is not a part of the standard statistical assumptions, it will have an influence on the assumptions of the analysis. For example, losing a subject just before the experiment can be critical if the necessary number of subjects have been estimated and divided

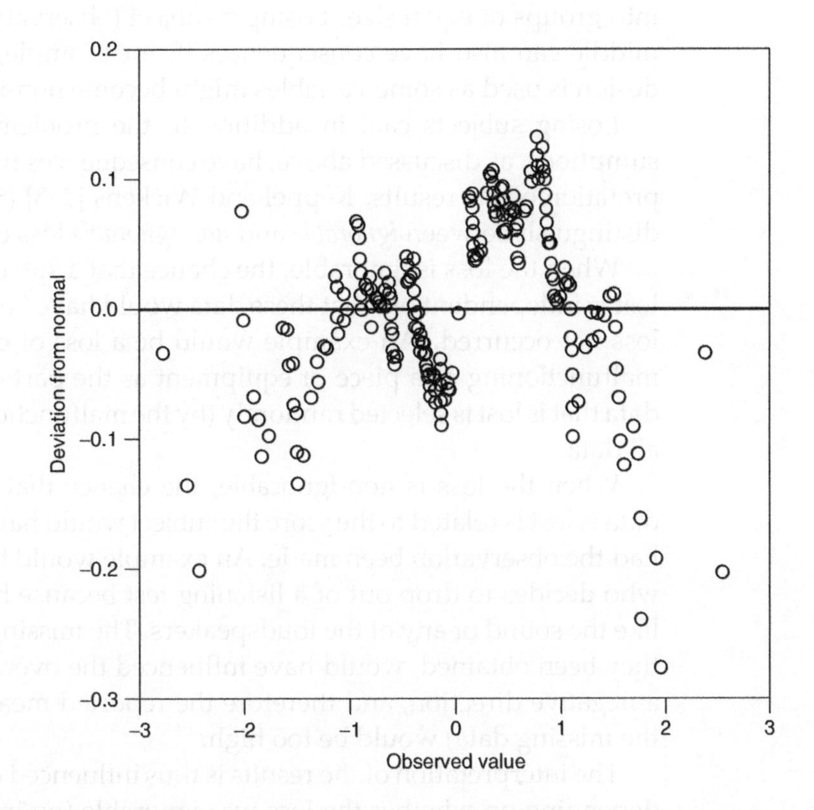

Figure 6.18: **Deviation between the observed Studentised residuals and the expected normal distribution.**

Distribution	Variance	Scale	Transformation
Binomial	$\mu \times (1 - \mu)$	interval (0, 1)	arcsine $\sqrt{}$
Poisson	μ	Positive	$\sqrt{}$
Gamma	μ^2	Positive	log
Inverse Gauss	μ^3	Positive	$1/\sqrt{}$

Table 6.19: **Distributions where the variance depends on the mean and associated calculations that will transform the distribution to a normal distribution.**

into groups of equal size. Losing a subject (observations) in the middle can also have consequences if, for example, a blocked design is used as some variables might become non-testable.

Losing subjects can, in addition to the problems with assumptions, as discussed above, have consequences for the interpretation of the results. Keppel and Wickens [253] (Section 7.2) distinguish between *ignorable* and *non-ignorable* loss of subjects.

When the loss is ignorable, the chance that a subject's data is lost is independent of what these data would have been, had the loss not occurred. An example would be a loss of data due to malfunctioning of a piece of equipment as the particular set of data that is lost is selected randomly (by the malfunction event) of all data.

When the loss is non-ignorable, the chance that a subject's data is lost is related to the score the subject would have obtained had the observation been made. An example would be a subject who decides to drop out of a listening test because he does not like the sound of any of the loudspeakers. The missing data, had they been obtained, would have influenced the overall mean in a negative direction, and therefore the reported mean (without the missing data) would be too high.

The interpretation of the results is thus influenced differently, depending on whether the loss was ignorable (no influence) or non-ignorable (a bias is introduced). The experimenter must thus be very careful when evaluating the consequences of losing subjects.

ANOVA model for more than one subject group The discussion above has been focussed on complete block within-subject designs, where all subjects are presented with, and assess, all stimuli in the experiment in completely randomised presentation order. However, in some situations it might be of interest to have several and different groups of subjects participate in the experiment. This was, for example, the situation when International Telecommunications Union (ITU) conducted verifications tests for the ITU model ([213]) to predict the perceived quality of low bit-rate audio codec systems. To increase the validity of the results, it was decided to conduct the same experiment in Japan, Germany and Sweden using local subject groups. The database generated at the three test sites has recently been used as verification for another series of experiments, and a description of the original experiment can be found in Bech *et al.* [38].

The experimental design used for this experiment was a so-called mixed design (see Section 6.1). The experimental variables were 'test site', 'item' (a piece of programme material en- and decoded by a certain codec algorithm) and 'subject'. At each test site, the local group of subjects evaluated all stimuli (items) in a completely randomised order, and the same selection of items were evaluated at each site. The experimental design was thus a combination of a within-subject design with the fixed variable 'item' and random variable 'subject' and a between-subject design with the fixed variable 'test site'. The experimental data can be analysed, for example, in SPSS using either the 'repeated measures' option under 'general linear model' or the 'mixed models' option. Such designs are not commonly encountered in listening tests, so the analysis will not be discussed in further detail here. The reader is referred to Bech *et al.* [38] for an example and to Keppel and Wickens [253] (Sections 19, 20 and 23) for other examples and a discussion of the statistical theory.

6.2.5 *Statistical checklist*

Altman *et al.* [13] present a set of statistical guidelines for contributors to medical journals and Gardner *et al.* [139] present a useful check list for checking the statistics applied in contributions to British Medical Journal (BMJ). The statistical guidelines and check list include a number of points that are relevant for contributions related to perceptual evaluation of sound, and these are referred–some in an adapted form–in the following:

Statistical design The reviewer should be able to answer affirmative to the following questions:

- Was the objective of the study sufficiently described?
- Was an appropriate experimental plan used to achieve the objective?
- Was a satisfactory description of the subjects, their background and selection criteria given?
- Were the treatments (stimuli) well defined?
- Was random allocation or another strategy used to allocate the stimuli to the subjects?

- If random allocation was used, was the method of randomisation described in detail?
- Was a pre-study calculation of required sample size reported?

Analysis and presentation The reviewer should be able to answer affirmative to the following questions:

- Were all statistical procedures described or referenced?
- Were the statistical analyses applied correctly?
- Was the statistical material presented satisfactory?
- Were the confidence intervals given for the main results?
- Were the statistical inferences justified?

PART II

Technical considerations

CHAPTER

7

Electroacoustic considerations

In this chapter, a number of different electroacoustic considerations associated with conducting listening tests will be discussed. The issue of listening spaces will be studied in detail, specifically in relation to several commonly used standards for listening rooms. Other types of listening spaces will also be dealt with in addition to some important practical considerations associated with such spaces. We will also provide a short section on the requirements associated with the placement of loudspeakers and listeners in listening spaces, once again referring to essential standards on the topic. A brief section that considers how the inclusion of accompanying pictures may interact with the acoustic conditions is also provided. Lastly, details pertaining to the selection and measurement of electrical devices required for listening tests will be presented.

Perceptual Audio Evaluation – Theory, Method and Application Søren Bech and Nick Zacharov

There are a number of different types of listening spaces that can be considered for listening tests. Most of these are acoustically and physically well controlled in nature and provide a stable and comfortable environment for subjects to perform listening tests. This is the primary category to be considered here, although a short mention of other suitable listening spaces will be provided. Naturally, ecologically motivated spaces can be used in any test and these are also well justified if representative of a particular usage situation, for example. This latter category is more difficult to characterise as almost any environment can be categorised as ecologically motivated, such as a domestic living room, a car, a concert hall, and so on, and thus will only be considered in general terms.

7.1 Listening rooms

The concept of a listening room has been developed in order to create reference spaces within which certain types of listening tests can be performed over loudspeakers. For headphone testing, the influence of the room is practically insignificant [112] and, as a result, the acoustic characteristics of the listening space may be less stringent. It is important when performing listening tests, as in all scientific work, that the results from one laboratory be comparable with those from another laboratory. For this to occur, the test facilities and methodologies at each laboratory must be similar. For listening tests, the acoustic properties of the acoustic space are known to play an important role [29, 333, 440]. As with concert hall acoustics, the size, shape and treatment of the listening room will strongly influence the characteristics of the sound field. It is thus desirable to attempt to regulate the acoustic properties of the listening testing facilities such that the inter-laboratory differences are minimised.

The primary intention of a listening room is to provide controlled acoustic characteristics with respect to room reverberation, background noise, and so on. The specific performance requirements are presented in detail in the following section. Additionally, a listening room must also fulfil numerous other requirements associated with listening tests, that will enable stimuli to be presented in a controlled manner and subjective data to be collected from subjects. In terms of acoustic and vibration characteristics, the room also needs to be well isolated from the surrounding environment to ensure that sound (e.g. office noise) and vibration (e.g. from ventilation machinery, adjacent road, etc.) do not adversely degrade the soundfield in the room. The rooms must be comfortable and livable for subjects for an extended period of time. Lastly, suitable cabling, lighting and provision for special equipment and computers must be ensured. All of these factors and more must be taken into consideration, ideally at the design stage, and will be discussed in further detail in Section 7.1.4.

When planning to build a listening room, it is wise to be aware of all the requirements that need to be taken into consideration. Many of the requirements are outlined in a number of standards or recommendations that are discussed in the following sections, with key characteristics of which are summarised in Table 7.2. It is advisable to contact an acoustic consultant with expertise in this field and, more specifically, with experience in building such rooms. Such a consultant may provide advise with the assistance of several experts in matters related to the design of electrical and ventilation systems and perhaps on vibration issues. It is important for all issues to be well designed at the outset, where possible, as later modification can be very complex and costly, if not near impossible without a partial or complete reconstruction of the space.

7.1.1 *IEC 60268-13 listening rooms*

One of the earlier technical reports associated with the specification of a listening room was provided by the International Electrotechnical Commission (IEC), originally developed in 1985 as IEC 268-13 and then updated in 1998 as IEC 60268-13 [184]. The technical report applies to listening tests of loudspeakers in domestic environments, with monophonic, stereophonic and multichannel configurations and the associated audio electronics. The recommendation also provides guidance for audio evaluation performed on television systems, that is, with audio-visual content.

The primary aim of the IEC listening room is to provide a stable and controlled listening environment that approximates 'a residential listening environment in the geographical region where the test results are intended'. One of the problems with this statement is that little guidance is provided for such regional differences and it is not stated for which geographical region the specified room is designed. Guidance is provided in order to avoid major acoustic defects that might adversely affect the room's performance. This is of particular importance with respect to the perception of reverberation, timbre, spaciousness and background noise.

To ensure a suitable distribution of low-frequency standing waves, the aspect ratio of a listening room, IEC 60268-13 recommends that the room dimensions be based upon the following three criteria:

$$(w/h) \leq (l/h) \leq (4.5(w/h) - 4) \tag{7.1}$$

$$(w/h) < 3 \tag{7.2}$$

$$(l/h) < 3 \tag{7.3}$$

where l is the room length, h is the room height and w is the room width.

Additionally, it is advised that the floor area for monophonic and two-channel stereophonic reproduction lie in the range 25–40 m^2 and in the range 30–45 m^2 for multichannel reproduction. The presented reference room is of dimensions $l = 7.0\,\text{m}$, $w = 5.3\,\text{m}$ and $h = 2.7\,\text{m}$.

In addition to the aspect ratio, which influences the distribution of standing waves in the room, the reverberation time should be carefully controlled. It is advisable that the average reverberation time T_m lies in the range 0.3–0.6 s in the frequency range 200–4000 Hz. The reverberation time should be measured in accordance with ISO 3382 [196] in 1/3 octave bands without subjects being present. In the cases where a more stringent specification is required, it is advisable that the reverberations lie within the bounds of the mask illustrated in Figure 7.1 with an average reverberation of 0.4 s.

The room should be normally furnished, as a domestic living room, providing typical absorption and diffusion of the sound field. The floor should be absorbent and the ceiling reflective in nature. Comfortable low back chairs should be provided for the comfort of the subjects and also to reduce reflection or obstruction of the sound near the subject's head. Optically opaque curtains should be provided to hide speakers under test. Such curtains should change the measured room response by less than 0.3 dB in all 1/3 octave bands below 10 kHz.

In order that the room be suitable for critical listening, the background noise level should be carefully controlled and not exceed NR15

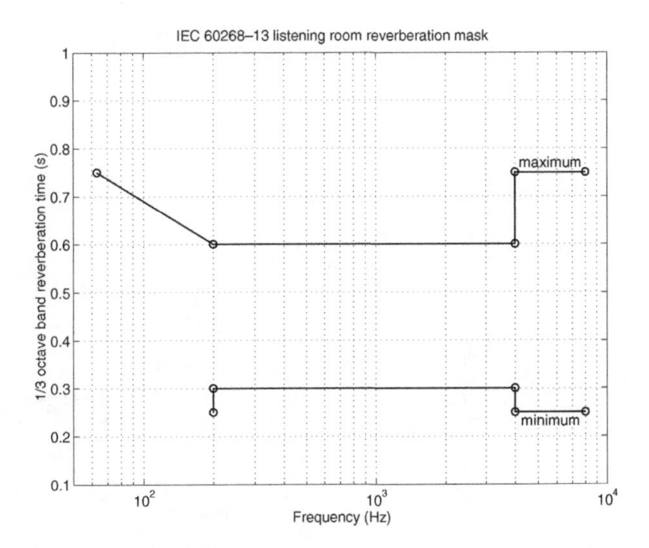

Figure 7.1: IEC 60268-13 listening room reverberation mask (adapted from [184]). Reproduced by permission of IEC.

(approximately 20–25 dBA), details of which can be found from [55] (see Figure 7.2). These performance figures are defined to be measured with the ventilation running, other equipment on and with no subjects present. An example of an IEC listening room can be found at Bang and Olufsen, Denmark, the primary characteristics of which are presented in Figure 7.3.

7.1.2 ITU-R Recommendation BS.1116-1 listening rooms

The ITU-R Recommendation BS.1116-1 [207] was developed for the assessment of small impairments in audio, including multichannel sound systems. Part of the recommendation provides the specification for the listening conditions including the characteristics for a reference listening room. This specification is not vastly different from that defined in IEC 60268-13; however, there are sufficient differences to justify an explanation here. At the time of writing, the majority of listening rooms were designed and built according to ITU-R Recommendation BS.1116-1. The reader should also be aware that there are slight differences between the room requirements specified within ITU-R Recommendation BS.1116 and the revised edition ITU-R

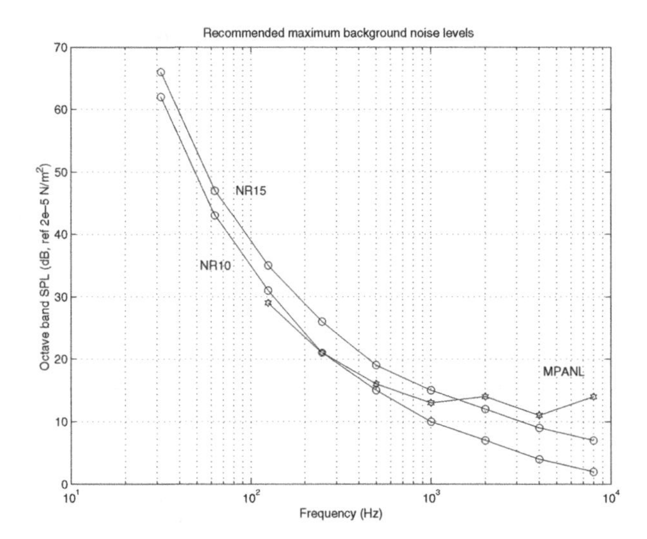

Figure 7.2: The recommended maximum background noise levels according to ISO Noise Rating (NR) curves, details of which can be found from [55] and maximum permissible ambient noise levels (MPANL), ears not covered, according to ANSI S3.1-1999 [14].

Recommendation BS.1116-1 and certain rooms have been designed according to the earlier recommendation.

The primary differences compared to IEC 60268-13 include

- shorter reverberation time with tighter tolerances;
- lower noise level;
- specification for early reflected sound;
- specification for operational room response;
- lack of furnishings.

The room should be symmetrical about the vertical mid-plane with a rectangular or trapezoidal[40] shape. The floor area for monophonic and two-channel stereophonic reproduction should lie in the range 20–60 and 30–70 m^2 for multichannel reproduction.

Uniform low-frequency standing wave distribution is ensured by using the following equations to specify the aspect ratio of the room.

$$1.1(w/h) \leq (l/h) \leq (4.5(w/h) - 4) \tag{7.4}$$

[40]It is not clear from the recommendation how a trapezoidal-shaped room would meet the requirements of Equation 7.4.

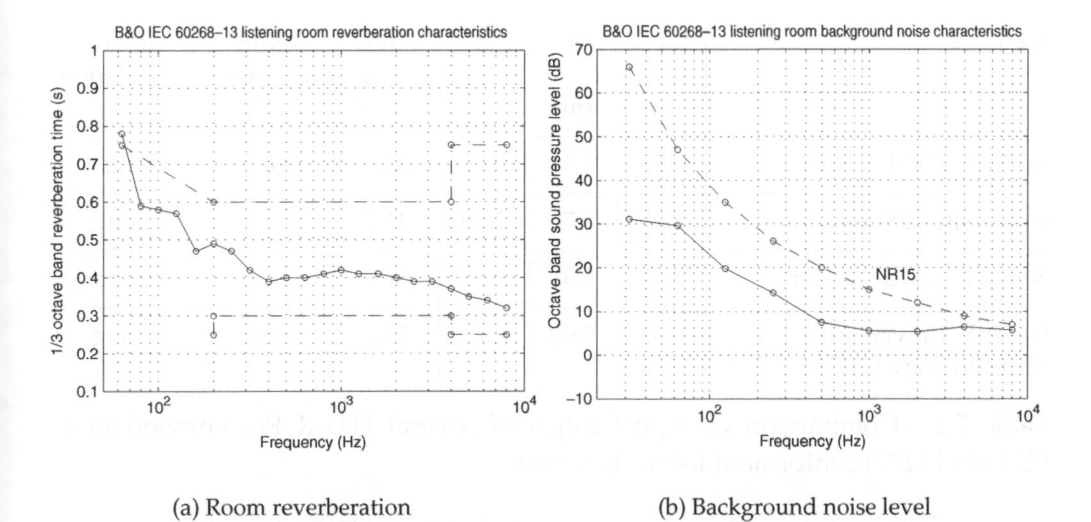

(a) Room reverberation

(b) Background noise level

(c) Photograph of interior

Figure 7.3: Characteristics of the B&O IEC 60268-13 [184] listening room. Reproduced by permission of Bang & Olufsen AV Mediacenter 2005.

Location	l (m)	w (m)	h (m)	Floor area (m^2)	Room volume (m^3)	Room orientation
BBC R&D [441]	6.76	4.94	2.70	33.4	90.2	Wide
University of Surrey	7.35	5.70	2.50	41.9	104.7	Long
Nokia Research Center	5.83	5.31	2.70	31.0	83.6	Long
Helsinki University of Technology [247]	6.35	5.60	2.70	35.6	96.0	Long

Table 7.1: Comparison of aspect ratios of several ITU-R Recommendation BS.1116-1 [207] conformant listening rooms.

where l is the room length, w is the room width[41] and h is the room height. Additionally, the conditions of Equations 7.2 and 7.3 must also be met.

For further information, this specification and other details regarding optimal dimension ratios for small listening rooms can be found in Walker [437–440]. This specification has led to a number of listening rooms being designed around the world, all meeting the requirements of the recommendation. Table 7.1 provides a comparison of how these rooms may differ in nature.

The average room reverberation, defined as T_m over the range 200–4000 Hz, is defined in Equation 7.5.

$$T_m = 0.25(V/V_0)^{1/3} \tag{7.5}$$

where V is the room volume and V_0 is a reference volume of $100\,\text{m}^3$. Additionally, the tolerance limit within which T_m must lie over the range 63–8000 Hz is illustrated in Figure 7.4. Note that the mask is relative to T_m.

Additionally, the recommendation provides some details of the early and late reflection characteristics, which should control the

[41] It is not made very clear in the ITU-R Recommendation BS.1116-1 what is meant by l and w. As this equation is only aimed at ensuring the correct modal distribution at low frequencies, the orientation of the room is unimportant. This can be seen in the BBC R&D listening room, which is employed as a *wide* room [442], the dimensions of which are found in Table 7.1. This matter is more clearly specified in the EBU's 3276 [112] interpretation of this equation, discussed in Section 7.1.3, which states that l is the length of the longest side of the room and w is the the length of the shortest side, irrespective of room orientation.

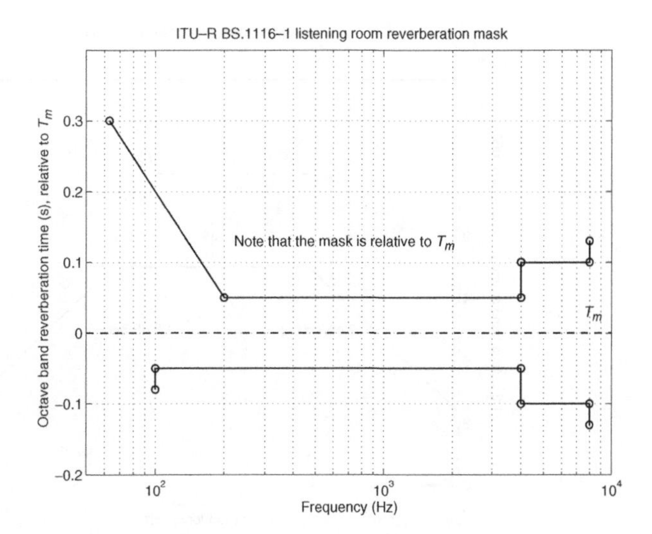

Figure 7.4: ITU-R Recommendation BS.1116-1 listening room reverberation mask (adapted from [207]). Reproduced with the kind permission of ITU.

anomalies associated with flutter echoes, tonal colouration, spatial imaging, and so on. The specification for early reflection requires that, in the range 1–8 kHz, all boundary reflections arriving within 15 ms in the listening area from the monitor loudspeakers be attenuated by at least 10 dB relative to the direct sound. Close attention is required to fulfil this aspect of the specification. Walker [441] provides specific details on this matter, based on a study of the patterns of the reflections from boundaries within the listening area. It has also been observed by the authors that carpeting of the floor is essential to achieve this specification with respect to the contribution of the floor reflection.

The operational room response characteristic is defined as the target 1/3 octave band sound pressure level measured at the reference listening position from each reproduction loudspeaker. The measurement is performed using pink noise over the range 50 Hz to 16 kHz and should lie within the mask illustrated in Figure 7.5. L_m is stated to be the average sound pressure level, without further definition.

The background noise is specified to preferably not exceed ISO noise rating, NR10, and should under no circumstance exceed NR15, details of which can be found from [55]. Figure 7.2 presents the recommended maximum background noise levels for listening rooms.

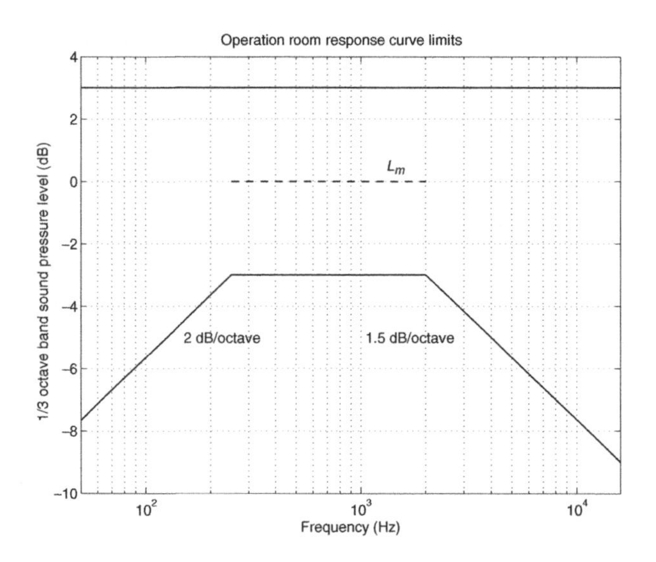

Figure 7.5: ITU-R Recommendation BS.1116-1 listening room operational room response mask (adapted from [207]). Reproduced with the kind permission of ITU.

An example of an ITU-R Recommendation BS.1116-1 listening room can be found at the Nokia Research Center, Tampere. Some details of this room are provided in Figure 7.6.

7.1.3 EBU 3276 listening rooms

The European Broadcast Union provides two technical documents on the topic of listening conditions for the assessment of sound programme material for monophonic and two-channel stereophonic sound [112] and an associated supplement for multichannel sound [113]. The former specifies in detail the requirements associated with the listening conditions for two primary applications quoted as follows.

Reference listening rooms Listening rooms used for critical assessment and selection of programme material for inclusion in a sound or television broadcast.

High-quality sound control rooms Sound control rooms used for the critical assessment of sound quality as a part of the sound or television broadcast production process.

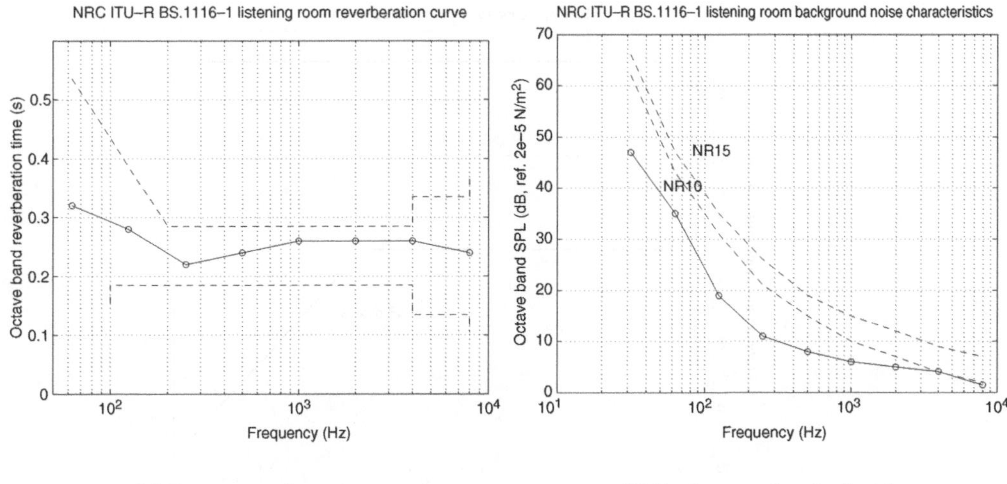

(a) Room reverberation

(b) Background noise level

(c) Photograph of interior

Figure 7.6: Characteristics of the Nokia Research Center ITU-R Recommendation BS.1116-1 [207] listening room. Reproduced by permission of Nokia Corporation.

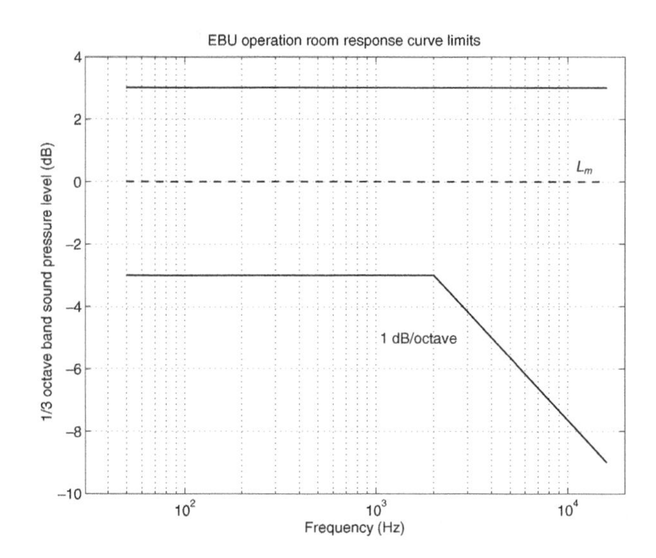

Figure 7.7: EBU 3276 listening room operational room response mask (adapted from [112]). Reproduced by permission of the European Broadcast Union.

The primary intention of these controlled listening spaces is to consider 'what the programme will really sound like to the listener at home' [112].

The listening room specification is based upon [207] with some additional details, which we shall present here. For the remaining details, the reader is referred to Section 7.1.2 and directly to the standard.

Regarding the listening room dimensions, Equations 7.2–7.4 must apply. l is the larger dimension of the floor plan irrespective of orientation, w is the shorter dimension of the floor plan irrespective of orientation and h is the room height. All dimensions are measured between the internal structural surfaces. Additionally, the ratios of l, w and h should differ by more than $\pm 5\,\%$ The floor area for a reference listening room should be greater than $40\,\mathrm{m}^2$ and the volume less than $300\,\mathrm{m}^3$.

The average reverberation time requirements follow those in ITU-R Recommendation BS.1116-1 as presented in Figure 7.4.

The operational room response for all loudspeakers measured at the reference listening position is defined in Figure 7.7 and it is measured in 1/3 octave bands using pink noise. L_m is defined as the mean 1/3 octave band sound pressure level measured over 200 Hz to 4 kHz.

7.1.4 General characteristics

Under ideal circumstances, listening rooms are designed and built into a new building. In this ideal case, there is a lot of scope for ensuring that an ideal laboratory space is built. Bearing this in mind, there are a lot of details that are of value in listening rooms that are not always clearly defined in the literature.

Some characteristics that are not explicitly mentioned within the standards include the following:

Ventilation It is important that the listening space is well ventilated and kept at a comfortable temperature, humidity, and so on. When specifying or designing the ventilation system, many factors should be considered. The first of these is to establish the heat sources in the room and their associated thermal output. Take into account the intended occupancy of the room (minimum–maximum) and also the thermal output of any devices in the room, for example, displays, amplifiers, lighting, and so on. After reckoning the heat sources in the room and estimating the target temperature and tolerance, the ventilation system should be specified so as to meet the requirements regarding the background noise. This is usually a technically challenging task. Ensure that the designer considers the matter of acoustic cross talk between adjacent spaces also, as this may affect the background noise levels. It is advisable to ensure that a heating, ventilation and air conditioning (HVAC) engineer is involved in the project to ensure a suitable design, as this is a complex and costly matter.

Computers It is quite common to employ computer equipment for the presentation of audio stimuli and/or the gathering of subjective data. Computers are generally noisy and it is advised that they be located outside the listening room. In this case, provision should be made for all the cabling of all control devices (mouse, keyboard) and display devices.

Displays devices If data projectors are used, it is also advised that these be mounted outside the listening room or in a sound isolated box. This latter solution may require attention to vibration isolation and ventilation of the projector. In each of these cases, provision should be made for easy, robust and clean signal cabling, either using a patching scheme or using wireless technology where possible. In the special case that a film projector is required, special attention is required to ensure that a suitable projection room is provided with sufficient ventilation, sound and vibration isolation and access to the projector(s). If a display is used in the listening room, ensure that it is of the appropriate size for the required viewing distance (see Section 7.5). Also verify that it does not produce excessive background noise.

Cabling Following on from the previous point, typically large numbers of audio signal cables will be required in the listening room in addition to the computer cabling. It is good to consider what the facility will be used for in future and try to design a cabling solution that will be future proof. Use of patch panels with different types of connections, for example, balanced line level, and so on, is a fairly common method. Try also to bear in mind the other types of connections that might be needed in future for video images, control signals, and so on. It is also wise to provide a cable trap solution for those cases that cannot be handled by the fixed patch installation.

Lighting It is important that the room is well illuminated to ensure that subjects are comfortable, able to read instructions, and see and use the stimulus presentation or data collection system. Additionally, if there is a desire to render the sound sources invisible, a combination of well located lighting and curtains can perform this function well. In the case of tests with accompanying pictures, there may be some additional lighting constraints associated with the lighting level, diffuseness and colour temperature. Guidance on such matters can be gained from ITU-R Recommendation BT.500-11 [216] and is briefly discussed in Section 7.5.

What is around the room In addition to designing the listening room, care should be taken to understand the sound and vibration environment adjacent to the room in order to limit problems. Before designing and building the room, consider employing a consultant to study the sound and vibration levels around the space. The existence of a nearby road, a railway, or an underground train system may require specific efforts to achieve the required background noise level in the listening room. Also, do not underestimate the existence of a photocopy room, corridor or fire door adjacent to the listening room space. This may require improved sound isolation or a special design of the floor or ceiling to achieve the required sound isolation. Consult an acoustician if in doubt.

Electrical requirements It is good to consider whether or not special earthing is required for the audio equipment. In many test facilities, several different mains are provided for different types of equipment. This requires careful design and installation to ensure that clean and noise-free power supplies are available for audio equipment. A more detailed discussion of electrical requirements with regard to perceptual evaluation can be found in Section 7.7.

Furnishings Select the furnishings according to the standard followed. ITU-R Recommendation BS.1116-1 [207] tends towards a sparse room while in IEC 60268-13 [184] the recommendation is towards a 'normally' furnished room. Bear in mind that the furnishings (carpets, chairs, cupboards) may affect the acoustics of the room and so need to be carefully selected and located. Ensure that chairs are of the appropriate height so that the subject's ears are at a height of approximately 1.2 m. Also select chairs that do not have high back that might affect the sound field near the head (unless this is an experimental factor). Select chairs that are comfortable and practical to improve listening comfort and ease during listening tests.

Curtains Curtains may be needed to hide the loudspeaker systems under test. In such cases, make sure the curtain rails will complement the planned loudspeaker configurations. The authors' experience is that acoustically transparent curtains can be obtained, providing 1/3 octave band deviation of less than 0.2–0.5 dB. IEC 60268-13 [184] recommends that curtains introduce less than 0.3 dB deviation in the operational room response in any 1/3 octave band below 10 kHz.

Subject's comfort While this is perhaps not the first thing that comes to our minds when designing a listening room, this should not be

	IEC 60268-13 [184]	ITU-R BS.1116-1 [207]	EBU 3276 [112, 113]
Application	Listening tests of loudspeakers in domestic environments	Subjective assessment of small impairment	Critical assessment and selection of programme material
Basis	–	Expansion of IEC 60268-13	Similar to ITU-R BS.1116-1
Floor area (m^2)	1–2 channel: 25–40 Multichannel: 30–45	1–2 channel: 20–60 Multichannel: 30–70	>40
Room volume (m^3)	–	–	<300
Room shape	–	–	–
Aspect ratio	$(w/h) \leq (l/h) \leq (4.5(w/h) - 4)$ $l/h < 3$ $w/h < 3$	$1.1(w/h) \leq (l/h) \leq (4.5(w/h) - 4)$ $l/h < 3$ $w/h < 3$	$1.1(w/h) \leq (l/h) \leq (4.5(w/h) - 4)$ $l/h < 3$ $w/h < 3$
Reverberation time (s)	$T_m = 0.3$–0.6 (1/3 Octave, 200–4000 Hz) See mask in Figure 7.1	$T_m = 0.25(V/V_0)^{1/3}$ (200–4000 Hz) Where $V_0 = 100\,m^3$. See mask in Figure 7.4	$T_m = 0.25(V/V_0)^{1/3}$ (200–4000 Hz) Where $V_0 = 100\,m^3$. See mask in Figure 7.4
Early energy	–	10 dB attenuation of early reflections (15 ms, 1–8 kHz)	10 dB attenuation of early reflections (15 ms, 1–8 kHz)
Late energy	–	–	–
Background noise level	NR15	NR10, NR15 max.	NR10, NR15 max.
Loudspeaker issues	1–2 channel and multichannel	1–2 channel and multichannel See operational room response mask in Figure 7.5	1–2 channel and multichannel See operational room response mask in Figure 7.7
Headphone issues	–	Diffuse-field frequency response according to ITU-R BS.708 [204]	Meet frequency response requirements of ITU-R BS.708 [204] and otherwise meet requirements of IEC 60581-10 [181]. (Applied to mono and stereo only)
Listener issues	–	–	–

Table 7.2: Overview of the primary difference between the various listening room standards and recommendations.

	AES 20 [5]	N 12-A [411]
Application	Listening tests of studio and high-quality loudspeakers	Reference listening room for listening tests
Basis	–	–
Floor area (m^2)	>20	60 ± 10
Room volume (m^3)	50–120	–
Room shape	Rectangular advised	Rectangular or trapezoidal
Aspect ratio	$h > 2.1$ m	$l : w = 1.25 - 1.45$ $w : h = 1.10 - 1.90$ $l : h \leq 1.90$ or ≥ 2.10
Reverberation time (s)	0.45 (mid-range)	$T_m = T_0(S/S_0)^{1/2}$ (200–2500 Hz) where $S_0 = 60\,m^2$ and $T_0 = 0.35$ s
Early energy	–	15 dB attenuation of early reflections (10 ms, >400 Hz)
Late energy	Suppress flutter echoes	Sufficient diffusion over listening area to avoid flutter echoes
Background noise level	35 dBA and 50 dBC	NR10 or 15 dBA
Loudspeaker issues	–	Refer to Recommendation N 12-B for loudspeaker requirements 1–2 channel only
Headphone issues	–	–
Listener issues	–	Capacity: 6–10 listeners

Table 7.2: Continued.

overlooked. The acoustic characteristics of the test room are vital, but thereafter it is important that the space is not unpleasant for subjects to spend a significant period of time. As the subjects will provide the majority of the data, it is advisable to invest in their comfort at this stage. Where possible, make an effort to ensure that the subjects are put at ease by the environment and are provided easy access to the basic required amenities.

7.2 Listening booths

In certain applications such as headphone listening, telecommunication headset testing or audiometry, a compact listening space is an interesting proposition. In such cases, the absolute size and shape of the listening space is not vital, as the room's reverberation and sound field do not contribute significantly to the listening experience. As a result, compact listening spaces or booths can be created with well-controlled acoustics and a smaller footprint.

For audiometry purposes, the maximum permissible background noise levels are well defined and quite strict. Maximum permissible ambient noise levels (MPANL) for audiometric test rooms are specified in ANSI S3.1 [14, 131]. The octave band MPANL for the 'ears not covered' case is presented in Figure 7.2 for 125–8000-Hz bandwidth audiometry. Listening booths for such purposes are commercially available.

Listening rooms are ideally intended for single occupancy when performing independent tests. This can be quite limiting in cases where large experiments need to be conducted or a large number of subjects need to participate in the test. In such cases, it is desirable to employ multiple listening spaces. However, owing to the size, complexity and cost of listening rooms, multiple listening rooms are a very rare occurrence.

For the purposes of telecommunication testing, the basic specifications of listening rooms are required, but without the need for a fully conformant listening room. Some of the basic requirements defined in ITU-T Recommendation P.800 [224] for single stimulus test include that the room volume should be within the range 30–120 m^3, background noise levels below 30 dBA and reverberation times below 500 ms.

In certain cases, telecommunication testing is performed using a telecommunication handset on one ear and with simulated background noise reproduced over loudspeakers. ITU-T Recommendation P.800 [224] provides some examples of simulated environmental noises, representing typical room noise (Hoth [170]) or internal vehicle noise. The reproduction levels for the former is 50 dBA. For the car noise case, sound pressure levels range from 75–90 dB for stationary through to simulated speeds of 110 km/h. In these higher reproduction level cases, care may be needed with respect to the sound isolation of listening booths, especially if multiple booths are adjacent.

A compact listening space has been developed by Kylliäinen *et al.* [265] to recreate the fundamental characteristics of a listening room in a small footprint (i.e. $<5\,\text{m}^2$) for the purposes of audiometry as well as telecommunication terminal testing. Specifically, the aims of the concept were to ensure

- controlled reverberation time;
- low background noise levels;
- high sound isolation;
- high vibration isolation;
- compact construction;
- comfortable listening for single occupancy.

Two aspects of the traditional listening room design need to be compromised in such a design. Firstly, owing to the aspect ratio and size limitations of a listening booth, it is not possible to replicate the structure of the reverberant field of a listening room. Secondly, owing to the the small internal dimensions of listening booths (e.g. $1.1 \times 1.4 \times 2.1$ m) only limited control of low-frequency reverberation is possible. As a result, the lower cut-off frequency for reverberation control must be approximately 100 Hz. The matter of listener comfort is one that should neither be overlooked nor underestimated. Typically, listening tests are performed within 20–60 min. While the latter time limit is not recommended owing to listener fatigue, such tests are also performed. In all the cases, the subjects should not become uncomfortable owing to lack of oxygen, excessive room temperature or humidity. In order to meet these comfort requirements, special efforts are required regarding the HVAC systems.

The design and implementation of six such listening booths has been performed and reported in [265]. In this manner, a technical solution was created to simulate the key characteristics of a listening room. Owing to the space and cost savings associated with the solution, several listening booths could be installed, allowing for parallel and independent listening tests to be performed. The key performance characteristics of these listening booths is presented in Figure 7.8 and summarised below.

- Reverberation time (<280 ms above 500 Hz)
- Low background noise level (< NR15)[42]
- High sound isolation: $R'_w > 57$ dB (between booth and adjacent laboratory space)[43]
- High sound isolation: $R'_w > 83$ dB (between adjacent listening booths)
- High vibration insulation.

7.3 Other spaces

While listening rooms provide a controlled acoustic environment for simulating key characteristics of domestic listening spaces, they are certainly not the only spaces that can be used for listening tests or perceptual evaluation. In order to perform well-controlled tests, certain key characteristics of the listening space should be controlled in order to ensure a stable acoustic environment. As can be seen from earlier in this chapter, typically these comprise of

- background noise level and spectrum,
- reverberation time characteristics, and
- minimisation of external disturbances (sound, vibration, etc.).

In the field of acoustics, several other acoustic test spaces also fulfil these characteristics, but do not simulate domestic listening. Anechoic chambers are usually very well controlled acoustics spaces that aim to eliminate the contribution of the room in the acoustic response and are usually very well sound and vibration isolated. ISO 3745 [197] provides guidance on the specification of anechoic chambers. Anechoic chambers are used when critical perceptual or psychometric tests are to be performed without the influence of room

[42] When used in conjunction with high-frequency audiometry headphones with high sound isolation characteristics, MPANL characteristics specified in ANSI S3.1 [14] are met.

[43] Where R'_w is the weighted sound reduction index measured according to ISO 140-4 [198] and ISO 717-1 [195].

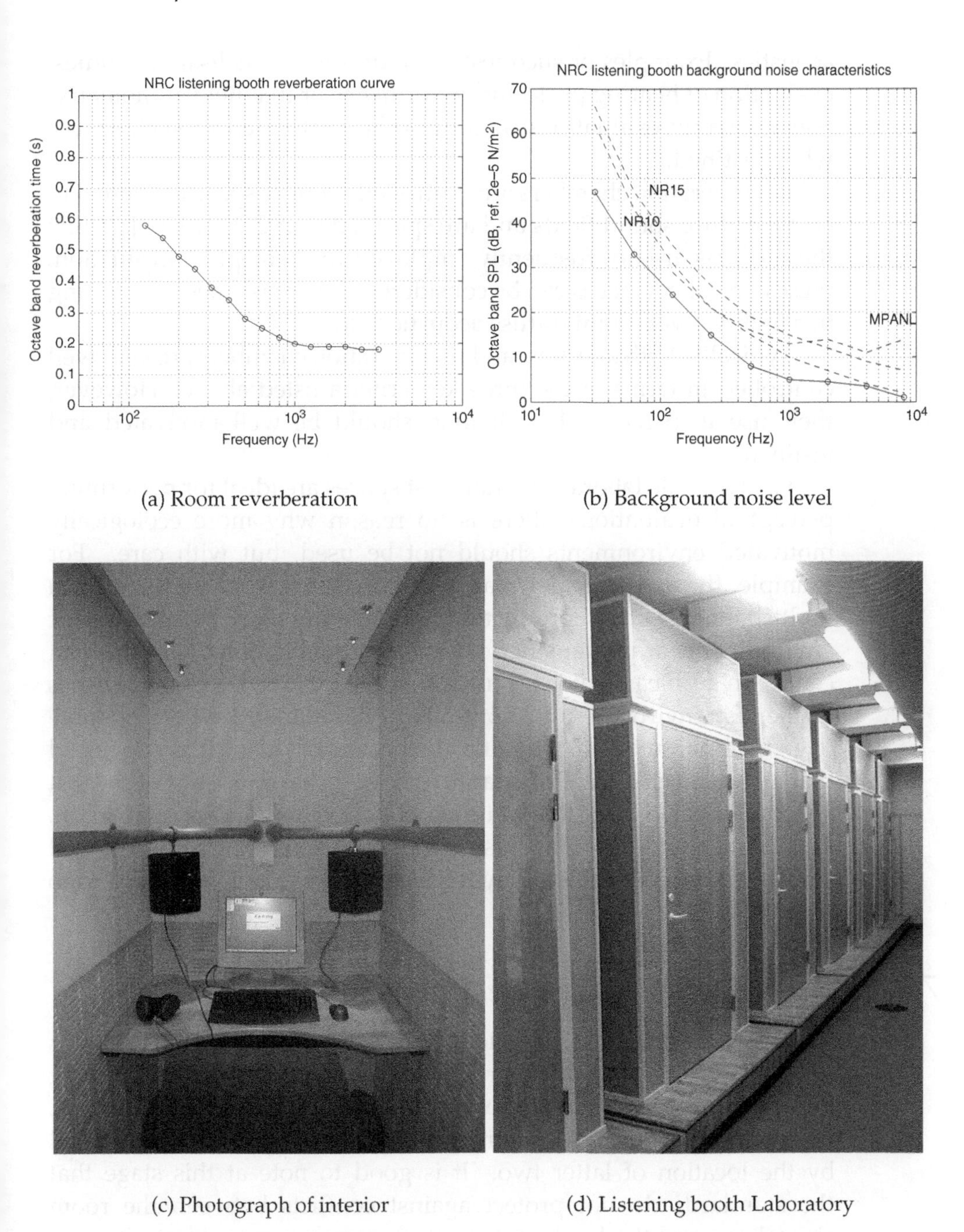

(a) Room reverberation

(b) Background noise level

(c) Photograph of interior

(d) Listening booth Laboratory

Figure 7.8: Characteristics and photographs of a custom-designed listening booth at Nokia Research Center and described in [265]. Reproduced by permission of Nokia Corporation.

acoustics. Examples of such tests include sound localisation studies, evaluation of hearing protector performance or free-field audiometry. Sound quality evaluations might also be performed in such spaces when justified.

Reverberation chambers are controlled acoustic spaces intended to create diffuse sound fields and are specified in ISO 3741 [200]. While these rooms are not frequently employed for perceptual evaluation, these controlled spaces can be considered if there is a need for testing in a highly reverberant diffuse acoustic field.

Typically, both anechoic and reverberation chambers are very well controlled in terms of the three key criteria listed above. However, their use in perceptual evaluation should be well motivated and justified.

Lastly, while laboratory grade test spaces are ideal for performing perceptual evaluations, there is no reason why more ecologically motivated environments should not be used, but with care. For example, the experimenter may considered it reasonable to perform multichannel sound reproduction tests in a number of real domestic environments or perhaps perform speech intelligibility tests in a call centre. In such cases, it is essential that the extraneous characteristics of the test environment be controlled or minimised where possible (e.g. external noise or vibration disturbance). Also, it is suggested that an attempt is made to quantify the key characteristics of the test space as much as possible in terms of background noise, and so on. Employing such environments has not been common, but when well controlled and quantified they may provide valuable alternatives to traditional test spaces.

7.4 Listener and loudspeaker positioning

Once the test space has been defined, it is time to define the locations of the listener(s) and the reproduction loudspeakers. The listening experience depends strongly upon the interaction between the room, the loudspeaker and the listening position and is significantly modified by the location of latter two. It is good to note at this stage that the standards do not protect against conflicts between the room size/shape and the loudspeaker set-up requirements. Please beware in order to avoid problems.

7.4.1 Monophonic reproduction

The standardised approach to monophonic listening is simply described in Figure 7.9(a), where a single loudspeaker is located directly in front of a listener at a minimum distance of 2 m and at least 1 m from any reflecting surface. It is advised that the loudspeaker height be 1.2 m as measured from the loudspeaker's listening axis, which is defined by the loudspeaker manufacturer (see Figure 7.10, for example). The reference listening position is highlighted in each case, though it is well advised to also test other listening locations, providing a more generically representative view of overall performance[44]. While this configuration is ideal for testing, for example, audio codec performance over loudspeakers, it is not very representative of monophonic reproduction in rooms. A study of Bech [29], Olive *et al.* [333] or Zacharov [458] reveals that the absolute location of the source loudspeaker in the room has a significant effect on the temporal, spectral and spatial characteristics owing to the manner in which the loudspeaker is coupled into the room. This is especially important in the case of perceptual evaluation to compare the performance of different types of loudspeakers. In such cases, it may become very difficult to separate the contribution of the perceptual evaluation associated with the loudspeaker location and the loudspeaker type. To overcome this problem, the experimenter can design an experiment where all loudspeakers are assessed in all locations, but this is quite time consuming. An elaborate alternative developed by Olive et al. comprises a hydraulic automatic loudspeaker shuffler [332] that allows for rapid switching between different speakers (see Figure 7.11). The shuffler system is able to move loudspeakers (mono or stereo) to their required position within 2–3 s and overcome many of the problems associated with loudspeaker positioning, but allowing loudspeakers to be tested in identical locations.

The experimenter should carefully consider what is the purpose of the experiment in order to determine whether the loudspeakers' location is to be factored into the experiment or whether it is sufficient to adopt the standardised reference configuration.

---

[44]Guidance on the usage of the non-reference listening positions is not provided in the standards.

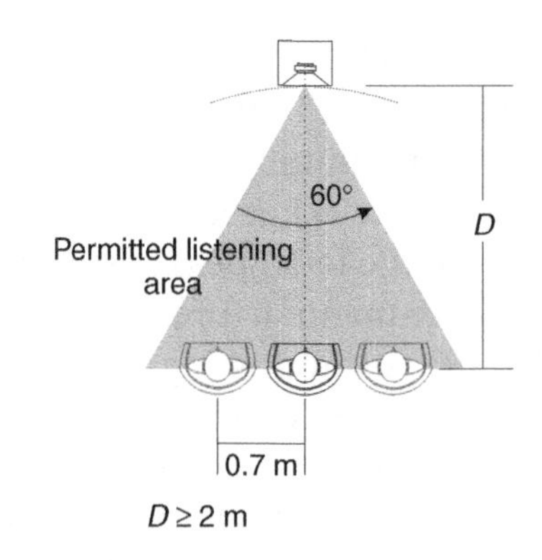

(a) Monophonic arrangement

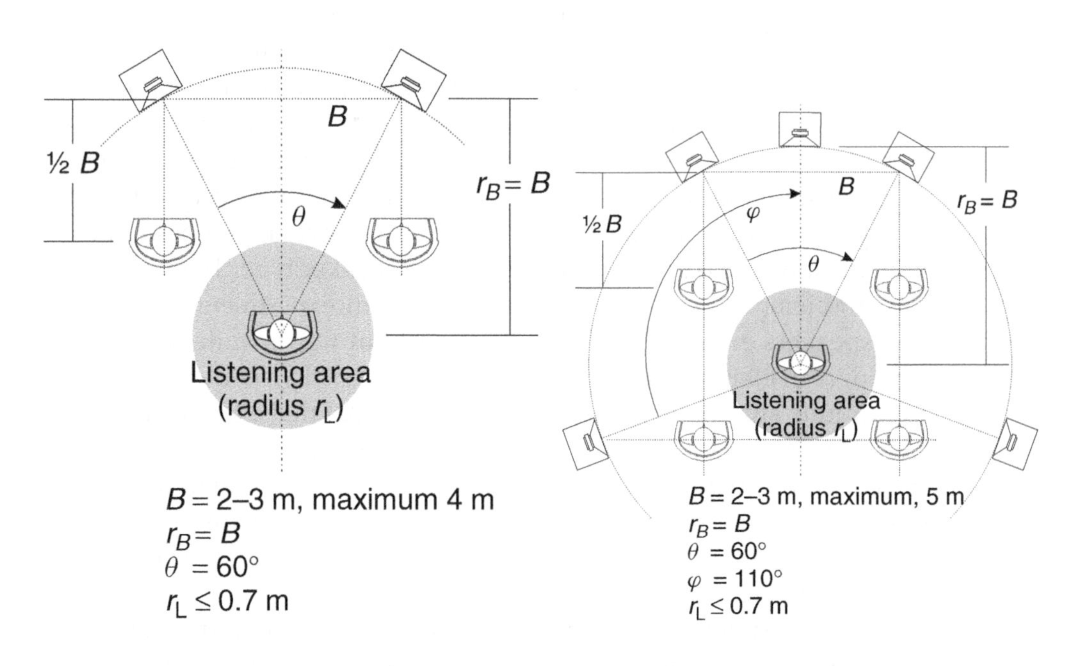

(b) Two-channel stereophonic arrangement

(c) Multichannel arrangement

Figure 7.9: Listener and loudspeaker arrangements (adapted from ITU-R Recommendation BS.1116-1 [207]). Shaded area represents the recommended listening area. The recommended or reference listening position is highlighted. Reproduced with the kind permission of ITU.

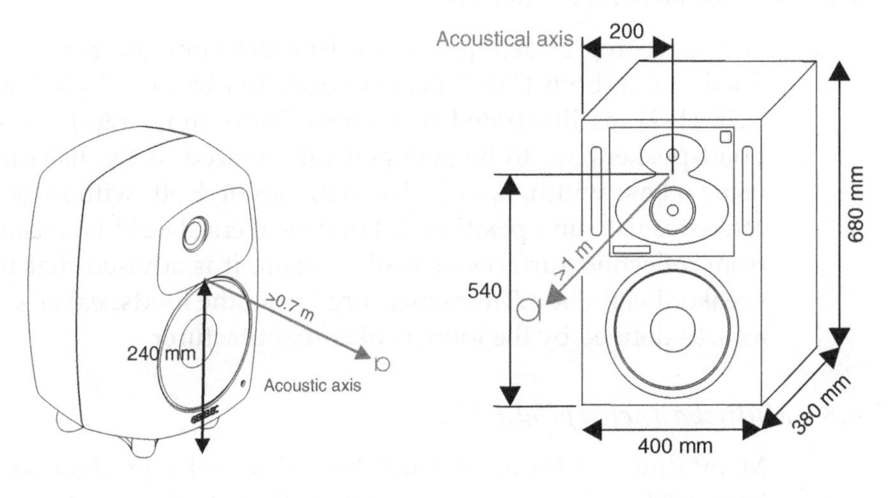

(a) Two-way loudspeaker reference axis

(b) Three-way loudspeaker reference axis

Figure 7.10: Example loudspeaker reference axis definitions. Reproduced by permission of Genelec Oy.

Figure 7.11: An automatic speaker shuffler [332]. Reproduced by kind permission of Sean Olive.

7.4.2 *Stereophonic reproduction*

For stereo reproduction, the loudspeaker configuration is well defined within both ITU-R Recommendation BS.1116-1 [207] and EBU 3276 [112], as illustrated in Figures 7.9(b) and 7.12(a) respectively. Loudspeakers are to be symmetrically placed in the listening room using a base width, B, of 2–4 m at angles of $\pm 30^\circ$ with respect to the reference listening position[45]. Loudspeakers should be located >1 m from reflecting surfaces or walls. Again, it is advised that the loudspeaker height be 1.2 m as measured from the loudspeaker's listening axis, as defined by the loudspeaker manufacturer.

7.4.3 *Multichannel reproduction*

Many different forms of multichannel sound reproduction systems have evolved in recent years and continue to do so. The most commonly encountered multichannel sound reproduction configuration employs five channels in a 3/2 configuration, that is, with three front and two surround loudspeakers and may be supplemented by a subwoofer (discussed in the following section). The common and standardised layout for such systems was originally defined in ITU-R Recommendation BS.775-1 [205] as illustrated with listening positions in Figures 7.9(c) and 7.12(b). Loudspeakers are located at 0°, $\pm 30^\circ$ and $\pm 110^\circ$. The angle of the surround speaker varies from standard to standard as illustrated in Table 7.3. It is advised that loudspeakers should all be placed on the reference circle, of radius r_b or b, for multichannel systems (see Figures 7.9(c) and 7.12(b) respectively). ITU-R Recommendation BS.1116-1 states that all speakers should be located more than 1 m away from any wall or reflecting surface. However, in many listening rooms, this becomes quite problematic with surround speakers. Loudspeaker height is defined as 1.2 m with respect to the loudspeaker axis with the exception of ITU-R Recommendation BS.775-1 [205], where the surround speakers maybe elevated (see Table 7.3).

ITU-R Recommendation BS.775-1 also presents a number of alternative surround loudspeaker configurations. However, these are not presented here, as many of these configurations have not found widespread acceptance or usage in recent times. The interested reader is referred for further information to the recommendation

[45] Also commonly referred to as the base or listening angle.

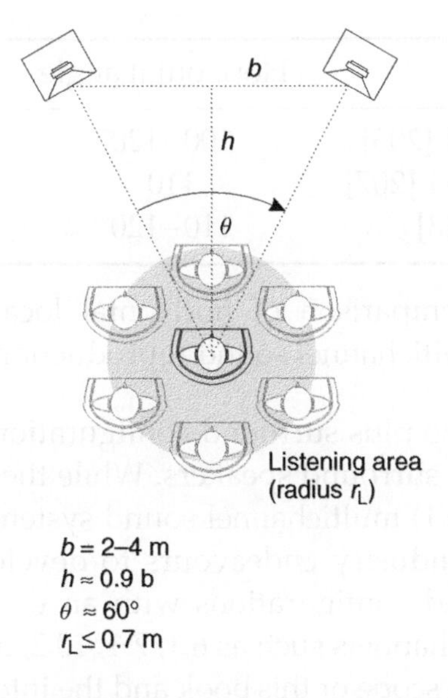

(a) Two-channel stereophonic arrangement

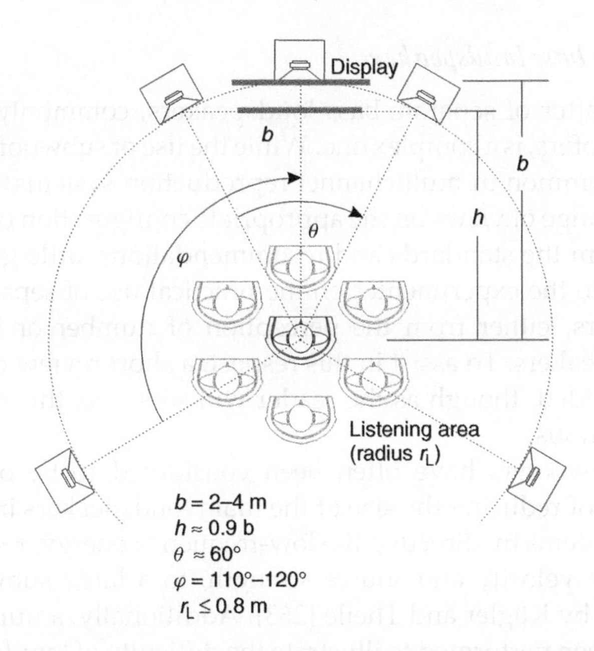

(b) Multichannel arrangement, with optional display locations

Figure 7.12: Listener and loudspeaker arrangements (adapted from EBU 3276 [112] and [113]). Shaded area represents the recommended listening area. The recommended or reference listening position is highlighted. Reproduced by permission of the European Broadcast Union.

	Horizontal angle	Height	Inclination
ITU-R BS.775-1 [205]	100–120°	≥1.2 m	0–15° down
ITU-R BS.1116-1 [207]	110°	1.2 m	–
EBU 3276-1 [113]	110–120°	1.2 m	<10°

Table 7.3: Comparison of horizontal location of surround loudspeaker for multichannel sound reproduction.

on mono/stereo plus surround configurations or arrangements with more than two surround speakers. While the commonly encountered five channel (5.1) multichannel sound system continues to be widely adopted, the industry endeavours to develop numerous new multichannel sound configurations with an ever increasing number of speakers and channels such as 6.1, 7.1, 10.2, and so on. These formats are beyond the scope of this book and the interested reader is referred to [166, 367] for an overview of related topics.

7.4.4 *Separate bass loudspeakers*

The matter of separate bass loudspeakers, commonly referred to as subwoofers, is a complex one. While the use of subwoofers has become very common in multichannel reproduction systems, there remain a wide range of views on the appropriate configuration of such systems.

From the standards and recommendations, little guidance is provided to the experimenter on the practical use of separate bass loudspeakers, either from the perception of number or the location of loudspeakers. To assist in this respect, a short review of the literature is provided, though as the reader will soon see, this matter is rather contentious.

Subwoofers have often been considered to be of interest as a means of reducing the size of the main loudspeakers in the reproduction systems by directing the low-frequency energy, requiring greater volume velocity and source strength, to a large subwoofer, as discussed by Kügler and Theile [263]. Additionally, a number of studies have been performed to illustrate the difficulty of low-frequency localisation. Borenius [60], for example, illustrated that in listening rooms below 200 Hz there is very little directional information and none at all below 100 Hz. This provides additional flexibility with respect to the absolute location of the subwoofer, which from a directional perception perspective is of little relevance.

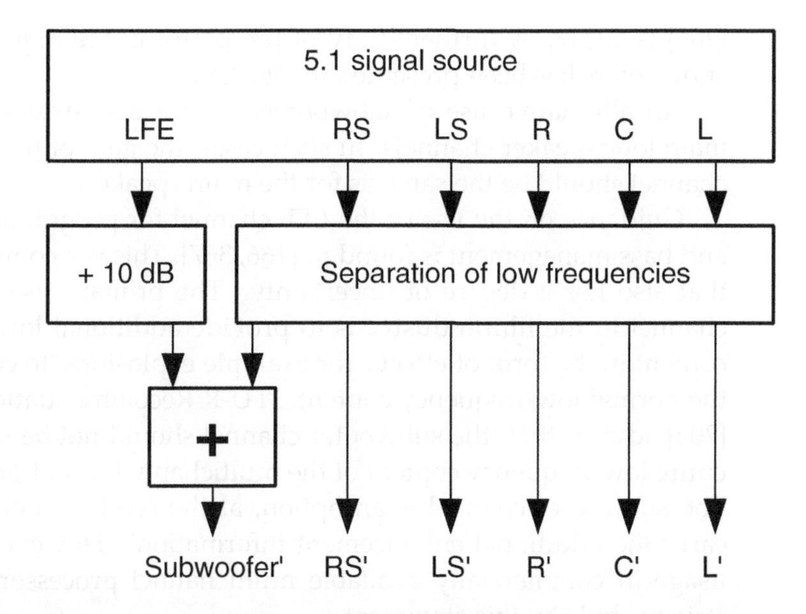

Figure 7.13: Derivation of combined subwoofer and LFE signal [113]. Reproduced by permission of the European Broadcast Union.

The human hearing system is known to be less sensitive at low frequencies, as can be seen from the equal loudness contours illustrated in Figure 8.2 [302]. As a result, the source strength at low frequencies needs to be stronger to balance the perceived level at higher frequencies. Studies by Fielder and Benjamin [122] considered the low-frequency content of commonly encountered programme material and concluded that there is often significant energy as low as the 12–16-Hz range.

Combining the above research observations has led to a specific set of performance requirements. This has resulted in the specification of a dedicated subwoofer channel now called *low frequency effects* (LFE) in cinemas. For domestic multichannel sound reproduction systems, the subwoofer commonly performs two roles: extending the frequency range of each of the principal channels downwards in frequency, and reproduction of the LFE channel (0.1) with 10 dB of in-band gain compared to the in-band level of any one of the main channels (see Figure 7.13 for a block diagram of a bass management system). The crossover frequency to the main channels should occur at a maximum of 120 Hz and typically in the range 80–120 Hz. The lower cut-off of subwoofer channels as defined in ITU-R Recommendation BS.775-1

[205] is 20 Hz. A further study of the performance requirements of subwoofers has been presented in [36, 339].

An alternative use of subwoofers is as a bass extension for the main loudspeaker channels. In such cases, the gain of the subwoofer channel should be the same as for the main speakers.

Guidance on the use of the LFE channel for programme creation and bass management is found in [166, 367]. This is a complex matter that also has a degree of uncertainty. The primary use of the LFE channel in the film industry is to provide additional low-frequency content in the form of effects, for example explosions, to complement the normal low-frequency content. ITU-R Recommendation BS.775-1 [205] advises that 'the subwoofer channel should not be used for the entire low-frequency content of the multichannel sound presentation. The subwoofer channel is an option, at the receiver, and thus only carry the additional enhancement information'. However, common usage in commercially available multichannel processors does not strictly abide by this statement.

The number of subwoofers has become a matter of discussion in recent years, with several perspectives on the subject becoming apparent. While all of the standards and recommendations refer to a single low-frequency loudspeaker, researchers and industry alike have been studying the use of multiple subwoofers. From the perspective of smooth room coupling of the loudspeaker(s) into the room, there is still significant debate. Holman (T. Holman, personal communication, 2005) in 1993 experimented by feeding the single subwoofer channel to more than one subwoofer placed differently with respect the standing wave pattern of the room and found this could lead to a smoothing of the response. Nousaine [327] states that a subwoofer located at a corner produces a more even response over multiple listening positions compared to any combination of upto five individual low-frequency sources located at typical multichannel loudspeaker locations. Salava also supports the location of a subwoofer in a corner in [371]. Zacharov *et al.* [462] illustrated the apparent smoothing in the spectral response with multiple low-frequency loudspeakers compared to a single subwoofer. However, the perceptual benefit in informal listening test was not clear. Recent studies by Welti [444] consider the number of subwoofers required to smooth out the excitation of standing waves modes through both simulation and measurement. The study considered a large number of subwoofers ($N = 5000$) in a theoretical case to illustrate the potential benefit. A more practical outcome of the work of Welti was the conclusion that two or four subwoofers placed at the mid-point of opposing walls could provide a smooth excitation of the room modes. The interested reader may study the significant spectral difference associated with different locations of subwoofer(s) in [263, 370, 444, 458, 462].

Griesinger [151–153] researched the mechanisms for the perception of spatial impression and envelopment. Griesinger [152] suggested that, by placing loudspeakers on either side of a listener, it would be possible to create fluctuations in the low-frequency interaural time delay (ITD) that could increase spatial impression. Work performed by Martens [279] and Martens *et al.* [280] has shown the potential benefits in the use of multiple and distributed decorrelated subwoofers located laterally on either side of the listening position. In both of these studies, benefits have been found in terms of enhancing the perceived spaciousness and envelopment of the sound field. This concept has been incorporated in Holman's 10.2 multichannel concept, which has 10 main channels and two subwoofer channels [168], and also briefly discussed in [166, 367].

One effect that is not visible in the literature is the coupling of the subwoofer(s) and the main loudspeaker systems at the crossover frequency. If a single subwoofer is located equidistant from the main speakers, a single phase correction control of the subwoofer signal can ensure that the subwoofer is approximately in phase with all the main speakers, ensuring a smooth spectral transition at the crossover frequency. This type of control is provided by certain manufacturers of active subwoofers, the impact of which is illustrated in Figure 7.14. However, in practice, limited guidance is provided on phase alignment of subwoofers in practical locations, that is, where the subwoofer is non-equidistant from each main channel loudspeaker, located at the side or corner. With a crossover frequency of 80 Hz, the associated wavelength is approximately 4.3 m. In such cases, it is quite possible for a subwoofer to be in-phase with the nearest speakers and approximately out-of-phase with the furthest speakers, in typical domestic configurations. Clearly, this is not an ideal situation, and even careful phase adjustment of the subwoofer can only have a limited positive effect.

It quickly becomes apparent that there is no single agreed upon view regarding subwoofer number or location with respect to modal excitation or spatial impression. However, from the research to date,

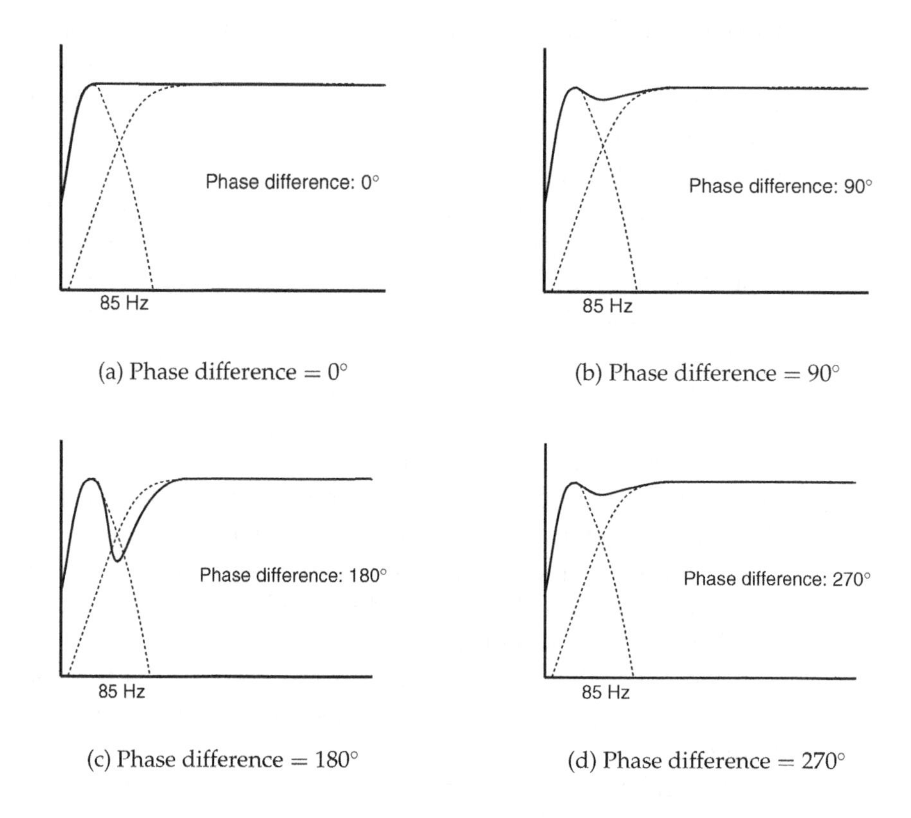

(a) Phase difference = 0°

(b) Phase difference = 90°

(c) Phase difference = 180°

(d) Phase difference = 270°

Figure 7.14: Illustration of influence of phase difference between subwoofer and main loudspeaker channels at the crossover frequency. Reproduced by permission of Genelec Oy.

there appears to be some merit in the use of multiple subwoofers, but this matter still requires further research.

7.4.5 *Listener position*

In all perceptual evaluations, it is essential to define the listening position. In some cases, this simply comprises the location of the subject while in a more detailed studies the absolute location and orientation of the head and ear may be specified. In certain types of perceptual studies, head tracking can be employed to provide real-time information regarding the absolute location and orientation of the head and or ears. For the purpose of this section, we will only consider listening matters associated with standardised listening room configurations. The interested reader should refer to Foxlin [130] for information on head tracking technologies.

In standard listening configurations, the reference listening position is usually specified as the ideal listening location in relation to the loudspeaker configuration (see Figures 7.12 and 7.9). It should be noted that none of the standards define the absolute location of the loudspeaker set-up/listening positions with respect to the listening room. Only the EBU [112] states that all listening positions should be >1.5 m from the side or back walls of the listening room. The experimenter should be aware that if the reference listening position coincides with the centre of the listening room, this may exaggerate standing wave effects at this location. Nonetheless, placing the loudspeaker and listening locations symmetrically in the listening room will ensure that the spatial sound image is symmetrical. In all of the recommendations discussed here, the reference listening position is the optimal central location. However, as highlighted in Figures 7.9 and 7.12, a listening area is defined as a number of 'suboptimal' listening locations. Guidance is not provided on how or when these locations should be employed. It is the authors' experience [459] that the use of suboptimal listening locations can provide significant information regarding the generalised performance of the systems under test. In practice, these off-centre locations may well be more representative of typical situations encountered in the domestic environment, which is rarely perfectly symmetrical.

In addition to the reference listening position, off-axis listening positions are defined in ITU-R Recommendation BS.1116-1 [207] and [112, 113] to illustrate suboptimal listening positions that are often encountered in domestic listening conditions. How these locations are employed by the experimenter will depend upon the experimental constraints. The practicalities of multiple listening locations is not always simple to resolve. To maintain subject independence and control of the test, only one person at a time 'should' ideally participate in a test. Another question is whether or not all subjects should perform the test in all listening locations. All of these issues should be considered as parts of the experimental design that are likely to be constrained by time, effort, cost and complexity of the test.

Regarding the listening axis, in most listening room cases the reference axis of the loudspeakers (see Figure 7.10) and the height of the ears are specified to be aligned at approximately 1.2 m above floor level.

7.5 Accompanying picture

In certain cases, the experimenter may wish to perform perceptual evaluation of audio in the presence of a picture in the form of a still image or video. This matter introduces a large number of additional constraints upon the experimental set-up, in order to ideally present both the audio and picture content. These matters are lightly discussed in a number of standards, but only a few provide concrete solutions to the potentially conflicting requirements of the room size, audio reproduction requirements, picture reproduction requirements and listening/viewing locations. In addition, when an accompanying picture is involved, matters of lighting and viewing conditions must be considered, as already briefly discussed in Section 7.1.4. For further details on this matter, the interested reader is referred to ITU-R Recommendation BT.500-11 [216].

For those experimenters planning to perform dominantly audio-visual experiments, it is advised to study the requirements of the listening/viewing room carefully in association with both the audio and video reproduction. It may or may not be possible to establish a set-up that meets all the standard requirements, depending largely on the dimension of the room and also the characteristics (size, resolutions, etc.) of the display or screen. In many cases, compromise solutions will have to be sought.

Appendix 5 of EBU 3276-1 [113] provides a realistic summary of the situation, stating that 'the addition of controlling viewing requirements to a listening arrangement may cause conflicts between the two sets of requirements'. It is typical for the ideal viewing distance to range from three to six times the screen height, depending on the display type, while listening distances remain in the range 2–4 m. Some potential guidance on this matter is presented in Table 7.4 based on viewing distance requirements for different display systems. Additionally, Appendix 5 of EBU 3276-1 [113] provides a summary of the relationship between some viewing and listening arrangements, based on the assumption of a 2 m listening circle. ITU-R Recommendation BS.775-1 [205] provides outline information on such configurations, but has little additional information.

Other problems encountered when there is an accompanying picture may include the conflict between the central loudspeaker and the location of the display. If using a projection screen for the image, a so-called acoustically transparent perforated screen type may be used.

Recommendation	System type	Display size	Viewing distance
ITU-T BT.1128 [208]	Conventional television system	≥ 22″	6H
ITU-T BT.1128 [208]	Critical view of conventional television system	–	4H (625 lines) 4H to 5H (525 lines)
ITU-T BT.1129-2 [214]	Standard definition digital television (SDTV) systems	≥ 22″	4H to 6H (ideal)
ITU-T BT.811 [206]	Enhanced Phase alternating line (PAL) and Sequential couleur avec memoire (SECAM) systems	≥ 22″ (16:9) ≥ 28″ (4:3)	4H to 6H (ideal)
ITU-T BT.710-4 [215]	High-definition television (HDTV)	55″ or ≥ 30″	3H

Table 7.4: Examples of typical viewing distances for different television display types. H is the display height.

However, these may degrade the quality of both the image and the sound, which may not be acceptable in all cases. If there is a plan to compare multiple displays, the literature provides little assistance on optimal configurations. Lastly, there is little guidance provided to the experimenter regarding testing with very large screens.

7.6 Commonly encountered problems

This section will highlight very briefly a few issues that we have encountered over the years. It is the authors aim here to provide observations, rather than solutions. Some of the complications discussed are associated with audio with accompanying picture installations, especially when employing multichannel sound reproduction, as already mentioned at the end of Section 7.5.

One of the primary aims of listening rooms and associated experimental configurations is to provide test environments that are well defined and controlled. In most cases, the intention is to approach the acoustics of the domestic listening environment. However, the authors observe that while this is achieved to a degree there are a few open issues.

- A question that may be asked is how well such rooms represent a domestic living space. A reference to Bech [29] or Olive *et al.* [334] illustrates some of the variations in perceived sound quality that can be found between a number of domestic listening spaces.

- Listening rooms are typically very symmetrical. Geometrically and acoustically this is very uncommon in the domestic listening environment, with the exception of custom-made home theatres. When combined with the symmetrical set-up of the loudspeaker system within the symmetrical room, certain characteristics emerge that are less common in domestic spaces. The first of these is that the spatial sound image maybe very ideal owing to the symmetrical lateral reflections. What is of greater concern is the effect at low frequencies. If the reference listening position is located in the centre of the room, this may lead to a significant lack of low-frequency energy due to standing waves.
- An important aspect that the standards appear to overlook is the influence of the orientation of a room, that is, wide or long rooms. While either can meet the requirements of the recommendations, it should be observed that there may exist differences in spatial characteristics due to the spatial and temporal distribution of lateral reflections, as illustrated by Martens and Zacharov [281]. The question thus arises as to whether results performed in similar rooms but of different orientations are similar.
- As already discussed in Section 7.4.4, the use of a separate bass loudspeaker or subwoofer as the low-frequency enhancement (LFE) channel can be quite problematic. Limited guidance is available regarding the use of such systems with respect to the number, location and configuration. Furthermore, as will be seen in the following chapter, limited guidance can be provided on the calibration of such systems. This is clearly an area in which further research is required.
- When measuring the performance of a listening room, the experimenter should be aware of the complexity associated with certain measurements. The correct measurement of reverberation time at very low and high frequencies requires attention to the measurement procedure to ensure valid results. The accurate measurement of background noise at low levels requires a degree of care. Ensure that the measurement equipment is suitable for the levels being measured, otherwise it becomes very easy to measure the noise floor of the measurement system, particularly at high frequencies.
- Lastly, the complex relationship of viewing and listening requirements intermixed with the listening room requirements can sometimes result in conflicting requirements. The interested reader is referred to Steinke [400] for an extensive analysis of viewing and listening requirements and conflicts between recommendations. It is

clear that in this respect the standards and guidelines do not always assist the experimenter to resolve these issues.

Whilst this list does not resolve these issues, it is hoped that at least the observations provoke some research to clarify these issues or help to avoid experimental complications.

7.7 Electrical considerations

Lastly in this section, we will address the issue of electrical considerations. For most listening tests amplifiers, digital-to-analogue converter (DAC), sound cards, and so on are needed. To ensure that these devices perform optimally, this short section provides brief guidance on aspects to be aware of when selecting and configuring the experimental set-up.

Amplifiers Power amplifiers and headphone amplifiers should be suitably selected to drive the transducers used in the experiment. It should be ensured that the amplifiers provide a suitable match for the transducers in terms of power output and also impedance matching. Ensure that the amplifier is not driven into clipping, nor that the speakers are overdriven, either of which can lead to undesired audible distortions. In practice, a well-selected amplifier will add no audible distortion to the sound reproduction, but this should be verified through listening and measurement. Guidance on the measurement and quantification of the performance of amplifiers can be found in IEC 60268-3 [185], Metzler [293], and Cabot [79]. For the specifics of measuring switch mode power amplifiers, the reader is referred to Hofer [164].

Digital-to-analogue converters DACs are commonly part of the audio reproduction chain and should be suitably specified and tested so that they function correctly. Improper synchronisation can be a problem with DACs, leading to clicks and other audible artifacts. Also ensure the selected DAC supports the digital input format (e.g. AES/EBU, S/PDIF, etc.) and also that the required sampling rates are supported.

General guidance on digital audio measurement techniques can be found from Dunn [108] and AES standard 17 [6] and specific guidance regarding DAC measurements are found in Dunn [109].

Sound cards When using a sound card, it is often believed that these devices will provide optimal audio output. However, this is not always the case. Not all sound cards are able to reproduce audio at arbitrary sample rates or bit rates. While many sound cards support the commonly encountered sampling rates, it is important to check whether these are handled at native sampling rates within the sound card or the sample rate is converted to a different sampling rate for processing. This process can lead to degradation in the audio signal and so the experimenter is advised to check this aspect and verify that the reproduction quality of the sound card is sufficient for the task. Signal-to-noise ratio (SNR) can also be an issue with certain cards. Again, check to ensure that there is sufficient SNR for the purpose of the experiment. For guidance on the measurement of such devices refer to [6, 7, 109].

Ground issues Electrical wiring and grounding can be a problematic issue if not correctly handled. Typically, grounding-related problems lead to, for example, noise, hum or crosstalk between channels. Any such problem can seriously compromise a listening test and this must be prevented at all costs. While a review of this topic is beyond the scope of this book, the interested reader is referred to an excellent collection of papers found in the Journal of the Audio Engineering Society [17, 121, 277, 311, 340, 445, 454] and also to the AES 48 standard [8].

CHAPTER

8

Calibration

This chapter addresses the complex and often underestimated topic of calibration. As will be explained later in this chapter, calibration is important to ensure that stimuli and systems are correctly calibrated enable enabling the experiment to be performed as expected. Differences in calibration can lead to undesired bias in the results.

The topic is divided into three main areas, namely, level, frequency and phase calibration or compensation. This chapter will concentrate on level calibration.

The calibration process can take many different forms depending on what the experimenter wishes to achieve. It may be applied to the aspects of the test stimuli and/or to the system(s) under evaluation. Here are a few forms of calibration the experimenter might wish to consider.

Perceptual Audio Evaluation – Theory, Method and Application Søren Bech and Nick Zacharov

Absolute level calibration Employed to ensure that the reproduction level is within the experimental and safe expected limits.

Peak normalisation Used to verify that test signals do not clip, leading to undesirable distortions.

Inter-system calibration Used to ensure that different reproduction systems (e.g. headphone or loudspeakers) are calibrated to the same reproduction level or frequency response.

Inter-channel calibration Employed to ensure that different channels in a multichannel reproduction system are calibrated to the same reproduction level.

Inter-stimulus calibration Used to ensure that all stimuli are calibrated to the same reproduction level.

There are two key aims of calibration. The first is to ensure that the experimental set-up is controlled and provides a neutral reproduction chain that faithfully reproduces the audio as was designed. For example, for music reproduction, the absolute aim may be to ensure that the signal heard by the listener is identical to that the music producer experienced when producing the recording. Another example would be to consider home theatre reproduction. In such a case, the signal received by the listener should be similar to that experienced in the cinema environment. The second aim is to ensure that the experimental set-up remains stable over time and can be repeated at a later date or at a different location.

Calibration can be performed in several ways. Most of the common approaches employ objective or instrumentation metrics to provide an objective measure of level that can be used for calibration. However, in some special and complex cases, the experimenter may establish that no suitable objective metric is available for the type of level of calibration sought. In such cases, the calibration can be performed through a listening test. Such methods may be applied for level, frequency and even phase calibration. However, performing calibration in this manner is a task in itself, requiring significant effort and as a result is seldom employed.

Whatever method the experimenter might have employed to perform the calibration, one last check should be made prior to running the experiment. The experimenter should listen to the system himself to check whether it is functioning as expected. He should ascertain, for example, whether the stimuli are reproduced at the same loudness, or whether the levels of all reproduction systems are similar. This is particularly important when using objective metrics for level alignment. The authors have come across several cases in which sub-optimal loudness alignment has occurred because of the selection and application of an unsuitable level calibration metric for the task at hand. The experimenter should also consider whether the reproduction levels are safe and comfortable for listeners (see Section 9.1.3 regarding noise exposure levels and associated ethical matters). If he notices that the calibration is not sufficient or appropriate, then perhaps the calibration should be revisited, using a more sophisticated calibration approach.

8.1 Level calibration

The importance of calibration of reproduction level has been well discussed in the literature. With respect to overall calibration and system reproduction level, Gabrielsson and Sjögren [137] established that the level of sound reproduction would influence certain perceptual attributes, including fullness, spaciousness, nearness, brightness and sharpness of a sound and may also affect clarity. These characteristics were further studied and clarified in Gabrielsson *et al.* [134]. Illényi and Korpássy [186] observed an association between the ranking of loudspeaker quality ranking and the calculated loudspeaker loudness. Bech [33] in a study of audio-visual interaction established that sound reproduction level could positively impact the perceived quality of reproduction of audio-visual content. In telecommunication testing, the influence of listening level has also been found to have a significant impact on the mean opinion scores as presented in [226] and summarised in Figure 8.1. In each of these studies, the impact of reproduction level on the perception of quality or qualities of sound are observed.

It becomes very clear from many of these studies that calibration and alignment of reproduction level are important to avoid uncontrolled bias effects. Naturally, this may not be the case for the experimenter wishing to study the effect of level and its interaction with other experimental factors. In such cases, the experimenter may well wish to measure and document the reproduction levels of the experiment for further analysis.

A large number of level metrics exist, as discussed in the following section, each with associated advantages and disadvantages.

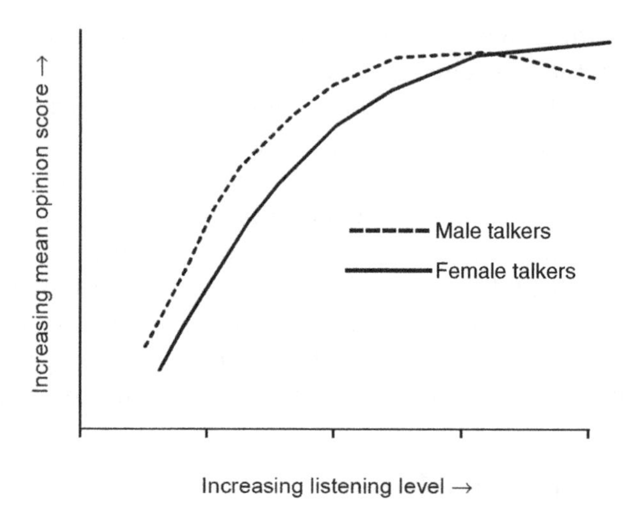

Figure 8.1: Mean opinion score as a function of listening level [226]. Reproduced with the kind permission of ITU.

The matter of establishing the most suitable metric for calibration of level has been discussed for many years. For example, Hellman and Zwicker [161] observed the misalignment between loudness and dB(A) measures. It can be seen from the equal loudness contours (see Figure 8.2) that the contours are both frequency and level dependent. The level calibration of speakers and stimuli for listening tests have been considered important and discussed by numerous researchers [134, 135, 137, 186, 421–424]. Aarts studied different means of evaluating the loudness of loudspeaker reproduction [2, 4] and subsequently compared the performance of these metrics for loudspeaker listening tests [3, 4]. More recently, the authors and associates have studied the correlation between subjective and objective level alignment metrics for the installation of multichannel sound reproduction systems [37, 40, 410, 458, 461, 463]. The matter of establishing the most suitable level calibration metric is still under discussion [384, 385, 390, 391] and, at the time of writing, also a work item within the International Telecommunications Union Radiocommunication (ITU-R) Sector under WP-6P, special rapporteur group (SRG-3). Details of some of this work are discussed further in Section 8.1.2. For an extensive and in-depth review of loudness models, the interested reader is referred to [384].

8.1.1 Level calibration methods

Level calibration is a generic term that can be applied in a number of contexts. Firstly, level calibration can be applied to either the reproduction system, for example, loudspeaker(s), or to the test stimuli. The calibration or alignment can then be performed in either the physical domain or the perceptual domain. For example, electrical calibration using voltage metrics only considers the physical characteristics of the signal. By comparison, loudness alignment of two acoustic signals will take account of human perception of the signals.

A large number of methods, metrics and models fall under the category of level calibration. These range from simple measures of voltage through to binaural loudness models for time-varying signals. All of them aim at providing an estimate of the level of the signal with different degrees of perceptual relevance and have their associated applications and limitations.

Ideally, level calibration would best be handled using the most accurate methods that would ensure the best accuracy in calibration and mostly closely approximate to a calibration performed through a listening test. To accurately model the hearing system a binaural, temporal loudness model would be required. However, such models of loudness are still in development, thus limiting the access to ideal metrics for loudness alignment. While level metrics are widely available in the form of voltage, sound pressure level (SPL) measures and loudness metrics are more complex and less accessible to the experimenter in standard test equipment.

A short description of the key level calibration metrics will now be presented. It is assumed that the reader is familiar with the basics of electrical and acoustic measurements and, where further information on such topics is needed, the reader is referred to [47, 55, 160, 293].

Intensity metrics

Intensity metrics are the starting point for calibration. Both voltage- and pressure-based metrics, such as SPL, are intensity-based metrics aimed at measuring the physical characteristics of the signal. These metrics do not aim to consider how the auditory system will interpret these signals and as a result do not describe well how the signals are perceived. Thus their association with the loudness is remote, but nonetheless provides some important means for physical calibration.

SPL is measured using a voltmeter connected to the output of a microphone. As a result, similar applications and limitations apply to both pressure and voltage metrics. Measures of peak electrical intensity such as peak voltage (V_{pk}) can be useful to prevent clipping and to ensure that the maximum dynamic range of the reproduction system is employed. Irrespective of the method of level calibration employed, it is a good practice to check whether the signal clips any part of the reproduction chain, which can cause audible distortions.

Electrical root mean square (RMS) level calibration is an effective means of calibrating and aligning signals that are spectrally and temporally similar.

Linear SPL is measured using a microphone connected to an RMS voltmeter. SPL measures (dB or dB(lin)) usually have an associated integration time [76]. The so-called *fast* setting commonly encountered provides a 100 ms integration time, which is associated with the integration time of the ear and the *slow* integration time is 1 s. The so-called peak SPL measured employes a time constant of 50 μs. In this respect, the measurement of SPL can differ from the measurement of RMS voltage. Additionally an SPL measurement averaged over a period of time is often possible with modern sound level meters and is referred to as an equivalent continuous sound level (L_{eq}) measurement. This allows for averaging of time-varying signals over a pre-defined period of time. These forms of SPL measurements are commonly available in sound level meters and often applied to the assessment of noise levels.

Speech activity level as defined in ITU-T Recommendation P.56, method B [220] is a particular form of intensity measurement employed in telecommunications. The basic premiss assumes that the speech signal is corrupted by noise, as is typically the case in telecommunications. As a result, it is not possible to obtain a good estimate of the speech level alone as the noise during speech pauses contributes to the level estimate. According to ITU-T Recommendation P.56, 'active speech level is measured by integrating a quantity proportional to instantaneous power over the whole aggregate of time during which speech in question is present (called active speech time), and then expressing the quotient, proportional to total energy divided by the active time, in decibels to the appropriate reference'. Measurement methods for speech activity level can be found from [220]. The active speech level can be used to estimate and calibrate the reproduction levels of speech samples with different speakers or sentences for example.

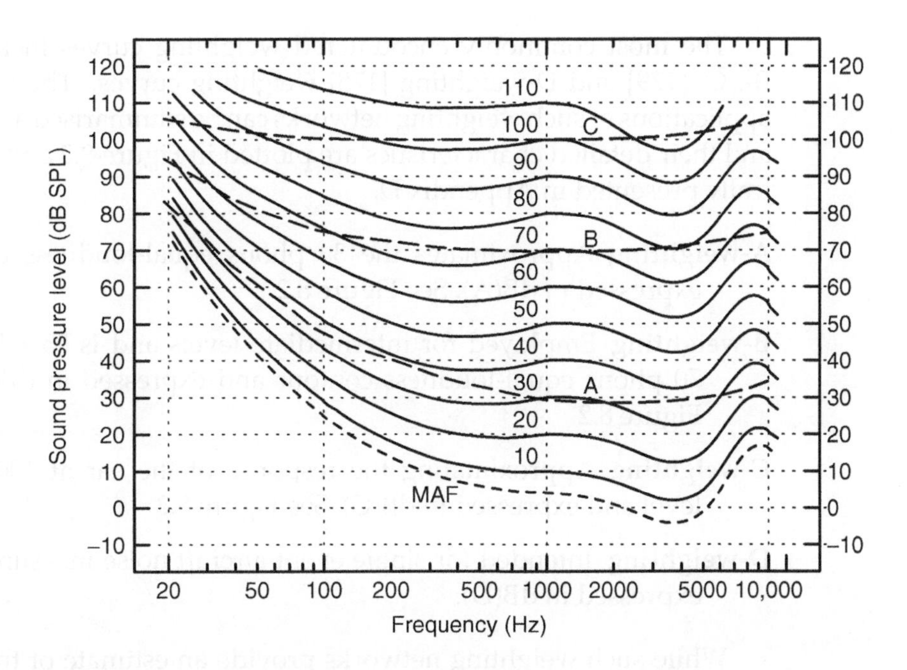

Figure 8.2: Equal-loudness contours (solid curve) and A-, B-, C-weighting curves (dotted lines). The A-, B- and C-weighting curves are approximately vertically aligned with the equal-loudness contour they are derived from. Reproduced by permission of B. C. J. Moore [302].

All of the metrics mentioned so far focus upon the measurement of the physical signal and do not provide links to auditory perception. Their application to calibration should thus be confined to the physical calibration of the signals, which may or may not bear any relevance to a perceptually motivated calibration, depending on the signal's spectral, temporal and level characteristics.

Spectral weighting is a simple means of improving the perceptual relevance of intensity-based metrics that can be applied to electrical or acoustic measurements. A range of such weighting curves has been created over the years, which is in common use in sound level meters and analysers. These spectral weighting functions are typically based upon an approximation to an equal-loudness contour at a given loudness level. For example, the A-weighting curve approximates the 30 phons[46] equal-loudness contour (see Figure 8.2).

[46]Definition of the phon is provided is Section 8.1.1.

The most commonly encountered weighting curves include A-, B-, C- [179] and D-weighting [178] weighting curves. The principle applications of such weighting networks can be summarised as follows and their detailed characteristics are plotted in Figure 8.2 and numerically presented in Appendix D.

A-weighting Approximates the 30 phons equal-loudness contour, expressed in dB(A). See Figure 8.2

B-weighting Employed for intermediate levels and is based on the 70 phons equal-loudness contour and expressed in dB(B). See Figure 8.2

C-weighting Approximating the response of the ear at 100 phons level and expressed in dB(C). See Figure 8.2

D-weighting Intended for single event aircraft noise measurements, expressed in dB(D).

While such weighting networks provide an estimate of the loudness of a test signal, they are still based upon a measure of physical magnitude and as a result only provide a limited estimation of the perceived loudness for signals reproduced near the intended level of the associated weighting curve. As such these metrics are unable to take into consideration the full spectral intensity and temporal dependence of hearing system and thus do not accurately estimate the true loudness of signals. The application of such metrics to the calibration or alignment of complex or significantly dissimilar stimuli can lead to poor or misleading results.

Loudness metrics

Loudness is defined as the 'attribute of the auditory sensation in terms of which sounds can be ordered on a scale extending from quiet to loud' [302]. Loudness, as can be observed from the equal-loudness contours (see Figure 8.2), is level and frequency dependent and also has a temporal dependence [119]. A loudness metric provides an estimate of the signal's loudness based upon a perceptual model of the human hearing system.

Many such auditory models have been developed for the purpose of hearing research and also for audio coding, as found, for example, in [159, 336, 386]. A number of models have been developed specifically for the accurate prediction of loudness as will be discussed shortly.

A generalised example of one such auditory model can be found in ITU-R Recommendation BS.1387 [213] and is illustrated in Figure 8.3, where one of the model's outputs is an estimate of loudness.

The unit of loudness can be reported in sones or phons. The 'loudness level (in phons) of a tone at any frequency is taken as the level (in dB SPL) of the 1 kHz tone to which it sounds equal in loudness' [345]. The sone is 'arbitrarily defined as the loudness of a 1 kHz sine wave at 40 dB SPL' [302]. As a perceptual quantity, loudness cannot be directly measured but can be estimated through listening tests or through the use of models developed to predict the subjective impression of loudness. This section will consider the latter case.

A number of loudness models have been developed over the years that can be applied for calibration purposes. Broadly speaking, these fall into two application categories, namely, models for steady-state and time-varying signals.

Several steady-state loudness models exist. The Stevens model (Mark VI) was one of the earliest developed [402, 405], employing octave band SPL measurements in diffuse fields. This model provides a single loudness figure in sones (OD) or phons (OD), where O refers to octave band measurement and D refers to diffuse-field measurements. The method is based upon a look up table approach, which is documented within the ISO 532 (method A) standard [188].

The Zwicker loudness [472, 476, 478] is also defined as part of ISO 532 (method B) [188] and provides a means for assessing one-third octave band SPL measurements from either diffuse (GD) or free fields (GF) and is reported in sones or phons. G refers to group (critical band or Frequenzgruppen), F refers to free field and D refers to diffuse field. For free-field measurements it is assumed that the sound source is located directly in front of the listener. Computational models for Zwicker loudness have been developed in Fortran and Basic and published in [473, 477].

A revision of the Zwicker model has been proposed by Moore *et al.* [305, 307] and is available in [306], the outline structure of which is presented in Figure 8.4. Real-time versions of the Moore model have been reported in [407, 429] and provide simple means for real-time calibration. This type of real-time functionality is highly desirable when performing the calibration of sound reproduction systems and speeds the process significantly.

The Stevens, Zwicker and Moore models are all based on a monaural input signal that is either a diffuse or free field. Additionally,

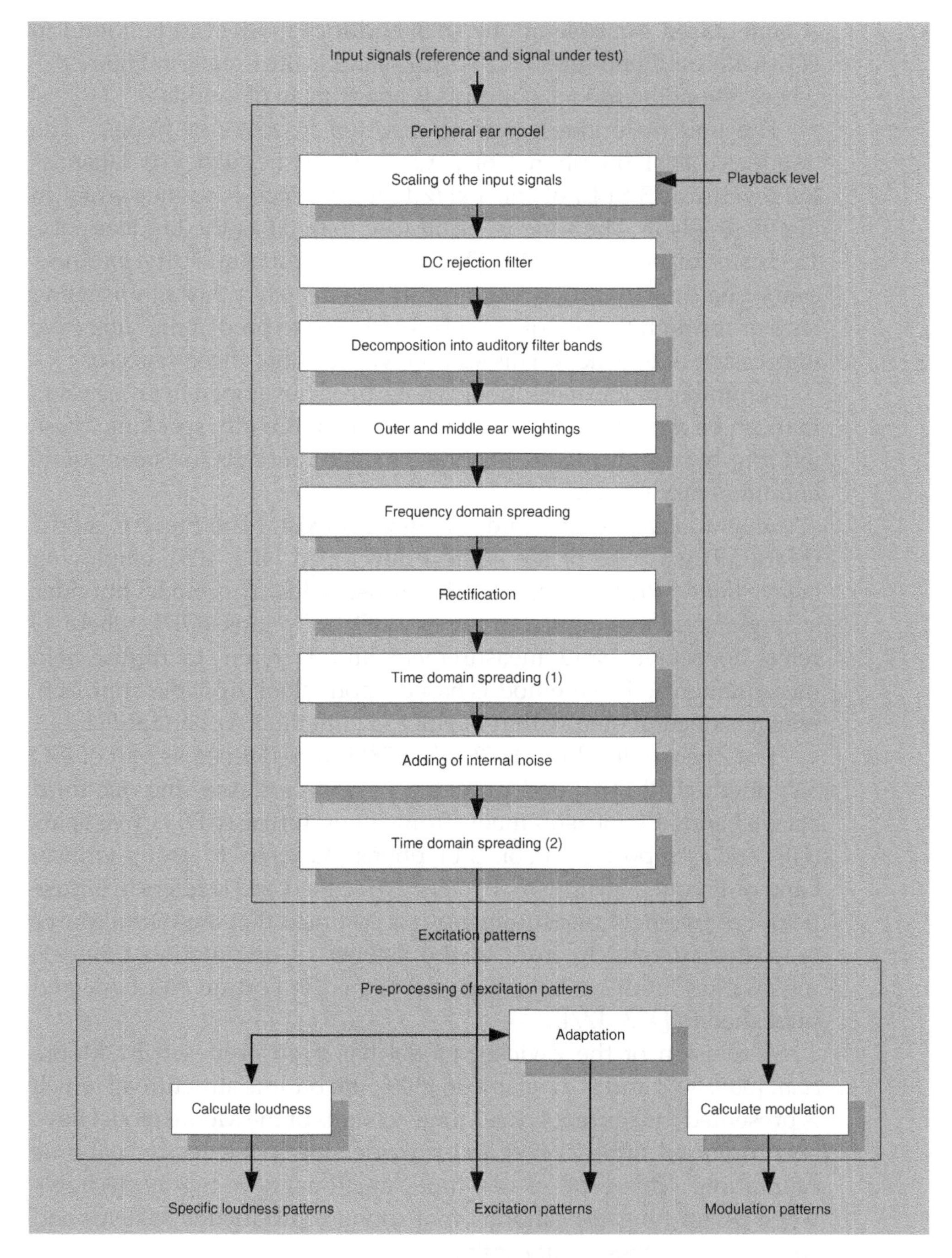

Figure 8.3: Peripheral ear model and pre-processing of excitation patterns for the filter bank–based part of the model, from ITU-R Recommendation BS.1387 [213]. Reproduced with the kind permission of ITU.

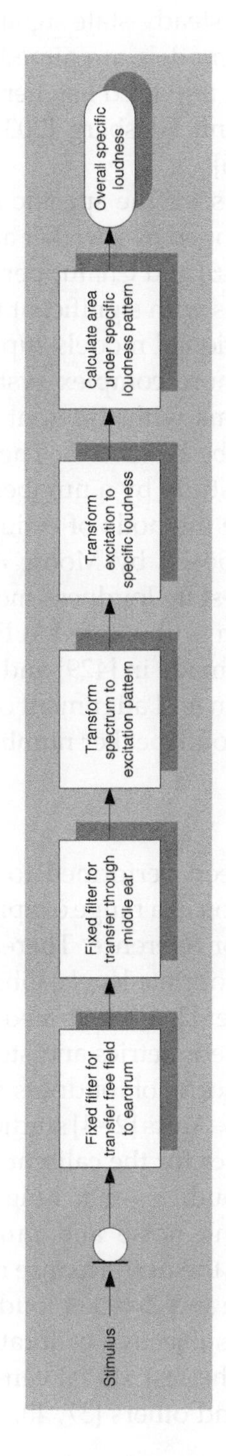

Figure 8.4: Block diagram of the sequence of stages of Moore *et al.* steady-state loudness model, adapted from [305].

they are all intended for steady-state signals and for modelling normal hearing. A number of models are aimed at modelling the loudness associated with hearing impaired listeners. These include the models developed by Moore and Glasberg [303], available from [304] and Chalupper and Fastl [86].

A number of models for the specific application to time-varying signals have been developed by Zwicker and associates [120, 474, 475], Glasberg and Moore [146] and Chalupper and Fastl [86]. Such models are well suited to signals with significant temporal variations.

All of the aforementioned models support monaural input. In the case of calibration of more complex systems, such as multichannel sound system or systems with different directivity loudspeakers, a binaural model would be beneficial. The topic of binaural loudness estimates is still under study by a number of researchers [91, 92, 364, 382, 383, 389, 469]. The methods of estimating binaural loudness of signals have been proposed by Moore *et al.* [305], based upon an extension of the steady-state loudness model [305, 306]. The concept applied in this approach is illustrated in Figure 8.5. An application of this approach has been made in [429] and successfully applied by the authors to the calibration and alignment of loudspeaker reproduction systems with different loudspeaker numbers and locations [467].

8.1.2 Level metric selection

Several studies have been performed to establish the most suitable metric for level calibration in a range of applications. A sample of these is summarised below for reference. Thereafter a short practical guide for selection of the appropriate level calibration method is provided.

Hellman and Zwicker [161] illustrated the difference in the performance of SPL A-weighted metrics and steady-state loudness models. In this paper, the superiority of loudness metrics is presented for several realistic signal cases. Aarts [2–4] studied the suitability of different kinds of loudness metrics for the calibration of loudspeaker listening tests. In this study, loudness matching was performed through a listening test using pink noise and music signals. These results were compared against the performance of several level metrics. The B-weighted L_{eq} measure and Zwicker loudness model (ISO 532B) [188] were found to predict subjective calibration best. Bech [32] further illustrated the impact the test signal can have on loudspeaker level calibration. Zacharov and others [37, 40, 410, 460, 461, 463] studied a

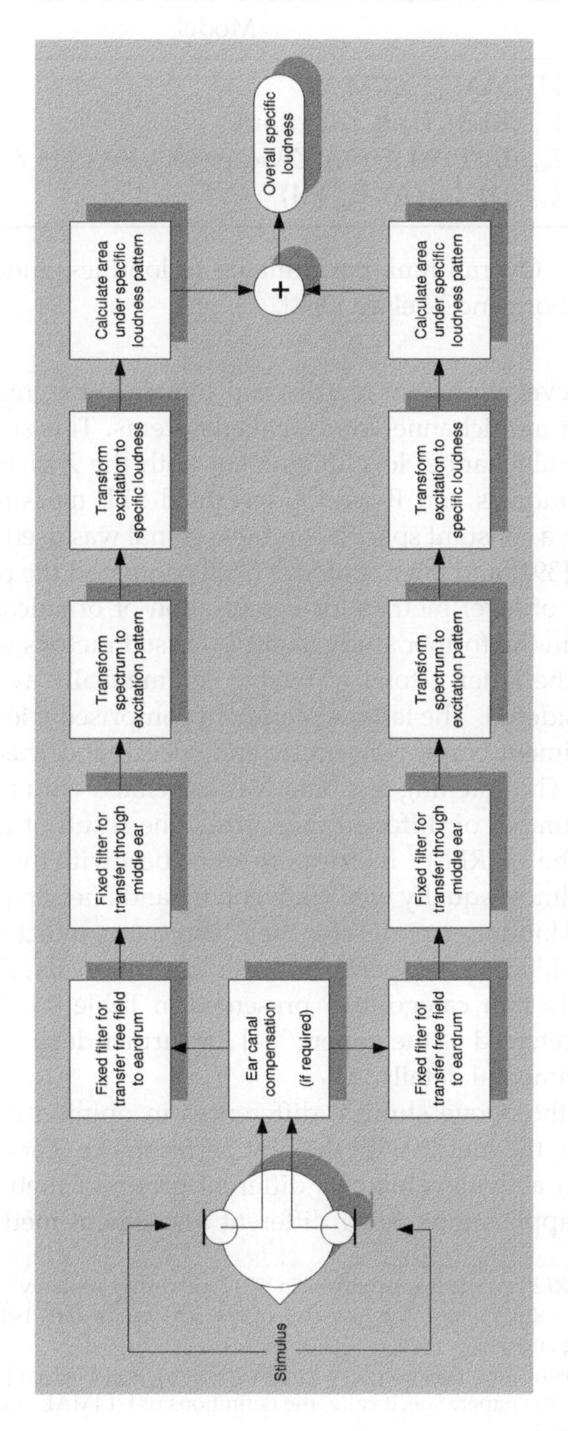

Figure 8.5: Binaural expansion of block diagram of the sequence of stages of Moore *et al.* steady-state loudness model, adapted from [305] and applied in [429, 469].

Rank	Model
1	HEIMAL, LARM
2	L_{eq} (RLB), L_{eq} (C), L_{eq} (Lin)
3	L_{eq} (B), PPM (50 %), Zwicker-ISO, Zwicker & Fastl (95 %)
4	L_{eq} (D), L_{eq} (A), L_{eq} (M)

Table 8.1: Overall rank performance of loudness models according to Skovenborg and Nielsen [384].

range of level alignment metrics and signals in the inter-channel calibration of multichannel loudspeaker systems. This study concluded that for multichannel level alignment both the Zwicker and Moore loudness models and B- and C-weighted SPL measures performed well when a constant specific loudness signal was used. Soulodre and Norcross [391] and later Soulodre [390] compared the performance of a number of level metrics for equalisation of broadcast programme material. In the former study, eight L_{eq} based metrics were evaluated while in the latter a total of twelve commercially available metrics were considered. The latter experiment comprised a loudness matching experiment employing music and speech and was performed in five sites. The listening test data was correlated with the level metric using a number of different measures. The result of the experiment was that the $L_{eq}(RLB)$[47] metric performed best with the L_{eq} metric performing almost equally well. Skovenborg and Nielsen [384] evaluated a range of loudness models for their 'ability to predict or measure the subjective loudness of speech or music'. Additionally, the results were ranked into four categories[48] presented in Table 8.1. The interested reader is referred to the paper [384] for further details on the metrics and experimental details.

From the above studies, differences in opinion are apparent in identifying the metric with the best performance. Each of the studies mentioned above evaluates a different group of metrics for slightly different applications using different assessment methods and, as a

[47] The $L_{eq}(RLB)$ metric was proposed in [391] and comprises a revised low-frequency B-weighting (RLB) where the low end of the new curve falls between the B- and C-weighting curves.

[48] The classification rule employed by Skovenborg and Nielsen [384] is not clearly presented in the paper. Specifically, the definitions of HEIMAL and LARM have not been published.

result, it appears difficult to draw firm conclusions. However, it is generally found that the B-weighted SPL and loudness metrics are often found to perform well. Several of the studies discussed above [384, 390, 391] have been performed in collaboration with ITU-R WP-6P, SRG-3. While this study evaluated time-varying programme material, temporal loudness models (e.g. [86, 146]) are not known to have been assessed as part of the study and as a result their relative performance for such applications has not been evaluated.

As already discussed and presented above, different metrics have different applications and associated complexities. In some cases, simple intensity metrics may perform as well as temporal loudness metrics. For example, regarding the calibration of samples of coded speech, which are time-varying in nature, a time-varying loudness metric might be selected. However, as typically signals with similar temporal content are aligned, for example, the same speech encoded with different codecs, a simpler level metric may well suffice. However, in other cases, the use of more complex loudness metrics are quite important to ensure a perceptually meaningful calibration, for example, when aligning several sound reproduction systems each with a different number of loudspeaker channels in different locations.

It is thus advised that the experimenter carefully considers the nature of the calibration to be performed in terms of the nature of the signal, reproduction systems and acoustics to identify which is the best suited calibration metric to apply. To assist the novice Figure 8.6 provides a rough guidance on this matter. The reader should be aware that this is neither a comprehensive nor an exhaustive selection guide, but rather an overview based on the authors' experience.

From the flow diagram shown in Figure 8.6, if the experimenter establishes that a loudness metric is best suited, there are then a number of factors to be considered with respect to the configuration of the loudness metric. These are mostly associated with the nature of the sources and the nature of the acoustic field in which the sources are located. Figure 8.7 provides an outline of some of the cases that may be encountered. It is most common to calculate loudness based on a microphone measurement and apply either the diffuse- or the free-field correction. The diffuse-field setting assumes the sources to be located in a diffuse field, while the free-field setting assumes the source to be located at 0° azimuth and 0° elevation compared to the listener. In more complex situations, such as with multiple sound sources or when sources of different directivities are used, there may be a need to

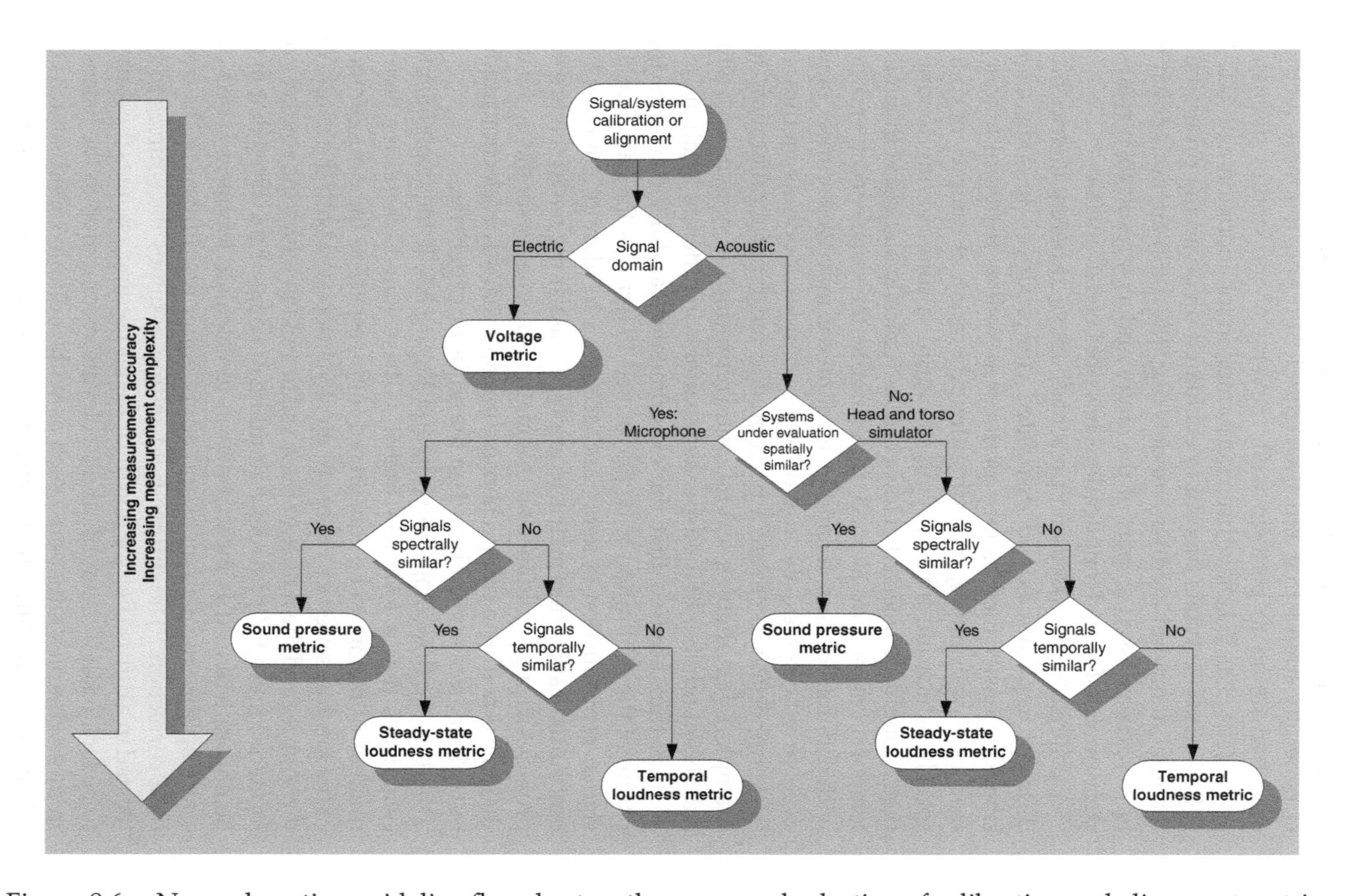

Figure 8.6: Non-exhaustive guideline flowchart on the usage and selection of calibration and alignment metrics.

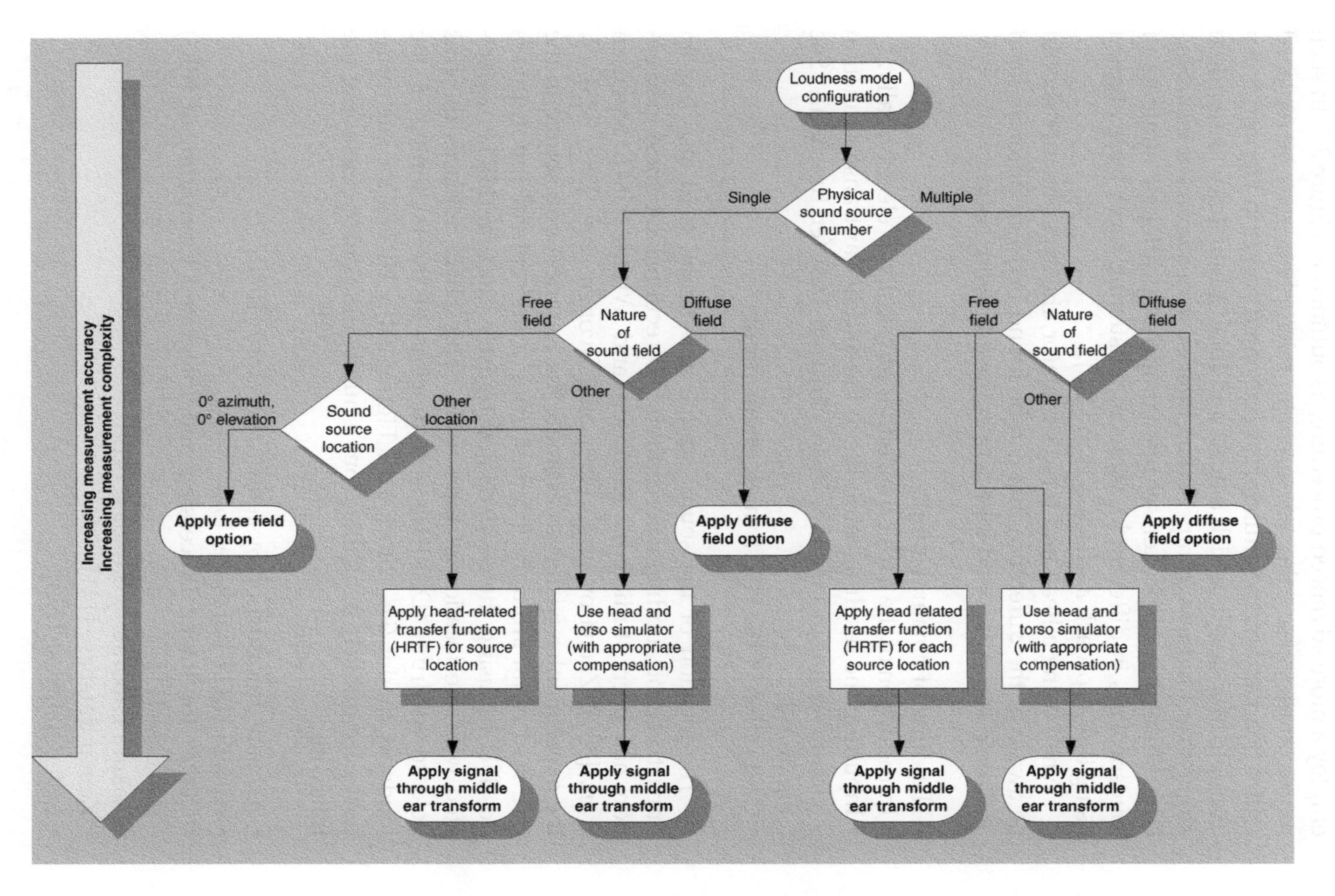

Figure 8.7: Non-exhaustive guideline flowchart on the usage of sound field options with loudness metrics.

employ a more advanced procedure, as outlined in Figure 8.7. In such more complex cases, the measurement microphone may be replaced with a head and torso simulator, as presented earlier in Figure 8.5. An example of a case in which a head and torso simulator was required aligned for loudness, comprised an inter-system alignment task. Several different loudspeaker set-ups comprising different numbers of loudspeakers in different locations in a listening room were aligned using this approach as discussed in [467].

When using a head and torso simulator, the experimenter should be aware of the difference that can be found between different head and torso simulators and take care of the following issues:

- Handling of binaural input.
- Introduction of the each canal, either physically or electronically.
- Inputting the resulting stimulus to the middle ear transfer function (circumventing acoustic field to eardrum transfer function). See Figure 8.4.
- Calibration of the HATS microphone inputs.

The location of the sound source in a sound field also has an impact on the loudness perception. Directional loudness perception has been studied by Robinson and Whittle [364], Sørensen *et al.* [389], Zacharov *et al.* [469] and Sivonen *et al.* [382, 383]. If the experimenter wishes to estimate the loudness of a source in the free field at an arbitrary location using a microphone input, this is possible by using the head related transfer function (HRTF) associated with the location of the source. This HRTF replaces the free field to eardrum (DRP) transfer function at the beginning of the loudness model (i.e. block 1 of Figure 8.4). HRTFs for such purposes are publicly available from a number of sources including the following:

- Acoustic Information Systems Laboratory [101]
- AUDIS HRTF databases [57]
- Center for Image Processing and Integrated Computing (CIPIC) database [11]
- Itakura Laboratory database [414]
- Kemar database [141]
- Listen database [273].

Author	Noise level	Programme type	Programme level (re. 20 μPa)
Ventry [435]		Speech	49.3 dB(A)
Pearson [337]	See Figure 8.8	Speech	see Figure 8.8
Beattie [25]	85 dB	Speech	90.9 dB
	100 dB	Speech	100.3 dB
Coren [98]		Speech	See Table 8.3
Benjamin [45]		Speech (TV)	57.7 dB
		Speech (home theatre)	64.8 dB
Mathers [284]		Music	83.5 dB(A)
Airo [10]		Music	69.0 dB
	72 dB	Music	85.0 dB

Table 8.2: A summary of preferred or the most comfortable listening level for different types of programmes under different noise conditions.

8.1.3 Preferred listening levels

When absolute SPL calibration is required, once the appropriate level calibration method has been selected, the experimenter can progress to define the target level. Most comfortable (listening) levels (MCL) provide an excellent means to establish the optimal sound reproduction level. The standards and literature provide some guidance in this respect. The MCL have been defined for different types of programme material and cases, which can provide some guidance to the experimenter. A summary of the research for different kinds of programme items is provided in Table 8.2.

Pearson, Bennett and Fidell [337] studied the influence of moderate noise levels on preferred listening levels for speech. Their study considered speech heard from a television and conversational speech levels. Beattie, Zentil and Svihovec [25] considered the MCL for speech in white background noise for moderate and high SPLs. The results for speech heard from a television [25] and from Beattie's study are both plotted in Figure 8.8. The interested reader is referred to [25] for further details on conversational speech levels in different background environments and levels. The reader interested in knowing about how preferred listening level of music is influenced by bandwidth may refer to Condamines [96].

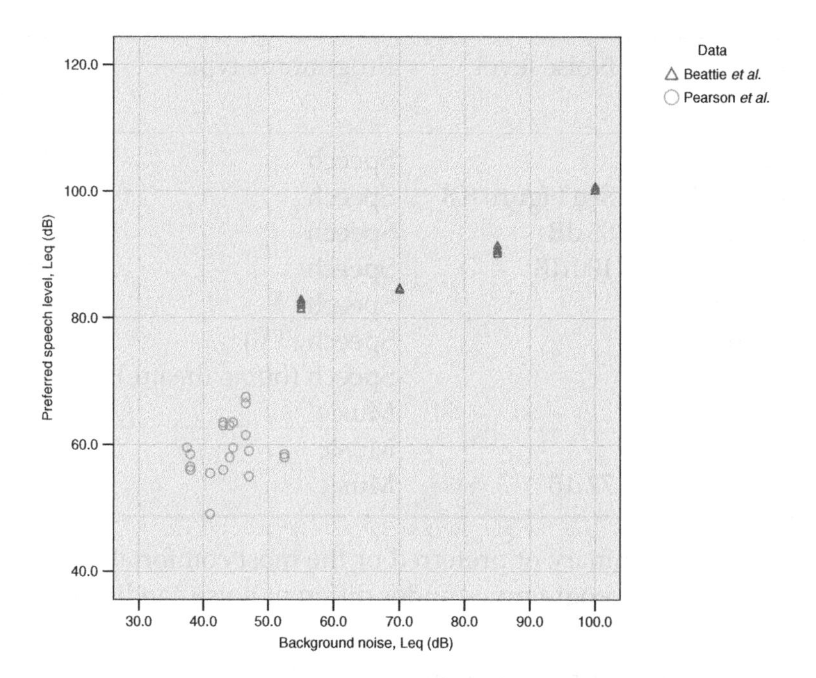

Figure 8.8: Preferred speech levels (SPL) as a function of background noise level (SPL) as heard from a television, according to Pearson, Bennett and Fidell [337] and in background white noise, according to Beattie, Zentil and Svihovec [25].

Coren performed a large study of the hearing comfort level of speech for different age groups using a random sampling of 799 subjects [98]. The results of this study are presented in Table 8.3 and can be approximated by the following equation:

$$C = 48.16\,e^{0.007A} \tag{8.1}$$

where C is the hearing comfort level in dB (SPL) and A is the age in years.

It is clear from the research presented that most comfortable levels depend on a number of factors including signal bandwidth, signal type, noise properties (level, bandwidth, type, etc.) and also the listening context (conversation, entertainment, etc.). The experimenter should consider these factors to establish the suitable MCL to which his experiment needs to be calibrated.

The term 'most comfortable listening' is mostly applied in listening tests as a general level for calibration of all stimuli for all subjects.

Age (years)	Hearing comfort level (dB SPL (Re. 20 μPa))
15	53.5
20	55.2
25	56.9
30	58.6
35	60.5
40	62.5
45	64.5
50	66.6
55	68.9
60	71.2
65	73.6
70	76.2
75	78.9
80	81.7
85	84.6
90	87.6

Table 8.3: Most comfortable listening levels (MCL) of speech as a function of age, according to Coren [98], using a random sampling of 799 subjects.

However, the most comfortable listening level can be left for adjustment by each subject at the beginning of the experiment. This should ensure that all subjects are comfortable, but may also lead to some bias in the results. In such cases, the experimenter is advised to record the levels set by each listener and to use this as a covariate during analysis.

8.1.4 Reference reproduction levels

In the previous section, we have studied the preferred or most comfortable listening levels for different cases. Additionally, several of the standardisation bodies provide recommendations regarding reference reproduction levels in different situations. Depending on the experiment, either the MCL or the reference reproduction level should be chosen for the calibration.

For those readers interested in performing tests in motion picture theatre environment, the SMPTE [388] provides guidance on the recommended reference level of 85 dB(C) SPL for normal theatrical operation. The recommended reference level is defined as 'the spatially averaged SPL of a single channel of a theatrical sound system measured with a broadband pink noise at the reference electrical level as a stimulus'. Guidance on spatial averaging is provided in [387]. The reference electrical level is defined as 'the voltage measured by an average responding voltmeter of wideband pink noise using a measurement bandpass filter of 20 Hz to 22 kHz bandwidth when the test signal is at the reference recorded level'. The reference recorded level is defined as 'the level of pink noise equivalent to 50 % modulation on an analogue photographic soundtrack, or the equivalent level on a digital photographic soundtrack or a digital cinema (D-cinema) sound track (typically in each case 20 dB level 100 % modulation)'. Holman [166] notes that the reference level in digital media varies between standardisation organisations, for example, the reference level employed by SMPTE is −20 dBFS, while the EBU employs −18 dBFS and other organisations employ a level as high as −12 dBFS.

With respect to video game audio reproduction, Tuffy [428] has made a recommendation for reference reproduction levels. It is proposed that the recorded reference level be −20 dBFS and the playback reference level be 85 dB(C) SPL.

In telecommunication tests, the recommended reproduction level is less strictly defined. ITU-T Recommendation P.800 [224] suggests that the level be pre-defined and recorded for the whole experiment and provides details of what signals to use to measure the reference level. According to the CCITT handbook of telephonometry [84], two definitions are provided with respect to *optimum* listening levels. The preferred listening level is defined as 'the speech level that in a listening or conversation test is judged as preferred on a loudness preference scale' (See Table 4.3). Alternatively, the optimum listening level is defined as 'the speech level that in a listening conversation test corresponds to the highest opinion score on the quality scale'. It has been found that the level difference that can occur between these two cases is on average 5.7 dB [84]. For typical speech bandwidths[49] a noise level of 84 dB SPL will yield a maximum score on a listening effort scale, while 79 dB SPL will yield a maximum score on a loudness preference scale.

[49] Applicable to telephony bandwidths, that is, 300–3400 Hz.

As the CCITT handbook of telephonometry notes, there is an assertion that a listening test can be carried out at one level; however, there are several indications that this may not be representative of telephone usage in real situations. The preferred listening level is affected by the type and level of background noise, as already illustrated (see Figure 8.8). Additionally, it can be noted that there will be variations in the listening level during practical telephone usage. However, in practice, a single level of 79 dB SPL is often employed, for example, in practical codec evaluations within the telecommunications industry and in standardisation. Gleiss [147] provides further details of handset and hands-free preferred listening levels in telephony.

With respect to loudspeaker calibration in stereo or multichannel cases, ITU-R Recommendation BS.1116-1 [207] provide the following guidance. The reference listening level is defined as 'as a preferred listening level produced with a given measuring signal at the reference listening point. It characterises the acoustic gain of the reproduction channel in order to ensure the same SPL in different listening rooms for the same excerpt'. See Section 7.1.2 for details regarding the reference listening position.

The reference SPL per channel is defined as

$$L_{\text{ref}} = 85 - 10 \log n \pm 0.25 \qquad \text{dB(A), slow} \tag{8.2}$$

where n is the number of reproduction channels in the total set-up. To align to this requirement, each loudspeaker in the loudspeaker is fed a pink noise signal at a level of −18 dBFS for digital media. The gain of the amplifiers are then adjusted to meet the L_{ref} level, for example, 78 dB(A) SPL, slow for a five-channel reproduction system. The authors have experienced that this calibration level yields too high SPLs, by about 5–10 dB SPL, for subjects for typical programme material.

8.2 Loudspeaker calibration

So far we have considered calibration in general terms. All listening tests employ either loudspeakers or headphones for sound reproduction. A vital aspect of the calibration is to ensure that the loudspeakers are correctly calibrated, and this section will introduce the various aspects of this matter. Issues relating to headphones calibration will be presented in Section 8.3.

8.2.1 Level calibration

Having discussed how the level can be measured and what levels might be appropriate for the experiment, the experimenter should now define what aspects of the loudspeaker reproduction system require calibration. This issue can be described by three categories.

Absolute As described in the previous section, it may be desirable that the absolute reproduction level of each loudspeaker channel or system be calibrated to an absolute level. Depending on the metric selected, this may or may not result in equal loudness of reproduction of the stimuli or the systems under consideration.

Relative (inter-channel) When more than one channel is being employed for reproduction, there is a need to ensure that the channels are correctly calibrated. In the case when all channels are *identical* with respect to room coupling, this is almost a trivial exercise and application of the ITU-R Recommendation BS.1116-1 [207] approach is quite suitable. Once again, if very different bandwidth or directivities occur between the channels, a more elaborate calibration approach may need to be considered, as discussed in Section 8.1.2.

Relative (inter-system) When multiple reproduction systems are being evaluated, the experimenter may wish to calibrate the level between systems. Such calibration is typically performed following the inter-channel calibration and aims to ensure that the reproduction levels of all reproduction systems are aligned or absolutely calibrated. ITU-R Recommendation BS.1116-1 [207] provides one basic calibration approach, as described in Equation 8.2, which can be applied to similar systems. More elaborate loudness metrics can be applied for more complex cases as discussed in Section 8.1.2 and shown in Figures 8.6 and 8.7. In the case when the reproduction systems are very different in nature, that is, in terms of number of channels, geometric locations or directivity, a more elaborate measurement scheme may be adopted in an effort to evaluate such differences accurately. Such a case encountered by one of the authors comprised calibrating the levels of several multichannel systems with differing speaker locations and channel numbers. The calibration approach taken, reported in [467], was to employ a head and torso simulator to provide and estimate the binaural loudness [429] of each system.

An alternative, listening test based, approach applied to multichannel systems has been proposed by Choisel and Wickelmaier [89], which comprises an adaptive 2AFC procedure.

Lastly, it should be noted that while in this section we have discussed the different approaches to calibration of loudspeakers in the sense of absolute, inter- and intra-systems, these concepts are equally applicable to the calibration of stimuli.

Separate bass speaker

As presented in Section 7.4.4, the use of a separate bass or subwoofer loudspeaker is quite common in multichannel sound reproduction systems. The separate bass speakers interact in a complex manner with the room acoustic and main speaker channels. This introduces challenges to the user or experimenter in terms of how to calibrate them in terms of level, frequency and also phase characteristics. The adoption of subwoofers has been very rapid, leading to several million units being in use worldwide. However, as noted by Benjamin and Gannon [46] 'users of those systems usually have no means to optimally adjust the subwoofer level, either by objective means or subjectively'.

The use of subwoofers is often associated with the term bass management in the context of multichannel reproduction. Bass management typically refers to the way the controller directs bass to the available speakers and manages the bass SPL in the context of reproduction of multichannel content from DVD media, for example. The majority of research on this topic focuses on such applications either in the home environment or in the cinema environment. Typically, the subwoofer channel is used in one of two modes for multichannel reproduction, referred to either as low-frequency effects (LFE) channel or as a bass extension speaker. The bass management and calibration for these two modes differs in some respects.

LFE channel calibration The guidance provided by the SMPTE [388] for cinema type applications states that the subwoofer LFE channel should be calibrated with 10 dB gain above the wideband screen channels when measured using a real-time analyser (RTA), using a broadband pink noise signal.

Application	Main channel level dB(C) SPL, slow	Subwoofer LFE level dB(C) SPL, slow
Television mixing	78	82
Film sound, small room	83	87
Film sound, large room	85	89

Table 8.4: Main channel and subwoofer calibration levels for 5 channel systems according to Holman [169]. See the main text for test signal characteristics.

Holman provides practical guidance on the subwoofer level alignment and also on the main channel calibration in [167, 169]. Holman suggests using 120 Hz low pass filtered pink noise[50] reproduced at −20 dBFSrms and measuring the C-weighted sound pressure level. Recommended levels for both the main[51] and subwoofer LFE channel levels are presented in Table 8.4. It is vital to note that, in this case, when using band-limited noise, the gain difference between the main and subwoofer channels is only 4 dB, compared to the 10 dB discussed above, which employs broadband noise. For more specific details on how these calibration levels have been defined, the interested reader is referred to [167].

Bass extension calibration The SMPTE [388] suggests that, when using the subwoofer as a bass extension speaker, the level of the subwoofer channel should be the same as the wideband screen channels. Also refer to Section 8.1.4 regarding the SMPTE guidance on reference reproduction levels for subwoofers.

Regarding the phase alignment of subwoofers, little if anything is found in the literature, although essentially the same principles apply as with any multi-way speaker design. The difference here is that, as the subwoofer is an independent source, phase alignment is required, as the distance and relation to the main loudspeakers and room are arbitrary. The primary aim here is to ensure that the subwoofer and main speaker are in phase at the crossover frequency, to avoid a

[50] For example, using a 7th order elliptical filter.

[51] Measured using a 500–2000 Hz bandpass filtered pink noise signal reproduced at −20 dBFSrms.

dip in the frequency response (see Figure 7.14). The SMPTE [388] advises that the polarity of the subwoofer be evaluated with a pink noise signal fed to both the main and subwoofer speakers to establish which of these provides the smoothest frequency response at the measurement position using a RTA. Some manufacturers provide greater phase control of their loudspeakers and as a result a superior phase alignment should be possible (see Section 7.4.4 and Figure 7.14).

8.3 Headphone calibration

Headphone calibration is an involved process that requires a fair level of knowledge regarding the headphone type and also measurement technique. This section summarises the key points the experimenter should be aware of before performing level and/or frequency calibration of headphones. For further details on headphones and their measurements, the interested reader is referred to the excellent chapter by Poldy [349].

8.3.1 Headphone types

Four primary types of headphones are commonly encountered, with several sub-variants. These are described in ITU-T Recommendation P.57 [237] and IEC 60268-7 [183] and illustrated in Figure 8.9. The four categories are defined as follows:

Supra-aural earphones Earphones that are intended to rest upon the ridges of the concha cavity and have an external diameter (or maximum dimension) greater than 25 mm and less than 45 mm (see Figure 8.9(a)).

Circum-aural earphones Earphones that enclose the pinna and seat on the surrounding surface of the head. Contact with the head is normally maintained by compliant cushions. Circum-aural earphones may touch, but not significantly compress, the pinna (see Figure 8.9(b)).

Intra-concha earphones Earphones that are intended to rest within the concha cavity of the ear. They have an external diameter (or maximum dimension) of less than 25 mm but are not made so as to enter the ear canal (see Figure 8.9(c)).

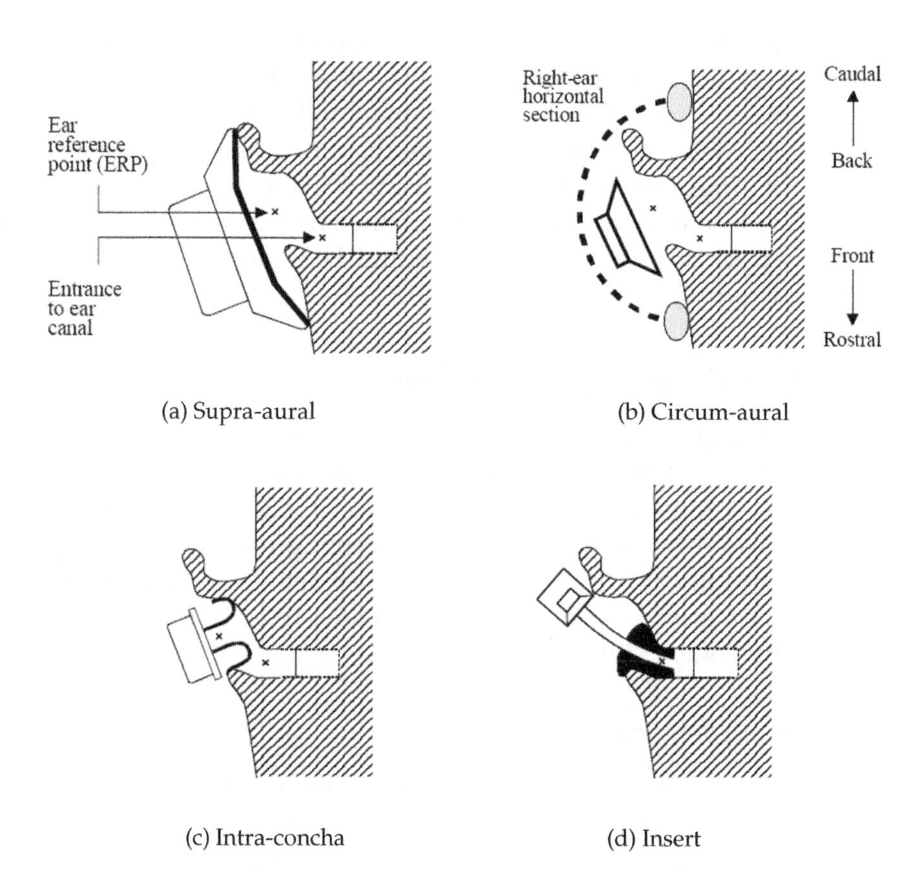

(a) Supra-aural (b) Circum-aural

(c) Intra-concha (d) Insert

Figure 8.9: Illustration of the relationship of the ear reference point (ERP) and entrance to ear canal point (EEP) for different types of headphones according to [237]. Reproduced with the kind permission of ITU.

Insert earphones Earphones that are intended to partially or completely enter the ear canal (see Figure 8.9(d)).

Møller *et al.* [298] defines so-called FEC headphones as those having a *f*ree air *e*quivalent *c*oupling to the ear. Such headphones have a coupling to the ear canal similar to the coupling to free air and as a result do not significantly interact with the response of the ear canal. Both intra-concha and insert headphones terminate in the ear canal and as a result modify the standing wave pattern in the ear canal.

8.3.2 Ear measurement points

Three measurement points are defined in ITU-T Recommendation P.57 [237] that are commonly discussed in headphone measurements, which will be used later in this chapter. The points are defined as follows and illustrated in Figure 8.9. Note that the DRP is not illustrated in this figure.

Ear reference point (ERP) A virtual point for geometric reference located at the entrance to the listener's ear, traditionally used for calculating telephonometric loudness ratings.

Entrance to ear canal point (EEP) A point located at the centre of the ear canal opening.

Eardrum reference point (DRP) A point located at the end of the ear canal, corresponding to the eardrum position.

Detailed discussion regarding the impact of the measurement point selection can be found in Møller *et al.* [299]. Møller notes that FEC headphones can be effectively measured either with blocked or open ear canal techniques.

8.3.3 Headphone measurement

The measurement of headphones can be performed by a number of methods including perceptual loudness balancing, or objectively with real ears and probe microphone or by employing coupler or artificial ears. All methods have associated benefits and limitations, depending on the domain of application. It would appear from the literature that the focus is placed on supra- and circum-aural headphones and limited guidance is provided regarding measurement and calibration of intra-concha and insert headphones.

Loudness balancing Loudness balancing aims to match the level of headphone reproduction to that of a source (e.g. loudspeaker) in a given sound field. The sound field often comprises either a free field or a diffuse field. Subjects listen to one-third octave, band-pass filtered pink noise from the loudspeaker source and then over headphones. The subjects then adjust the level of the headphone to match the loudspeaker's level using the method of adjustment. This process is iterated until the subjects find the matching loudness level in each frequency band. Typically, eight subjects are used for this process. Such methods are defined in detail in IEC 60268-7 [183] and have applications in both diffuse- and free-field measurements.

Real ear measurement Once loudness balancing has been performed, the experimenter may wish to measure the response of the headphone on the person's ear. This can be achieved by the use of a very small probe microphone, which is inserted into the subject's ear. Such methods discussed in IEC 60268-7 [183] and ITU-R Recommendation BS.708 [204] define in detail the physical characteristics of the microphone and its location in the ear to ensure meaningful and repeatable measurements.

Alternatively, methods may be used to establish the diffuse-field response target illustrated in Figure 8.14 and discussed later in Section 8.3.4, as developed by Spikofski [393]. According to ITU-R Recommendation BS.708 [204], 16 subjects should be measured to obtain a stable estimate of the diffuse-field frequency response.

Artificial ears and couplers Table 8.5 provides a summary of the different couplers and artificial ears commonly in use and numbered according to ITU-T Recommendation P.57. Details of the application of type 3.3 and 3.4 ears in head and torso simulators is provided in ITU-T Recommendation P.58 [223].

From Table 8.5 it can be see that not all artificial ears are suitable for all types of headphones nor are they intended for application to the full audio bandwidth. In these respects, artificial ears cannot replace the other methods discussed earlier. The authors have also encountered challenges using the type 3.3 ear for the measurement of insert headphones owing to the difficulty of properly fitting such devices into the ear canal.

It should also be noted that not all couplers and artificial ears measure at the DRP. As a result the measurements may not be comparable. Convention in parts of the industry has been to convert all measurements to be comparable with the type 1 ear, that is, to ERP. A mapping from DRP to ERP is provided in ITU-T Recommendation P.58 [223] and reproduced in Appendix E.

Ear type	Description	Reference point	Frequency range (Hz)	Measurement applications	Example
1	IEC 60 318 ear simulator [177]	ERP	100–4000		Brüel & Kjær 4153, GRAS RA0039
2	IEC 60 711 occluded ear simulator [180]	DRP	<8000		Brüel & Kjær 4157, GRAS RA0045
3.1	Concha bottom ear simulator	DRP	<8000	Intra-concha	–
3.2	Simplified pinna simulator	DRP	100–8000	Supra-concha, cirum-aural,	Brüel & Kjær 4195
3.3	IEC 60 959 pinna simulator [182]	DRP	100–8000	Supra-concha, cirum-aural, *intra-concha, insert*	Brüel & Kjær 4128C
3.4	Pinna simulator	DRP	<8000	Supra-concha, cirum-aural, *intra-concha, insert*	Head Acoustics HMS II.3

Table 8.5: Description of artificial ears defined in ITU-T Recommendation P.57 [237].

8.3.4 *Target frequency response*

As Poldy stated 'a palate of different frequency responses is available to cater for individual preferences' ([349], Section 14.4.1b) indicating that this topic is not a definite science. As already mentioned, the most commonly encountered target responses are the free-field and diffuse-field functions.

The diffuse-field response aims to match the response of the headphone measured at the DRP to that of a source in a diffuse field. An example of the diffuse-field response measured using a blocked ear canal method is presented in Figure 8.10[52,53]. This data is calculated using the CIPIC HRTF database[54] [12] and available from [11].

[52]The diffuse-field response is based on the average of the following angular locations: azimuths = [−80°, −65°, −55°, −45°−45°, 55°, 65°, 80°]; elevations = −45° to ∼230° every ∼5.625°).

[53]A 5 dB offset has been applied to the FF and DF data to provide the normally encountered 3 dB boost at 600 Hz and convergence to 0 dB at low frequency.

[54]This HRTF database is made available by CIPIC–Center for Image Processing and Integrated Computing, University of California, 1, Shields Avenue, Davis, CA 95616-8553, USA.

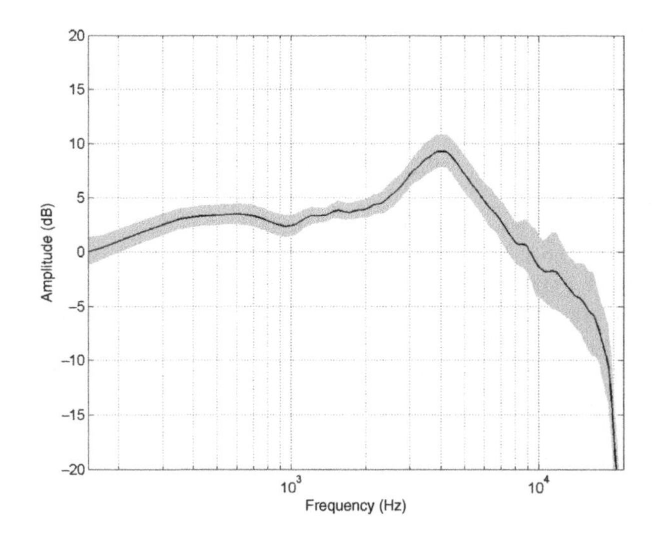

Figure 8.10: Average listener block ear canal diffuse-field frequency responses calculated from the CIPIC HRTF database (see footnote 54). Grey region illustrates the standard deviation of individual responses and the solid line illustrates the average diffuse-field response based upon 45 subjects.

The free-field response aims to match the response of the headphone measured at the DRP to that of a source located at (0° azimuth and 0° elevation) in a free or anechoic field. Example of the free-field response measured using a blocked ear canal method is presented in Figure 8.11 (see footnote 53).

A comparison of the average diffuse- and free-field blocked ear canal responses is presented in Figure 8.12. To compare the responses of blocked and open ear canal measurements, DF and FF responses are measured using a head and torso simulator (HATS) are presented in Figure 8.13. The reader should note that the roughness of the HATS response is due to fact that only a single HATS was being measured compared to the responses provided in Figure 8.12, which were based upon an average of 45 subjects. A clear difference in the first resonance can be seen between these measurements, which are associated with the presence of the ear canal resonance in open ear canal measurements.

The diffuse-field response is commonly supported as the target response for high-quality headphones, as presented by Theile [415], ITU-R Recommendation BS.708 [204] and also IEC 60268-7 [183].

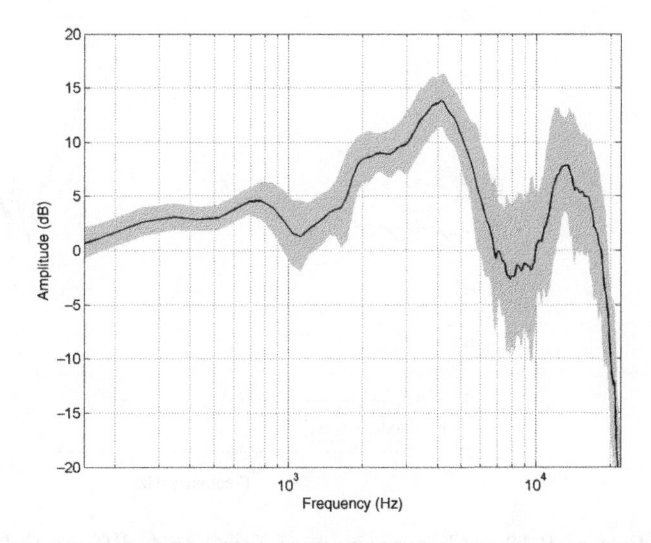

Figure 8.11: Average listener block ear canal free-field frequency responses calculated from the CIPIC HRTF database (see footnote 54). Grey region illustrates the standard deviation of individual responses and the solid line illustrates the average free-field response based upon 45 subjects.

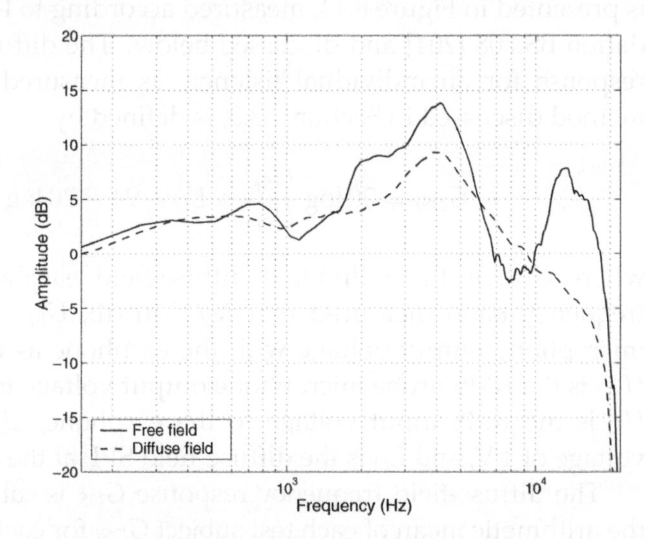

Figure 8.12: Comparison of blocked ear canal free- and diffuse-field responses, based upon an average of 45 subjects' HRTFs from the CIPIC HRTF database (see footnote 54).

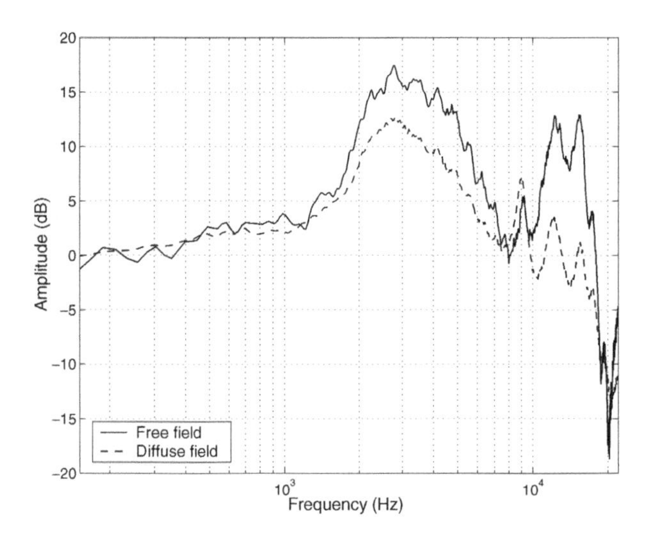

Figure 8.13: Comparison of free- and diffuse-field responses for a Brüel & Kjær 4128C head and torso simulator (open ear canal). Data measurement according to [359–361]. Reproduced by permission of Nokia Corporation.

Target response mask for the diffuse-field earphone response (G_{DS}) is presented in Figure 8.14, measured according to ITU-R Recommendation BS.708 [204] and discussed below. The diffuse-field earphone response for an individual listener, as measured using the direct method discussed in Section 8.3.3, is defined by

$$G_{DSi} = 20 \log \frac{U_{SK}}{U_{SD}} + L_D - 94 - 20 \log \frac{U_K}{U_0} \tag{8.3}$$

where G_{DSi} is the individual diffuse-field earphone response per frequency band measured re 1 Pa/V in dB, U_{SK} is the RMS probe microphone output voltage with the earphone as the sound source, U_{SD} is the RMS probe microphone output voltage in the diffuse field, U_K is the RMS input voltage to the earphone, U_0 is the reference voltage of 1 V, and L_D is the diffuse-field SPL at the reference point.

The diffuse-field frequency response G_{DS} is calculated by taking the arithmetic mean of each test subject G_{DSi} for each frequency band.

The less common free-field response is presented in IEC 60268-7 [183] and performance requirements and tolerance masks in IEC 60581-10 [181]. Alternative responses to the free or diffuse field have been discussed by a number of researchers [299, 348, 358, 455].

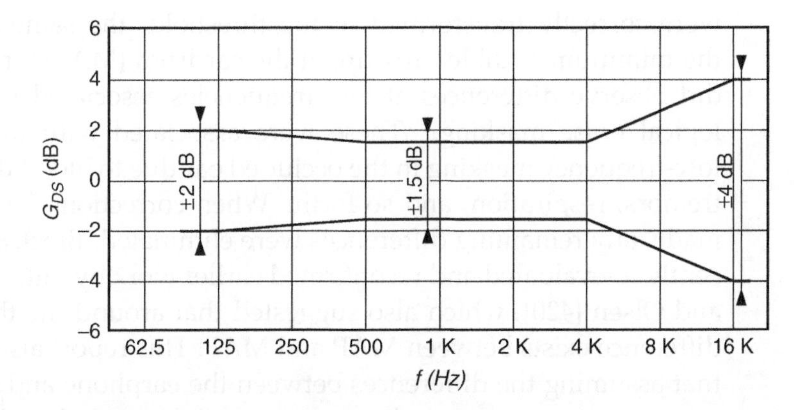

Figure 8.14: Tolerance mask for the diffuse-field frequency response of studio monitoring headphones [204]. Reproduced with the kind permission of ITU.

8.3.5 *Level calibration*

Once the experimenter has studied the ins and outs of headphone measurements in the above section, the issue of level or frequency calibration should be self evident.

Regarding the matching of left and right earphones, ITU-R Recommendation BS.708 [204] defines the acceptable tolerance as 1 dB in the range 100 Hz to 8 kHz and 2 dB from 10–16 kHz.

However, there still remains the issue of the illusive −6 dB, which has been observed between free-field loudspeaker and headphone reproduction.

The illusive −6 dB

For reference, we shortly present the matter of the apparent missing 6 dB. This topic has been discussed for some 70 years and considers the apparent difference of ∼6 dB between loudspeaker and headphone calibration around and above the threshold of audibility. It was originally studied by Sivian and White [381] and later by Munson and Wiener [312]. Both the groups found approximately a 6 dB difference in threshold and loudness judgements. A number of other authors also studied this topic and made similar observations.

To study this matter further, Killion [254] considered the available data for earphones and loudspeakers (referred to as sound field data). He observed that when the earphone and sound field data were correctly transformed to the threshold, the same estimate of the minimum audible pressure at the eardrum (MAPD) resulted. He did observe differences at low frequencies associated with physiological noise masking. These were associated with the increased low-frequency masking in the occluded ear due to blood flow, muscle tremors, respiration, and so forth. When corrections for these were made, any remaining differences were eliminated. Block *et al.* [58] recently re-evaluated and reconfirmed earlier work by Tillman, Johnson and Olsen [420], which also suggested that around the threshold no difference exists between MAP and MAF. This report also concluded that assuming the differences between the earphone and sound field measurement are correctly compensated for and the physiological factors are taken into consideration, no difference exists between the MAP and MAF. The interested reader is referred to Block *et al.* [58] for a thorough summary of the topic.

The topic of headphone calibration levels above the threshold still appears to be open as discussed briefly by Poldy [349]. Brixen and Søndergaard [68] have performed some experiments to study this issue further, from which it would appear that there exists approximately 5–9 dB difference between headphone and loudspeaker calibration levels, depending on the signal type and other factors. In practice, this means that if headphones are calibrated to the same physical level as a free-field source, the headphone will be perceived to sound quieter than the free-field source.

As a result of these studies, the experimenter should be aware that calibration of headphone levels may differ compared to loudspeaker calibrations, although concrete guidance on exact calibration levels is not presently available.

CHAPTER

9

Test planning, administration and reporting

This chapter, will overview the key practical issues that should be considered when planning, administering and reporting listening tests. This chapter, while very practical in nature, should help to minimise stress for both the experimenter and the subjects, with the aim of ensuring the quality of the collected and the reported listening test data.

9.1 Planning

9.1.1 Experimental planning

The design of the listening test will have an impact on the practicalities associated with running the experiment. On the basis of the number of dependent and independent variables and experimental paradigm selected, the size and duration of the test will be dictated. Frequently, the initial designs of the experiments prove to be unmanageable.

Perceptual Audio Evaluation – Theory, Method and Application Søren Bech and Nick Zacharov

Several design iterations may be required to achieve a design that balances the need to test the hypothesis and the size and duration of the task. This will be an iterative process until a sufficient balance is found.

The size of the test has an impact not only on the quality of the results (as discussed in Section 6.1) but also at a more practical level. In addition to the increased cost, running a large test will impose additional requirements on the experimental design. It should be borne in mind that subjects are prone to fatigue and boredom during such, sometimes tedious, tasks. If the subject is required to judge hundreds of samples, the experiment will need to be divided into several sessions, thus requiring additional designing. From practical experience, it has been observed that sessions of 20 minutes are quite a good idea and that several such sessions can be run during a day with occasional breaks. In practice, sessions of 30–40 minutes are still acceptable. To ensure that subjects can rest at ease, a suitable location should be provided for the breaks with some refreshments, and so on. While it is a common practice to test subjects for many hours per day with hundreds of samples, this is not advised. It is advised that listeners ideally participate in 1–2 hours of listening tests per day.

In the case that it is paramount to run a very large number of samples, the experimenter should consider using a particular experimental design method to ensure the quality of the results and avoid overloading subjects. In case the experimenter wishes to run a full factorial experiment but needs to split the experiment into multiple sessions, a blocked experimental design may be suitable. For very large experiments, it may be advisable to consider a fractional factorial design, such that each subject evaluates a partial stimulus set. Well-documented experimental design methods of this kind, such as incomplete balanced block designs, can be found in [63, 94, 300]. In very complex cases, a response surface method (RSM) experiment may need to be considered, for which the interested reader may refer [313]. Several excellent software packages are available for creating such experimental designs, for example, Design Expert[55] or DOE Kiss[56].

When designing an experiment, it is good to estimate the time required to complete all the preparations, run the experiment, analyse and report the results. More often than not, the inexperienced researcher will underestimate the effort required. The areas that will require significant effort include:

[55] `http://www.statease.com/`

[56] `http://www.does.org/airacad/doe_kiss.html`

- Experimental design and planning
- Subject selection
- Sample selection and processing
- Configuration and calibration of the experimental set-up
- Pilot study
- Main study
- Analysis
- Reporting.

Roughly speaking, about one-third of the time will be spent preparing the experiment, one-third for administrating the pilot and main experiment and the remaining one-third for analysis and reporting. Figure 9.1 illustrates the main steps in administrating the test.

Pilot experiments, also discussed in Chapter 3, aim to provide a final check for the experimenter to ensure the test will run as planned and deliver the sought data. Often the pilot study is a small subset of the main experiment performed with a few subjects. For this stage to be of value, the experimenter should plan to analyse the pilot study to establish whether refinements are needed for the main experiment on the basis of the findings. Time allocation for this to occur should be planned.

When estimating the time required for running the test, it is good to include some overheads. It is quite common for the experiment to take 2–3 times longer than the inexperienced experimenter initially expected. For example, the effort associated with escorting subjects to and from the test room, and so on, will often lead to an increase in the testing time by 1.5–2 times.

9.1.2 *Logistic considerations*

A few very practical but vitally important logistic topics are mentioned here that, if considered sufficiently well in advance, will ensure that the experiment designed can be performed within the planned schedule. If overlooked these, apparently trivial matters, can lead to significant delays in the research work.

The following five topics should be borne in mind when planning the test.

WHERE will the listening test be performed? The experimenter should ensure that a suitable test room (see Section 7.1) exists and is available for the duration of the configuration and

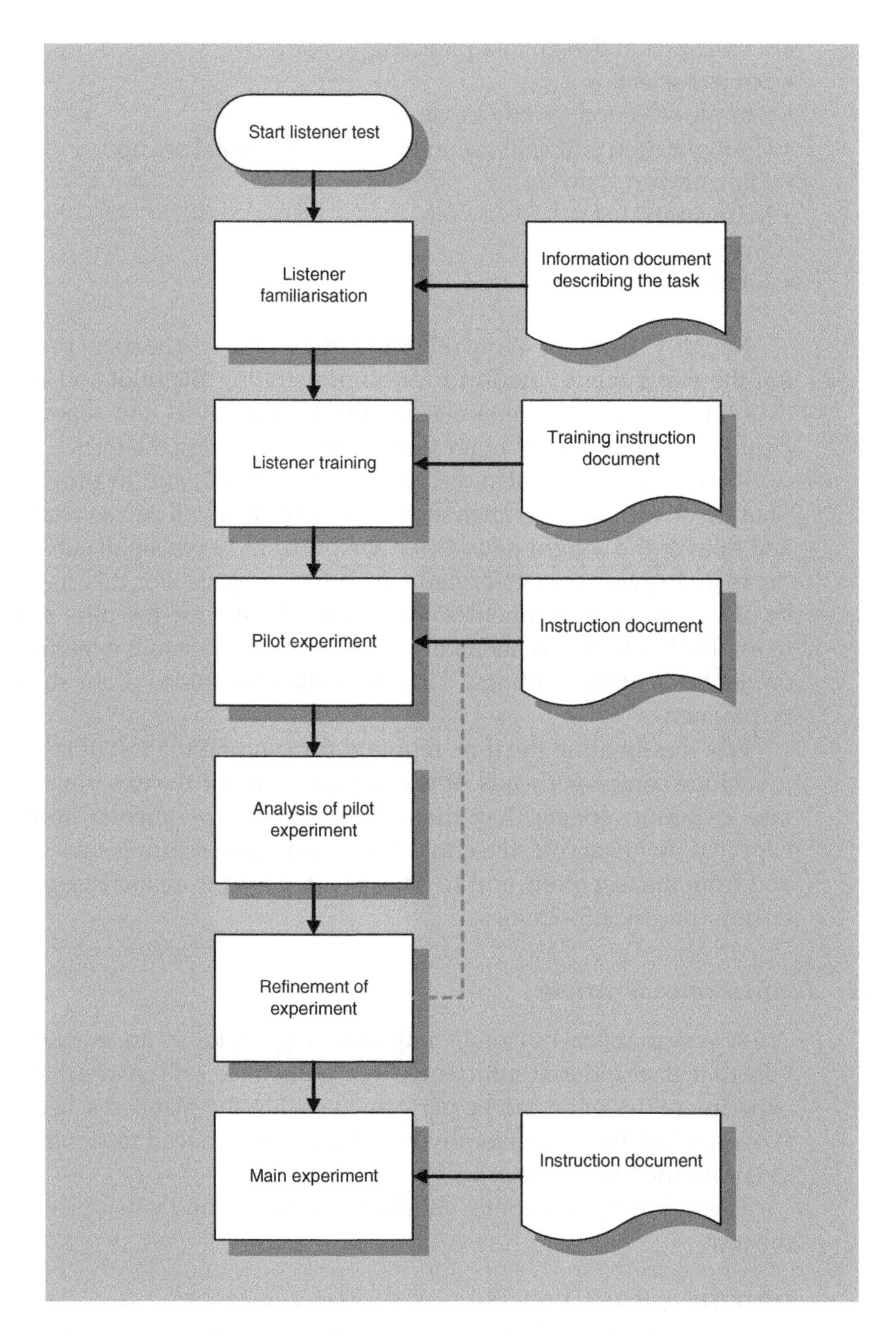

Figure 9.1: Flowchart of listening test administration stages.

administration of the test. Ensure that sufficient time is reserved for the set-up and calibration (see Chapter 8). Also verify that the test room meets the acoustic requirements as discussed in Chapter 7.1.

HOW shall the test be administered? Typically, listening tests take many hours or days to be completed. In practice, subjects will require some assistance to perform the test. Most commonly an administrator would be required to escort subjects, provide them with instructions on the task to be performed and ensure they can use the listening test system. Typically, this work is performed by the experimenter. However, in the case of descriptive analysis (DA) experiments, as described in Section 4.1.1, a neutral panel leader may be employed. Guidance on training of panel leaders is provided in ISO [202] with more general information provided in [201]. Administration of a test is a time-consuming task.

WHO shall perform the listening tests? When planning the experiment, the experimenter should have considered the type and number of subjects to be employed (see Section 5.4.1).

Irrespective of what category of subject is to be employed for the experiment, these subjects are to be found and evaluated for suitability. Some practical guidance on such matters is provided in Section 9.2.1. Scheduling of the experiment with the subjects should be performed well in advance.

In the case of using expert subjects, if a listening panel does not exist, additional overhead will be required for this work. Procedures for establishing such a panel have been presented in Section 5.4.2.

WHAT equipment is needed to set up and conduct the experiment? Once the experiment is planned, make a list of the required components and ensure that they are in working order, calibrated and available at the time of the experiment for the required duration.

WHEN will the test be performed? Ensure that it is planned that all of the above elements are available simultaneously to enable the experiment to run as expected. Bear in mind that ideally a small pilot study will precede the main experiment.

Considering these details sufficiently in advance will certainly ease the process of administrating a listening test.

9.1.3 Ethical considerations

When planning and performing listening tests, it is important to ensure the well being of the subjects. Issues relating to the general comfort and well being of the subjects have been discussed in Section 7.1.4. In this section, the matter of ethical codes of conduct will be briefly considered to ensure that subjects are well treated taking into account the safety aspects.

Thankfully, the topic of ethics and codes of conduct are well established in the field of psychology, to which reference will be made for guidance on good practice. An extensive code of conduct is provided by the America Psychological Society (APA), entitled 'Ethical principles of psychologists and code of conduct' [15]. This document is being revised continuously and updates are available online via `http://www.apa.org/ethics/code.html`. This document provides an extensive overview of many of the relevant ethical issues, which should be considered for listening tests. The key areas that are of interest to the audio experimenter and have to be considered include the following:

- Unfair discrimination
- Sexual harassment
- Other harassments
- Avoidance of harm
- Multiple relationships
- Conflict of interest
- Third-party requests for services
- Exploitative relationships
- Cooperation with other professionals
- Informed consent.

One of the vital topics in the field of audio is that relating to *Avoiding Harm*, which is defined in [15] in the following manner: 'Psychologists take reasonable steps to avoid harming their clients/patients, students, supervisees, research participants, organisational clients and others with whom they work, and to minimise harm where it is foreseeable and unavoidable'.

While in the field of audio the experimenter is often not a psychologist, the topic of avoiding harm is relevant in relation to noise/music/speech exposure. The experimenter must always ensure that subjects are not exposed to potentially damaging sound pressure levels. The Noise at Work Regulations [118] provides guidance in this respect. The exposure limit value based on a daily 8-hour noise exposure is defined as

$$L_{\mathrm{EX,8h}} = 87 \quad \mathrm{dB(A)} \tag{9.1}$$

where the time-weighted average of the noise exposure levels (reference 20 μPa) for a nominal 8-hour working day as defined in ISO 1999 [191] and the maximum peak sound pressure levels are defined as

$$L_{\mathrm{peak}} = 140 \quad \mathrm{dB(C)} \tag{9.2}$$

and the upper exposure action limit is defined as

$$L_{\mathrm{EX,8h}} = 85 \quad \mathrm{dB(A)} \tag{9.3}$$

and the lower exposure action limit is defined as

$$L_{\mathrm{EX,8h}} = 80 \quad \mathrm{dB(A)} \tag{9.4}$$

If the noise level exceeds the lower exposure level, individual hearing protection *should* be made available for subjects. If the upper exposure level is exceeded, individual hearing protection *shall* be used. As the use of hearing protection may invalidate the planned experiment, it is strongly advised to avoid such noise exposure levels, by either shortening the exposure time for subjects or reducing reproduction levels. Where the experimenter plans to perform listening tests at or near such reproduction levels, he is referred to [118] for guidance to ensure the well being of the subjects.

The experimenter should additionally ensure that the subjects are comfortable in their work. This is generally a good practice as happy subjects are more likely to be motivated to perform the task well. To this end, ensure that the experiment is designed so that it does not fatigue the subjects too much and that sufficient breaks are provided to avoid boredom.

Another good practice is to ensure that subjects have provided their consent to participate in the experiment. This requires the experimenter to inform the subject of the nature of the task, often without providing too much detail which could bias the subject and the experimental results. Examples of information sheets and consent forms are presented in Figures 9.2 and 9.3.

Department of Music and Sound Recording

Listening tests
General information sheet for participants

Thank you for agreeing to take part in one of our listening tests. Listening tests are an important part of our research into reproduced sound quality and we rely upon a committed panel of listeners to provide us with their responses to a range of different sounds.

The sorts of responses that you will typically be asked for fall into the following categories:

- Ratings of sound quality attributes on scales provided by the experimenter
- Descriptions of sound quality using your own terms or drawings
- Preference or 'liking' responses
- Judgements of differences and/or similarities between sounds

Sometimes sound reproduction will be accompanied by video pictures and you may be asked to undertake a task relating to the picture or the sound, such as playing a game, following an object or identifying features in the scene.

We may ask you to undergo a screening exercise prior to selecting you for a listening panel. This is normally to determine factors such as your consistency of judgement and your sensitivity to the attributes under investigation. All data relating to such screening tests is stored confidentially and anonymously. In some cases you will subsequently be trained in the identification and scaling of the sound quality attributes under investigation.

There are normally no right or wrong answers during the listening test proper – in other words it is not you that is under test. We are interested in your responses to the sounds that we present because of what they tell us about the way sound signals are perceived and/or described.

Health and Safety

Listening tests are structured in such a way as to allow for breaks so as to avoid listener fatigue. The precise time commitment required of you will be agreed beforehand.

Sound reproduction levels are controlled so as to be within safe limits as recommended by UK legislation.

Although we naturally hope for your continued commitment to the listening test, you are free to opt out of it at any time without giving a reason and without prejudice.

Figure 9.2: An example of a listening test–general information sheet for participants. Reproduced by permission of University of Surrey.

Department of Music and Sound Recording

Listening test consent form

This form is to be completed by any subject that agrees to take part in a listening test, before the test begins.

I the undersigned voluntarily agree to take part in the study on

______________________________ (project title)

I have read and understood the information sheet provided. I have been given a full explanation by the investigators of the nature, purpose, location and likely duration of the listening test, and of what I will be expected to do. I have been given the opportunity to ask questions on all aspects of the listening test and have understood the advice and information given as a result.

I agree to comply with any instructions given to me during the listening test and to cooperate fully with the investigators.

I understand that all documentation held on a volunteer is in the strictest confidence and complies with the Data Protection Act (1998). I agree that I will not seek to restrict the use of the results provided that my anonymity is preserved.

I understand that I am free to withdraw from the listening test at any time without needing to justify my decision and without prejudice.

*I acknowledge that in consideration for completing the listening test I shall receive the sum of £____.
I recognise the sum would be less, and at the discretion of the Principal Investigator, if I withdraw before completion of the study.

I confirm that I have read and understood the above and freely consent to participating in this study. I have been given adequate time to consider my participation and agree to comply with the instructions and restrictions of the study.

Name of volunteer ______________________
(block capitals)

Signed ______________________

Date ______________________

Name of witness ______________________
(block capitals)

Signed ______________________

Date ______________________

*Delete if not applicable

Figure 9.3: An example of a listening test consent form. Reproduced by permission of University of Surrey.

9.2 Administration

9.2.1 Subject matters

To ensure that the experiment runs as planned, the researcher should consider the subject-related matters well in advance. While the topic of subject selection has been covered in Chapter 5.4, a few minor logistical details need to be taken into consideration.

The first of these is the question of which population to select the subjects from and this applies to members of both listening panels and groups of untrained subjects. Panels may be created by using colleagues or recruiting from local universities, and so on, as discussed in Section 5.4.2. Finding untrained subjects can be more challenging as, typically, they are employed only once. A practical way to approach these matters is to contact local organisations, youth clubs and societies, which may be able to propose subjects. Posting information about such experiments in universities and music schools is also a good way to locate untrained subjects. Employment agencies may also be able to assist in locating suitable participants. Of course, there needs to be some appeal to participate in the experiment, which brings up the next topic.

Remuneration is often a vital key to enticing subjects to participate in an experiment. While in the short term it may be possible to coax colleagues to participate in a few experiments, this is not a sustainable mode in the long term. If the experimenter wishes to ensure that the experiment is complete in a timely fashion and that the subjects are diligent about their task, some kind of remuneration is recommended. If subjects are to be paid, ensure to check with the payroll department of your organisation on the specifics of paying individuals, as there may be taxation and related issues to be handled.

In some cases, depending on the nature of the work and organisation, subjects may need to pass a security check and even sign a non-disclosure agreement (NDA) prior to performing the test. Lastly, asking subjects to sign a consent form is a good practice so that all parties agree to the nature of the task (see the example of a consent form in Figure 9.3).

9.2.2 Subject familiarisation

Having established a group of subjects to perform the test, the next step is to familiarise the subjects with the task in hand. This can happen over a couple of sessions or just before the listening test, depending on the nature and complexity of the task. The main aim here is to ensure that the subjects are sufficiently well informed of the task and how it should be performed with ease. This small investment should help to motivate the subjects and reduce error in the data due to subjects misunderstanding the task.

The process of familiarisation should comprise several steps :

Task introduction If at all possible, gather all the subjects together prior to the commencement of the task and provide them with a broad overview of the experiment and the associated practical details (e.g. duration) and also illustrate how important this work is. A balance is needed between what information is needed to inform the subjects and what information is needed to motivate them while avoiding overly biasing them.

Written instruction Prior to the test, provide the subjects with written instructions of the task and details of how it is to be completed. Indicate the nature of the test and define the manner in which the items are to be scaled, for example, according to basic audio quality (BAQ) or timbre. If complex or multiple attributes are employed, for example as presented in Appendix B, additional training of the attribute usage may be required. These topics are discussed earlier in Sections 4.1.1 and 5.4.4. Where possible provide these instructions prior to the test so that subjects can read and consider them. Additionally, a copy of the instructions should be available throughout the duration of the test, to which the subjects can refer if needed. An example of instructions are provided in Figures 9.4 and 9.5.

Verbal instruction To support the written instructions, the experimenter should verbally review the instructions with each subject. Confirm with the subjects that they have understood the task and answer or clarify any issue the subjects may have.

User-interface familiarisation While the experimenter is probably quite familiar with the scales and user interfaces (UIs) to be used, the subjects are probably not. To ensure the quality of the results, spend some time to familiarise the subjects with the scales and the UI. The best way to do this is to allow subjects to try the UI prior to the test and ask for clarification on its usage when needed.

Instruction to listeners
Comparison category rating test.
'Evaluation of the influence of various environmental noises on the quality of different telephone systems'

In this experiment, you will hear pairs of speech samples that have been recorded through various experimental telephone equipment. You will listen to these samples through the telephone handset in front of you.

What you will hear is one pair of sentences, a short period of silence, and another pair of sentences. You will evaluate the quality of the second pair of sentences compared to the quality of the first pair of sentences.

You should listen carefully to each pair of samples. Then, when the green light is on, please record your opinion about the quality of the second sample relative to the quality of the first sample using the following scale:

The Quality of the Second Compared to the Quality of the First is:

- 3: Much Better
- 2: Better
- 1: Slightly Better
- 0: About the Same
- -1: Slightly Worse
- -2: Worse
- -3: Much Worse

You will have five seconds to record your answer by pushing the button corresponding to your choice. There will be a short pause before the presentation of next pair of sentences.

We will begin with a short practise session to familiarise you with the test procedure. The actual tests will take place during sessions of 10 to 15 minutes.

Figure 9.4: Example subject instructions for an ITU-T P.800 comparison category rating experiment [224]. Reproduced with the kind permission of ITU.

Dear listener

Welcome to the AES16 *virtual home theatre* round robin experiment. Please make yourself comfortable in the seat provided and read through these instructions carefully before starting the experiment. If you have any further questions, please don't hesitate to ask the person conducting the experiment.

The experiment consists of listening to a number of test samples labelled *A* to *G*. You can switch between these samples by clicking, with the mouse, on the labelled buttons. You are asked to listen to all the samples, at least partially, before you start grading. Rating the samples is performed by clicking on the pull down menu below the questions. Here, you will find a choice of grades from 1–7 to select from. You are asked to place the samples in rank order. Please consider 1 as the lowest (worst) rank and 7 as the highest (best) rank. If you find two or more samples very similar, you may select the same grade/rank.

You will be asked to grade the samples in terms of two questions:

- Rank order the samples by spatial sound quality (1 = lowest rank, 7 = highest rank)

  When evaluating the spatial sound quality, please consider all aspects of spatial sound reproduction. This might include the locatedness or localisation of the sound, how enveloping it is, it's naturalness and depth.

- Rank order the samples by timbral quality (1 = lowest rank, 7 = highest rank)

  When considering aspects of the timbre quality, please consider timbre as a measure of the tone colour. Timbre can be considered as the sensory attribute which allows two similar sounds of the same pitch and loudness to be different, for example a clarinet and a cello. Any audible distortions can also be considered as an aspect of the timbral quality.

When you have completed your ranking for each sample you can move onto the next set by clicking on the *Done* button. In total you will hear four sets of seven samples, to be rank ordered. Each set will consist of a different program item.

Please take your time to evaluate and grade all the samples. The test will require approximately 30 minutes for completion. You are kindly asked not to discuss your judgements with other delegates prior to the end of the conference.

Thank you for participating, have fun and good luck.

Figure 9.5: An example of subject instructions for a rank order listening test [464].

Figure 9.6: Example sample familiarisation graphical user interface.

Sample familiarisation Sample familiarisation aims to ensure that subjects are aware of the scope and range of stimuli to be experienced during the experiment. This will help subjects to employ the rating scale effectively and without constraint. The primary aim is to avoid, for example, the problem of the subject's wish to assign grade 11 [357] on a 10-point scale.

Several means can be applied to perform the sample familiarisation. The most basic is just to allow subjects to hear a number of samples that span the range of qualities to be evaluated. This can be performed manually or by using a graphical UI as presented in Figure 9.6.

Both the sample and user-interface familiarisation can be combined into a training experiment, where the subjects perform a subset of the experiment with a set of samples that span the range of qualities to be evaluated. In this manner, they hear the range of samples and can also become familiar with scaling their perception with the UI. This data should not be included in the analysis of results[57].

Another way to perform the familiarisation without the subject being aware is to include a number of dummy training items at the beginning of the listening test session. For example, the first 20 items of a test may be representative training items that are duplicated later in the test. The ratings of these items are not to be included in the analysis, but provide a simple means of blind familiarisation for the subject. The limitation of this method is that subjects do not have the chance to clarify any questions or issues they have regarding the experiment.

[57]This familiarisation approach is not commonly encountered; however, in the authors' experience this can quickly and easily familiarise subjects, leading to more reliable results.

Lastly, subjects like to be informed of what they have been involved in. Once the experiment has been completed, it is beneficial to inform subjects what the test was about and how the results will be used, for example, for standardisation, algorithm selection, and so on. This should only be done within the constraints of confidentiality permitted, but will help to maintain the subjects' motivation for future tests. Additionally, any feedback on the subjects' performance may interest the subjects and also encourage development through training.

9.2.3 *Listening test software*

When performing listening tests, audio stimuli are to be presented to subjects and their responses are to be collected. While these steps can be performed manually, this is a highly complex, time-consuming and very error prone approach. Nowadays, computer-based systems are available to automate stimulus presentation and/or data collection, avoiding most of the limitations associated with a manual procedure. Such software tools are highly desirable in listening test work to lighten the burden on the experimenter and also to provide a better control over the experiment. This latter aspect leads to a reduction in experimental error, as well as providing robustness. Additionally, using a computer-based system allows for similar experiments to be perfectly duplicated or repeated at different locations or times.

The following software provides means to collect subjects' responses in general, but have limited or no audio presentation functionality.

Fizz Generic commercial software for data collection.

Compusense Generic commercial software for data collection, with potential for presenting audio samples.

The following non-exhaustive list of software packages provides means to present audio stimuli and collect subject responses and include aspects of the audio. A generalised view on a software stimulus playback architecture for subjective audio testing is presented in Figure 9.7, based upon the GuineaPig system [175, 176] for example.

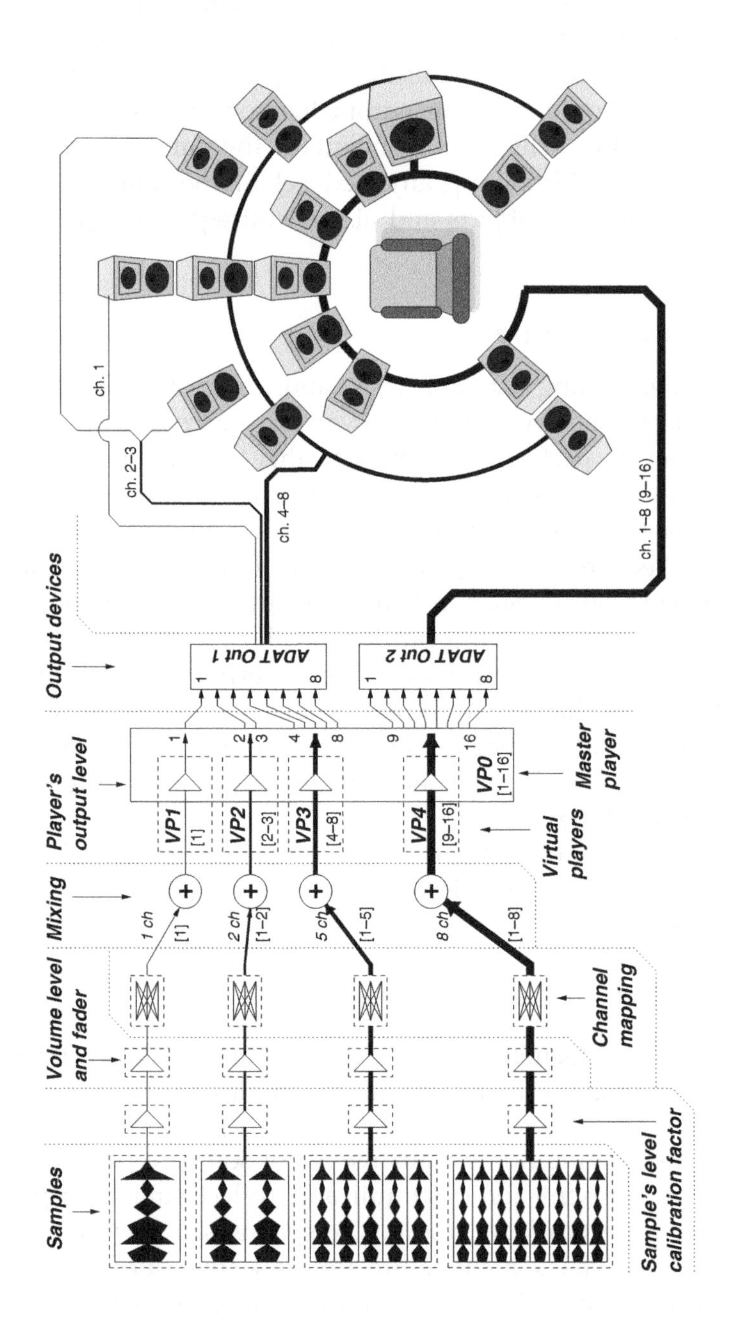

Figure 9.7: Block diagram of GuineaPig2 software–based multichannel listening test system [175, 176].

STEP[58] is a commercially available computer-controlled system for audio presentation and subjective evaluation of speech or audio quality running on Windows XP. [19].

Three test types are included:

- ITU-R Recommendation BS.1116-1 [207], triple stimulus, hidden reference tests.
- ITU-R Recommendation BS.1534, 'MUlti-stimulus test with Hidden Reference and Anchor (MUSHRA)' [217] tests.
- Two-comparison forced-choice tests with 5-point grading scale.

CRC-SEAQ is a commercial software System for Evaluation of Sound Quality, providing support for mono, stereo and multichannel reproduction using a range of windows systems [95]. Also provides access to an objective test module using the ITU-R PEAQ model [213] and a sample time alignment module.

Two test types are included:

- ITU-R Recommendation BS.1116-1 [207], triple stimulus, hidden reference tests,
- ITU-R Recommendation BS.1534, 'MUlti-Stimulus test with Hidden Reference and Anchor (MUSHRA)' [217] tests.

PCABX Dedicated software for performing ABX type tests under windows [262].

GuineaPig Proprietary generic software for audio-visual tests running on Linux[59]. The system provides highly scalable and easy listening test creation with or without image. Supports multichannel audio output. Further details on the application can be found in [175, 176].

Generic test types provided with this package include the following:

- Single stimulus, for example, for ACR type tests [224]
- Paired comparison
- Paired rating, for example, for DCR or CCR test types [224]
- ABX tests
- ITU-R Recommendation BS.1116-1 [207], triple stimulus, hidden reference tests,

[58] www.audioresearchlabs.com

[59] Previous generations of GuineaPig ran under Silicon Graphics IRIX.

- ITU-R Recommendation BS.1534, 'MUlti-Stimulus test with Hidden Reference and Anchor (MUSHRA)' [217] tests
- Two-alternative forced-choice tests [272]
- Method of adjustment tests [81]
- Single stimulus continuous quality evaluation [216]
- Intelligibility test: speech reception threshold (SRT) [433]
- Rank order tests.

Head Acoustics–SQuare is a commercially available jury testing software application.

Brüel & Kjær–Psychoacoustic Test Bench Software is a commercially available jury testing software application providing the means in

- the designing and setting up of jury tests;
- the conducting of jury tests;
- data collection and statistical processing;
- calculation of standard metrics on jury test samples;
- generation of combination metrics.

Both paired comparison and semantic differential tests are supported.

Lise Listening Interface for Sound Experiments is a package of routines aimed at listening, comparing and describing sounds. LISE runs under Matlab (5.3) on PC, Mac and Linux. Described in [362] and available via `http://cthta4.ta.chalmers.se/HomePages/vincent/doc/matlabRoutines.html`.

Example test types include the following:

- ABX tests
- Paired comparison
- Multiple comparison.

MUSHRA Software–Fraunhofer Institut Integrierte Schaltungen Commercially, available listening test software for performing and analysing MUSHRA [217] tests. The tool developed in tcl/tk can be run on MS Windows or Unix platforms.

MUSHRAM–A Matlab interface for MUSHRA listening tests A Matlab-based GUI for performing ITU-R Recommendation BS.1534, 'MUlti-Stimulus test with Hidden Reference and Anchor (MUSHRA)' [217] tests. Software is available from [436] under GNU General Public License (GPL).

9.3 Reporting

When finalising the reporting of an experiment, it is important to include all relevant information. While this is a slow and laborious process for the experimenter, this effort will provide a means for referring to the work in the long term. This is of interest when the work is to be continued, repeated or verified at a later date.

An excellent and very thorough guide to publication, reporting of results, authorship and associated topics is available in the 'Publication Manual of the American Psychological Association' [15]. Additionally, guidelines on ethics in reporting are discussed by Michaelson [294]. Guidance on technical scientific writing can be found in Wilkinson [450]. Some key elements from [15] regarding good reporting practices are summarised below.

When reporting an experiment, the author should aim to provide a sufficient level of detail to motivate the research, the applied research methods, analysis and final conclusions. Providing this information in a concise and effective manner is of particular importance for scientific publications. Reports of experiments should comprise the following key elements:

Introduction The introduction should present the problem under consideration and consider why it is of importance. The hypothesis should be defined and related to the experimental design method selected and to the research problem. The relevant literature and the related research should be reviewed.

Method The method section should consider all practical aspects relating to the experiment. Where possible, sufficient detail should be provided to ensure that the experiment can be reproduced or repeated. The experimental design should be presented in detail and specify the dependent and independent variables and any specific details (e.g. blocking, etc.). The experimental method should be defined with reference to any relevant documents (e.g. standards). The subjects' characteristics are to be defined, including details of the category of the subject chosen and any important demographic details. Additionally, details of the experimental set-up and apparatus are to be defined. Any relevant calibration information should also be provided.

Results This section should provide a summary of the data collection and the subsequent statistical analysis applied. Testing of statistical assumption should be presented and, if required, justification for the particular analysis method should be provided if it is considered uncommon. Testing of the null hypothesis is often the core of a listening test. Discussion on whether the hypothesis is supported or rejected should be presented with the associated interpretation of the implications of the results. Sufficient detail should be provided to support the key observations and conclusions (positive or negative) to be made. Where results are to be reported, for example, it is important to provide estimates of the significance of the results, commonly provided through the use of confidence intervals.

Discussion The discussion should evaluate and interpret the results and their implications. As the experiment often relates to hypothesis testing, it should be clearly stated whether or not the hypothesis is supported. An open discussion of the results can be presented, examining the results and any inferences to be made. The theoretical impact of these consequences may be discussed. Where possible, comparison to other related work should be presented, considering the implications of the comparison and the resulting conclusions. Lastly, it is a good practice to state the scope and limitation of the reported study.

References A list of references should provide background motivation for the work, support for the test methodology and also the results of related work for comparison when available.

When performing scientific work and reporting the results, it is vital to ensure that good scientific practices are employed. Guidelines to good scientific practice regarding the reporting of results are provided by the Finnish National Advisory Board on Research Ethics [317]. In this document, the key elements to good scientific practices suggest that researchers:

- follow modes of action endorsed by the research community, that is, integrity, meticulousness and accuracy in conducting research, in recording and presenting results, and in judging research and its results;
- apply ethically sustainable data-collection, research and evaluation methods conforming to scientific criteria and practise openness intrinsic to scientific knowledge in publishing their findings; and
- take due account of other researchers' work and achievements, respecting their work and giving due credit and weight to their achievements in carrying out their own research and publishing its results.

Further, it is in keeping with good scientific practice that:

- research is planned, conducted and reported in detail and according to the standards set for scientific knowledge;
- questions relating to the status, rights, co-authorship, liabilities and obligations of the members of a research team, right to research results and the preservation of material are determined and recorded in a manner acceptable to all parties before the research project starts or a researcher is recruited to the team;
- the sources of financing and other associations relevant to the conduct of research are made known to those participating in the research and reported when the findings are published; and
- good administrative practice and good personnel and financial management practices are observed.

Where possible these guidelines should be adhered to when performing and reporting results to maintain scientific standards.

PART III

Applications

CHAPTER

10 Commonly encountered experimental paradigms

In this chapter, a number of commonly encountered test paradigms will be reviewed briefly. The intention here is to provide the reader with key information regarding the applicability of methods and key sources of information.

10.1 Standards

10.1.1 ITU-T Recommendation P.800 methods

Absolute category rating

Test paradigm Absolute category rating (ACR) comprising a single stimulus paradigm

Perceptual Audio Evaluation – Theory, Method and Application Søren Bech and Nick Zacharov

Description Subjects are presented with a single stimulus at a time and asked to rate the listening on a 5-point categorical scale

Primary Application Listening-only test of analogue or digital telecommunication systems

Non-applicability Not intended for the assessment of echo-cancellers (see ITU-T Recommendation P.831 [231]), noise suppression algorithms (see ITU-T Recommendation P.835 [239]) or for conversational tests (See Annex A of ITU-T Recommendation P.800 [224]. Also not intended for assessment of audio codecs for which ITU-R Recommendation BS.1116-1 [207] and ITU-R Recommendation BS.1534 [217] are intended (see Sections 10.1.2 and 10.1.3 respectively)

Sources of information Annex B ITU-T Recommendation P.800 [224], ITU-T test plans, ITU-T telephonometry handbook [84]

Dependent variable(s) Listening quality or listening effort or loudness preference

Scale five-point category scale (see Table 4.3 and Figure 4.7(a)) [238]. Averaged results of listening effort and loudness are presented in units of MOS_{LE} and MOS_{LP} respectively.

Independent variable(s) System/codec type, speech sample, talker gender, sentence, listening level

Subject requirements Untrained or naïve subjects (see Section 5.4.1)

Number of subjects 24–36 (at least 12 subjects from either gender)

Stimuli Two to five speech stimuli comprising meaningful short sentences (length 2–3 s). Both male and female speech should be presented.

Technical requirements Reference condition should always be included in the test. Typically comprises the MNRU (modulated noise reference unit) [225] for a reference digital system. Other references may be used for example, for signal-to-noise ratio cases; refer to [226] Section 8.2.3.

Listening system should follow the requirements of the intermediate reference system (IRS) [219] or modified-IRS system

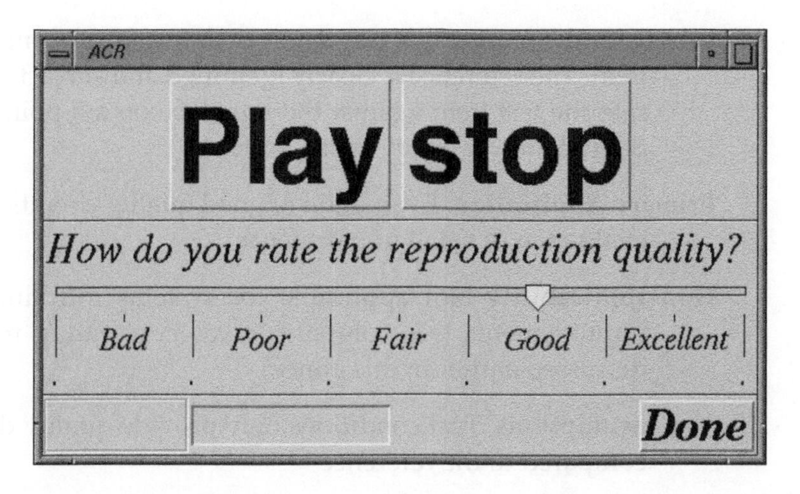

Figure 10.1: Example user interface for an ITU-T Recommendation P.800 [224] absolute category rating (ACR) listening test using GP2 software [175, 176].

(considered suitable for evaluation of all-digital connections using speech codecs, see [226] annex D).

Room requirements: 30–120 m^3, reverberation time <500 ms (ideally 200–300 ms), background noise level 30 dBA

Background noise should be reproduced using the Hoth [170] characteristics at a level of 50 dBA SPL (fast) discussed further in Section 7.2.

Significant details are provided in the standard regarding the requirements for recording of the speech samples. The interested readers are referred to the standard for all the details.

Analysis approach ANOVA based. Reports means (MOS). Confidence limits should be evaluated and significance tests performed using ANOVA techniques.

Example experiments See Barret *et al.* [23]

Example user interface See Figure 10.1

Degradation category rating

Test paradigm Degradation category rating (DCR) comprising a paired rating with fixed reference

Description Subjects are presented with a pair of samples for each item. The reference is clearly identified and subjects are asked to rate the test item against the reference on a 5-point categorical degradation category rating scale.

Primary Application Evaluation of good quality circuits. Suitable for evaluation of small impairments

Non-applicability Not applicable for systems introducing quality improvements (see comparison category rating method [224] discussed earlier in this annex)

Key assumptions Test conditions only provide quality degradations compared to the reference

Sources of information Annex D ITU-T Recommendation P.800 [224], ITU-T test plans, ITU-T telephonometry handbook [84]

Dependent variable(s) Degradation and annoyance

Scale Annoyance or 5-point degradation category rating scale (see Figure 4.7(b)).

Independent variable(s) Talker, speech samples, codec/system

Subject requirements Untrained or naïve subjects (see Section 5.4.1)

Number of subjects 24–32

Stimuli Speech samples comprising four talkers and two phonetically balanced sentences separated by a silent period of 0.5 s

Technical requirements The quality reference comprises a band-limited version of the original sample.

Test planning, administration and reporting Stimuli presented as pairs (A–B) or repeated pairs (A–B–A–B), where A is the quality reference and B the test stimuli. It is claimed that only one randomised presentation order is required! Some null pairs (e.g. A–A) are to be presented for at least one talker to check the quality of anchoring.

Analysis approach Sensitivities can be quantified by means of a statistical multiple comparison test. When a posteriori comparison of circuits is needed a Tukey Honestly Significant Difference (HSD) test can be applied effectively.

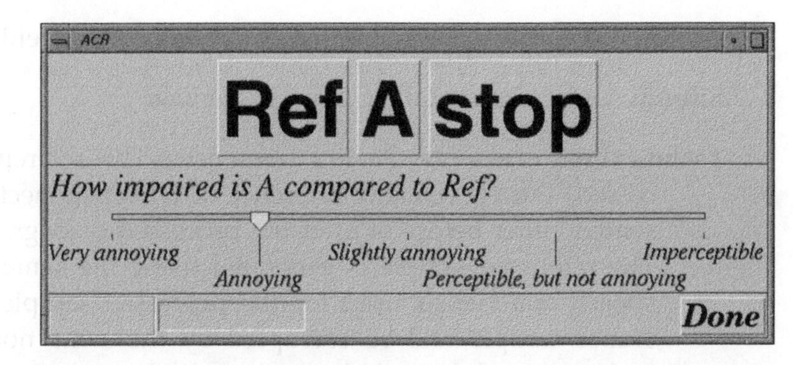

Figure 10.2: Example user interface for an ITU-T Recommendation P.800 [224] degradation category rating (DCR) listening test using GP2 software [175, 176].

Example experiments See 3GPP Technical Specification Group Services and System Aspects; AMR-WB Speech Codec Performance Characterisation [1]

Example user interface See Figure 10.2

Comparison category rating

Test paradigm Paired rating hidden reference

Description Subjects are presented with a pair of samples for each item. The hidden reference is identified and subjects are asked to rate the test items against each other on a 7-point categorical comparison category rating scale

Primary Application Applied to evaluations of systems that may improve or degrade speech quality compared to the reference

Sources of information Annex E ITU-T Recommendation P.800 [224], ITU-T test plans, ITU-T telephonometry handbook [84]

Dependent variable(s) Improvement or degradation

Scale 7-point comparison category rating scale (see Figure 4.7(c)). Subject is instructed to rate the second sample against the first.

Independent variable(s) Codec/system, background noise, talker, and speech sample

Subject requirements Untrained or naïve subjects (see Section 5.4.1)

Number of subjects 32 (minimum of 12 subjects from either gender)

Stimuli 8 samples per talker, 2 male, 2 female

Technical requirements Quality reference: The reference (unprocessed) sample (Quality reference or Direct connection) is presented either before or after the processed or degraded signal. The reference sample is generated using the same talker and speech material as used for the processed sample. This reference sample will be corrupted by the same noise (if any) and processed through the same preliminary processes, such as transmitter characteristic, logarithmic compounding, and so on. Thus, there will be a different quality reference for each of the test conditions.

MNRU (modulated noise reference unit) reference conditions should be included [225].

Null pairs, i.e. A-A sample comparisons, for each quality reference should be evaluated.

Test planning, administration and reporting

Analysis approach The first step is to recode the data relative to the reference. Apply ANOVA and report means (CMOS)

Example experiments See Barret *et al.* [23]

Example user interface See Figure 10.3

10.1.2 ITU-R Recommendation BS.1116-1

Test paradigm Double-blind triple stimulus with hidden reference.

Description The subject can at will switch between three stimuli 'A', 'B', and 'C'. The known reference is always available as the 'A' stimulus and the item (an en- and decoded version of the reference programme) and the reference are randomly, from trial to trial, assigned to either the 'B' or 'C' stimuli. The subject is asked to assess the impairments on 'B' compared to 'A', and 'C' compared to 'A', according to the continuous 5-grade impairment scale. One of the stimuli, 'B' or 'C', should be indiscernible from stimulus 'A'; the other one may reveal impairments. Any perceived differences between the reference and the other stimuli must be interpreted as an impairment.

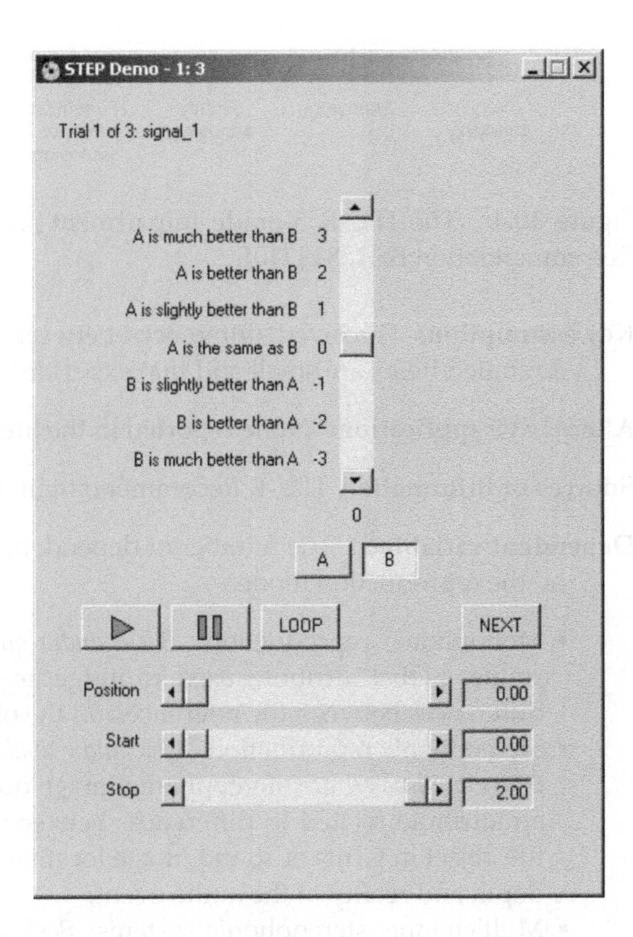

Figure 10.3: Example user interface for an ITU-R Recommendation P.800 [224] comparison category rating (CCR) listening test using STEP software. Reproduced by permission of S. R. Quackenbush, Audio Research Labs. (www.audioresearchlabs.com).

Primary Application Assessment of systems which introduce impairments so small as to be undetectable without rigorous control of the experimental conditions and appropriate statistical analysis. Applicable to monophonic, stereophonic and multichannel sound systems evaluations

Non-applicability To systems that introduce relatively large and easily detectable impairments, as it will lead to excessive expenditure of time and effort and may also lead to less reliable results than a simpler test. (see ITU-R Recommendation BS.1534 [217] and Section 10.1.3

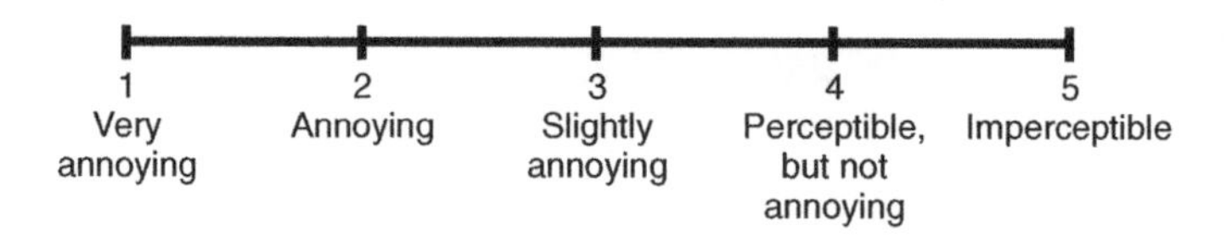

Figure 10.4: The ITU-R 5-grade impairment scale given in ITU-R Recommendation BS.1284 [210].

Key assumptions Perceived differences between the reference and degraded items are small and that expert listeners are applied.

Alternative applications None reported in the literature.

Sources of information ITU-R Recommendation BS.1116-1 [207]

Dependent variable(s) The number of dependent variables depends on the reproduction mode.

- Monophonic reproduction: *Basic audio quality* defined as a single, global attribute used to judge any and all detected differences between the reference and the object.
- Stereophonic reproduction: *Basic audio quality and Stereophonic image quality* where stereophonic image quality is defined as an attribute related to differences between the reference and the object in terms of sound image locations and sensations of depth and reality of the audio event,
- Multichannel stereophonic systems: *Basic audio quality, Front image quality, and Impression of surround quality* where front image quality is defined as an attribute related to the localisation of the frontal sound sources including stereophonic image quality and losses of definition. Impression of surround quality is defined as an attribute related to spatial impression, ambience or special directional surround effects.

Scale The continuous 5-grade impairment scale. The grading scale shall be treated as continuous with 'anchors' derived from the ITU-R 5-grade impairment scale given in Recommendation ITU-R Recommendation BS.1284 [210] and in Figure 10.4. It is recommended that the scale be used to a resolution of one decimal place.

Independent variable(s) Codec, programme and subjects. Note that the term 'item' is often used for a given combination of a programme and codec.

Subject requirements Expert subjects (see Section 5.4.1). Pre-screening of subjects is recommended and suggested methods include audiometric tests, degree of previous experience, performance in previous tests and elimination based on results in pilot tests. Post-screening of subjects is also recommended as it is very important to ensure that the reported impairments are the result of the codec and not an inability of the subjects to detect the differences. To ensure that the subjects can discriminate appropriately a procedure for evaluation of listener expertise is provided.

Number of subjects A minimum of 20 subjects is recommended.

Stimuli Only critical programme material that stresses the codecs under test should be used. Any programme that can be considered a potential broadcast material can be used; however, synthetic signals should be avoided. The artistic or intellectual content of the programme should neither be so attractive nor so disagreeable or wearisome that the subject is distracted from focusing on the detection of impairments. The selection of the programme material should be done by a group of skilled subjects and it is important that the attributes which are to be evaluated are precisely defined for the group before the selection process starts. The group should define the reproduction level, both absolute and relative between the programmes. A reasonable number of programmes is 1.5 times the number of codecs to be tested with a minimum of 5 programmes. The duration of each programme should be 10–25 s.

Technical requirements The standard has a detailed description of reproduction devices (loudspeaker or headphones), acoustic properties of the listening room, background noise, listening levels, loudspeaker and listener positions within the room. The reader is advised to consult the standard for further details.

Test planning, administration and reporting The standard includes specific instructions as to the planning (including randomisation, etc.), administration (instruction to the subjects) and content of the test report. The reader is advised to consult the standard for further details.

Analysis approach The basis for the statistical analysis is the difference between the grade for the object minus the grade given to

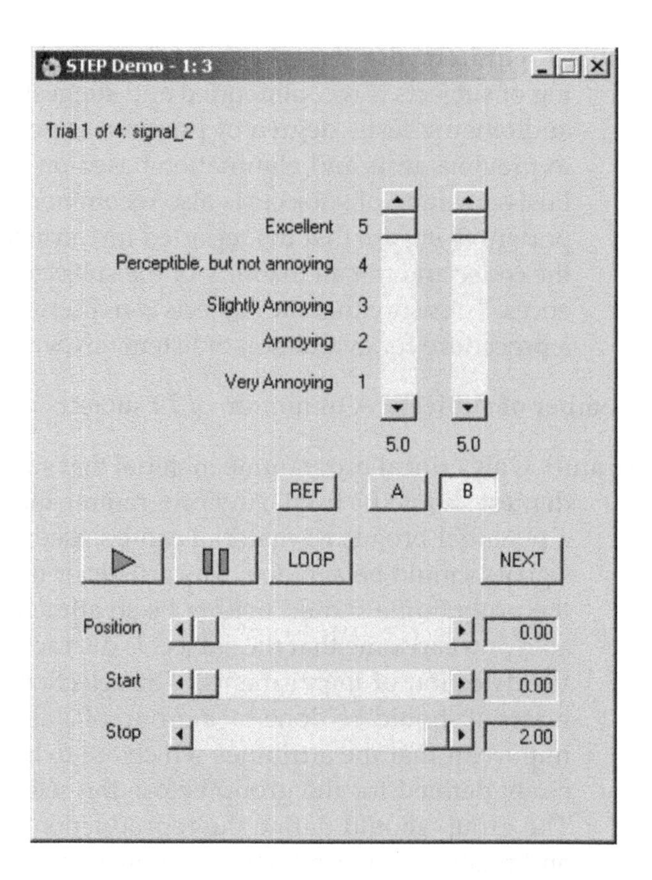

Figure 10.5: Example user interface for an ITU-R Recommendation BS.1116-1 [207] listening test using STEP software. Reproduced by permission of S. R. Quackenbush, Audio Research Labs. (`www.audioresearchlabs.com`).

the hidden reference. The subjects have been instructed to use the 5-point impairment scale as a continuous scale with a resolution of 0.1, which means that data can be consider quantitative and a standard ANOVA can be applied. It is very important to note that unless the grading scale can be shown to be linear comparisons of difference, grades can only be made on the basis of rank order. Both the absolute grade as well as the difference grade should be reported with appropriate confidence intervals.

Example experiments See Kirby [255].

Example user interface See Figure 10.5.

10.1.3 ITU-R Recommendation BS.1534-1

Test paradigm Double-blind multi-stimulus with hidden reference and hidden anchors.

Description The subject is allowed to switch at will between the reference signal and any of the systems under test. In each trial, the subject is presented with the reference version, the versions of the test signal, one hidden reference, and one hidden anchor. A maximum of 15 signals should be included in a trial.

Primary Application Assessment of intermediate audio quality. The standard is intended to provide a reliable and repeatable measure of audio quality of the system, which normally would be in the lower half of the 5-point impairment scale used in ITU-R Recommendation BS.1116-1 (see Section 10.1.2).

Non-applicability Assessment of systems that introduce impairments so small as to be undetectable without rigorous control of the experimental conditions and appropriate statistical analysis.

Key assumptions That the systems under test will introduce significant impairments.

Alternative applications None reported in the literature.

Sources of information ITU-R Recommendation BS.1534-1 [217]

Dependent variable(s) Basic audio quality, stereophonic image quality, front image quality, impression of surround quality. For definitions of attributes, see ITU-R Recommendation BS.1116 (see Section 10.1.2).

Scale 0–100 point continuous quality scale (CQS) divided into five equal intervals (see Figure 4.7(f)).

Independent variable(s) Systems, programmes, subjects.

Subject requirements Subjects should be experienced listeners in the sense that they should have experience in listening to sound in a critical manner. To ensure this, it is recommended to apply both pre- and post-screening of subjects. The pre-screening requirements include having experience in critical listening and normal hearing according to ISO 389 [199]. Post-screening

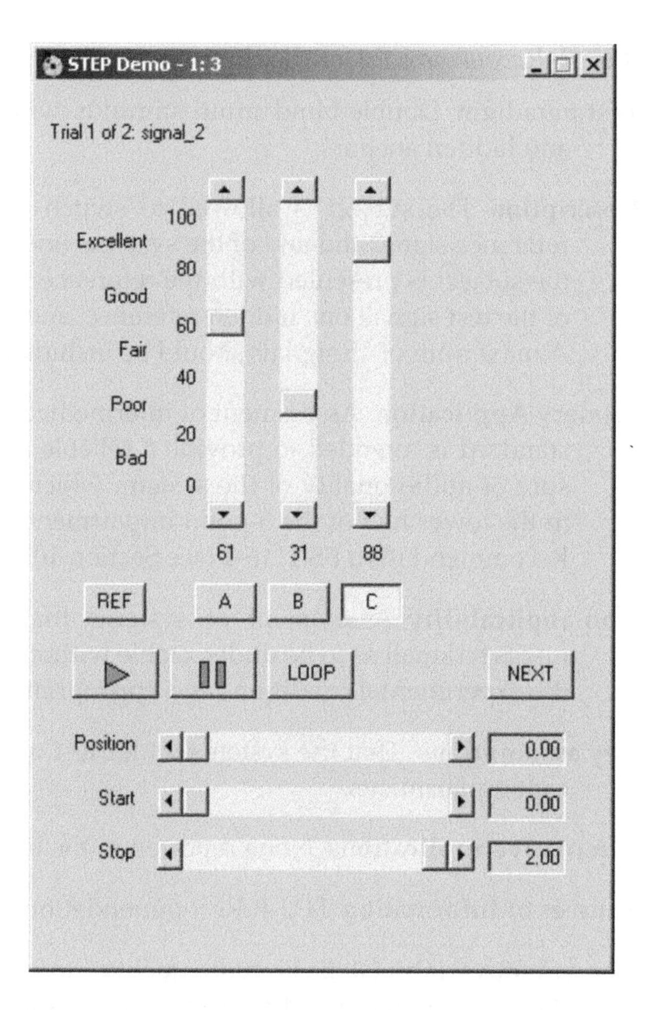

Figure 10.6: Example user interface for an ITU-R Recommendation BS.1534-1 [217] listening test using STEP software. Reproduced by permission of S. R. Quackenbush, Audio Research Labs (`www.audioresearchlabs.com`).

criteria include the ability to make consistent repeated ratings of the degree of deviation from the mean grading of all subjects.

Number of subjects A minimum of 20 subjects is recommended.

Stimuli The requirements are as described for ITU-R Recommendation BS.1116-1 (see Section 10.1.2). In addition to the selected programmes, an anchor should be included. It is recommended that this be a low-pass filtered version of the reference (unprocessed) signal. The bandwidth should be 3.5 kHz. Depending on the systems under test, other anchors can be introduced and a number of suggestions are provided in the standard.

Technical requirements The requirements are as described for ITU-R Recommendation BS.1116-1.

Test planning, administration and reporting The requirements are as described for ITU-R Recommendation BS.1116-1 10.1.2.

Analysis approach The scores are assumed to be quantitative and the standard statistical analysis should be conducted. Results are reported as mean values and associated confidence intervals.

Example experiments See Soulodre and Lavoie [392].

Example user interface See Figure 10.6.

PART IV

Appendices

APPENDIX

Standards and Recommendations

THIS appendix will provide the reader with a non-exhaustive short guide to the scope and content of key standards relating to perceptual evaluation of audio and the associated key topics discussed in earlier chapters. The reader should be aware that standardisation is a continuous process and revisions to standards and recommendations are not uncommon. As a result, the information provided herein is accurate at the time of writing, but the interested reader should check with the standardisation organisation whether revisions have been made to any particular recommendation. The chapter is organised in the alphabetic order of the standardisation bodies.

Perceptual Audio Evaluation – Theory, Method and Application Søren Bech and Nick Zacharov

A.1 Audio Engineering Society

AES20 AES recommended practice for professional audio–Subjective evaluation of loudspeakers [5].

This standard is a set of recommendations for the subjective evaluation of high-performance loudspeaker systems. It is believed that, for certain audio components, including loudspeakers, subjective evaluation is a necessary adjunct to objective measurements. The strong influence of listening conditions, programme material, and individual evaluators is recognised. This document seeks, therefore, to assist in avoiding testing errors rather than to attempt to establish a correct procedure.

A.2 American National Standards Institute

ANSI S3.1-1999 (R2003) Maximum permissible ambient noise levels for audiometric test rooms.

Specifies maximum permissible ambient noise levels (MPANLs) allowed in an audiometric test room that produce negligible masking (<2 dB) of test signals presented at reference equivalent threshold levels specified in ANSI S3.6-1996. The MPANLs are specified from 125–8000 Hz in octave and one-third octave band intervals for two audiometric testing conditions (ears covered and ears not covered) and for three test frequency ranges (125–8000 Hz, 250–8000 Hz, and 500–8000 Hz).

ANSI S3.2-1989 (R1999) Method for measuring the intelligibility of speech over communications system.

Provides a guide for measuring the intelligibility of speech over communications systems. Three alternative English word lists are provided for use in the measurement of monosyllabic word intelligibility. The method is intended for application to room and outdoor acoustics and for all types of communication systems (e.g. public address, radio, telephony, etc.).

A.3 European Broadcasting Union

Tech 3276 Listening conditions for the assessment of sound programme material: monophonic and two-channel stereophonic [112].

This recommendation considers in detail the technical specification of two types of listening spaces:

Reference listening rooms: Listening rooms used for the critical assessment and selection of programme material for inclusion in a sound or television broadcaster's programme output.

High-quality sound control rooms: Sound control rooms used for the critical assessment of sound quality as a part of the sound or television broadcast production process.

Significant details are provided on the room acoustic specification as well as the performance requirements of reference monitor loudspeakers.

Tech 3276: Supplement 1 Listening conditions for the assessment of sound programme material: multichannel sound [113].

The main part of Technical Document 3276, 'Listening conditions for the assessment of sound programme material: monophonic and two-channel stereophonic' does not include any recommendations for sound systems or programme material using more than two channels. The purpose of this supplement is to add requirements that relate specifically to the use of additional channels. In this context, 'multichannel' implies the use of more reproduction channels than the two used by the well-established, two-channel stereophonic system. The scope of this supplement is limited to the multichannel sound system described in ITU-R Recommendation BS.775-1, which uses five, full-bandwidth main loudspeakers usually arranged on the perimeter of a circle centred on the main listening position (5.0), with an optional 'low-frequency extension' channel (5.1). In order to align this document with established cinema practice, this '0.1' channel will be described in this document as the low-frequency effects (LFE) channel, which more clearly describes the established function of this channel. In ITU-R Recommendation BS.775-1, the code '3/2' (three front channels / two surround channels) is used for this loudspeaker arrangement. In addition, the listening levels in this document are essentially aligned with those in the SMPTE Recommended Practice RP200. For ease of cross-referencing, this supplement is arranged with the same layout and section headings as the main document, even though in many cases there are no additional or altered requirements. The recommendations given in this supplement should be taken as additional or alternative to those in the main part of Tech 3276. For clarity, and in order to make this supplement more readable without a great deal of cross-referencing, some parts of the text of the main part of Technical Document 3276 have been repeated or summarised.

Tech 3286 Assessment methods for the subjective evaluation of the quality of sound programme material–Music [111].

This document gives details of the method recommended by the EBU for the evaluation of the quality of sound programme material. The recommended listening conditions are given in EBU Recommendation R22 and EBU document Tech 3276: Listening conditions for the assessment of sound programme material.

The method described in this document was developed for the assessment of the quality of 'classical music' programmes. This includes symphonic music, orchestral music, choral music, opera, chamber music and solo performances. The method and many of the parameters may also be applied to the assessment of other types of music where the source consists essentially of a live acoustical performance taking place in a real space. It does not apply to music sources that are mainly electronic in origin or to speech or drama productions.

Tech 3286: Supplement 1 Assessment methods for the subjective evaluation of the quality of sound programme material–Multichannel [114].

A.4 International Electrotechnical Commission

60268-13 Sound system equipment–Part 13: Listening tests on loudspeakers [184].

The recommendations in this report apply to loudspeakers intended for residential systems and environments. Although, the procedures are specifically designed for loudspeakers that have separate sound system components for monophonic, two-channel stereophonic or multichannel stereophonic reproduction, they may also be applied to other devices such as complete audio and television receivers with integrated sound.

This report gives recommendations for establishing, conducting and evaluating listening tests on loudspeakers. Although various aspects are under continuing discussion, this report is intended to give general guidelines.

Certain aspects may vary–in particular, the listening room may reflect regional construction practices. The tests described in this report are to be performed in a room, the size and acoustical properties of which are similar to those of an 'average' living room. Specific recommendations about the room size, acoustical properties, arrangements of loudspeakers and listeners, and environmental conditions are given.

This technical report describes experimental procedures, including recommendations on the choice of programme material and the processing and presentation of the final data.

A.5 The International Telecommunications Union standards

To assist the reader in selecting the suitable ITU recommendation to follow, a graphical summary of the ITU recommendations is provided in Figures 1.3 and 1.4.

A.5.1 Telecommunications Standardisation Sector

The ITU Telecommunications Standardisation Sector focuses upon applications relating to telecommunications. As a result, the sector has focused upon the following aspects:

- Telephony bandwidth
  - Narrowband: 300–3400 Hz
  - Wideband: 150–7000 Hz
- Speech, listening and conversational, quality

As time passes, the frequency bandwidth of communication audio widens. As a result, it can be expected that there will be an increasing overlap between the application domains covered by the Telecommunications and Radiocommunication Standardisation Sector in future.

Handbook of telephonometry [84]

This handbook provides a thorough guide to recommendations within the ITU 1992. The aims of the handbook are to describe

- the main methods of subjective measurement,
- objective measurement techniques,
- calculation techniques for estimating transmission performance.

For those performing perceptual evaluations in the field of telecommunication, this handbook provides excellent background information for many of the commonly employed methods.

Recommendation P.84 Subjective listening test method for evaluating digital circuit multiplication and packetised voice systems [221].

This recommendation focuses on the preparation of test material for the assessment of speech quality in Digital Circuit Multiplication Equipment (DCME) or packetised voice systems. The text details speech recording, simulation of sample loads on DCME and stimulus processing. Listening test methodologies are in accordance with ITU-T Recommendation P.800 (annex B–ACR).

Recommendation P.85 A method for subjective performance assessment of the quality of speech voice output devices [222].

Various services providing vocal answers related to telephone directory inquiries, weather forecast, mail order, and so on, are now available to PSTN users using vocal servers. As the speech messages are produced by machines, they may suffer some impairment.

In this recommendation, a method is defined for subjective performance assessment of the quality of speech of voice output devices. This method allows the comparison of several systems between them. It will be useful for system designers and service providers for checking the quality of their products.

This method is of the listening test type. Messages are presented aurally to subjects. The subjects express their opinions on one or more rating scales after having answered specific questions on the information contained in the messages. The results are measures of the perceived quality in several aspects, which makes it possible to compare the effectiveness of different speech synthesis systems.

Recommendation P.563 Single-ended method for objective speech quality assessment in narrowband telephony applications [243].

This recommendation describes an objective single-ended method for predicting the subjective quality of 3.1 kHz (narrowband) telephony applications. This recommendation presents a high-level description of the method and advice on how to use it. An ANSI-C reference implementation, described in Annex A, is provided in a separate file and forms an integral part of this recommendation. A conformance testing procedure is also specified in Annex A to allow a user to validate that an alternative implementation of the model is correct.

Recommendation P.800 Methods for subjective determination of transmission quality [224].

This recommendation describes methods and procedures for conducting subjective evaluations of transmission quality. The main revision encompassed by this version is the addition of an annex describing the comparison category rating (CCR) procedure. Other modifications have been made to align this recommendation with the recent revision of Recommendation P.830.

- Annex A: Conversation-opinion test.

  This annex specifies in detail the requirements for performing the more complex conversation-opinion tests, which involve real-time processing of stimuli. The acoustic requirements of the listening and transmission rooms are clearly specified in addition to the background noise requirements. The subject population to be selected from is also strictly defined and a 5-point opinion scale is employed for rating performance.
- Annex B: Absolute category rating (ACR).

  This annex describes in detail a common perceptual evaluation method employed in telecommunications audio. Details are provided regarding the acoustic requirements for both recording of stimuli and for the perceptual evaluation phase. The single stimulus paradigm is described with a 5-point listening-quality scale employed (see Figure 4.7(a)). The listening effort or loudness-preference scale may also be used (see Table 4.3). Examples of instructions for subjects are also provided.

- Annex C: Quantal-response detectability tests.

  This is a methodology to test the detectability of a sound (or echo). The method employs a 3-point categorical scale and audiometrically screened subjects. Unfortunately, the standard does not provide much information on the precise details of this methodology.
- Annex D: Degradation category rating (DCR).

  The paired comparison, fixed reference method is introduced, which is considered more sensitive than the ACR method. The reference condition is defined such that it is superior compared to the test stimuli. A 5-point degradation category rating scale is employed (see Table 4.3 and Figure 4.7(b)).
- Annex E: Comparison category rating (CCR).

  This annex introduces the CCR method, which is similar to the DCR method. The method employs a hidden reference sample and a 7-point CCR scale, as illustrated in Figure 4.7(c).

Recommendation P.800.1 Mean opinion score (MOS) terminology [238].

This recommendation provides a terminology that shall be used in conjunction with speech quality expressions in terms of mean opinion score (MOS). The new terminology is motivated by the intention to avoid misunderstanding as to whether specific values of MOS are related to listening quality or conversational quality, and whether they originate from subjective tests, from objective models or from network planning models.

Recommendation P.830 Subjective performance assessment of telephone-band and wideband digital codecs [226].

This revised recommendation describes methods and procedures for conducting subjective performance evaluations of digital speech codecs. Revisions encompassed by this version of the recommendation are to include new information that reflects current practices in subjective evaluation of digital codecs, including an expanded section on creating source recordings and addition of two annexes. One annex describes an implementation of a PCM codec (A-/m-law) that generates one quantisation distortion unit (qdu) of distortion to input signals. The other new annex describes the modified IRS transmit and receive characteristics. These characteristics are recommended as the transmit and receive responses to be used in situations where the codec being tested is intended for use in fully digital circuits.

Recommendation P.831 Subjective performance evaluation of network echo cancellers [226].

This recommendation describes methods and procedures for conducting subjective performance evaluations of network echo cancellers. The deployment of digital technology in the public switched telephone network (PSTN) has had numerous advantages for users of the network as well as for network operators. These new technologies come at the price of increased transmission time, which increases the likelihood that any echo impairment will be annoying to voice users of the network. Hence, the deployment of echo cancellers in the network is widespread. Recommendations G.165 and G.168 define certain instrumental tests that must be met to ensure a minimum performance of an echo canceller. However, there has been some concern that those tests do not address fully the echo cancellation needs of voice users of the network. Subjective testing is a commonly used method of assessing the performance of digital devices, including digital speech codecs and digital circuit multiplication equipment (DCME). This recommendation defines natural extensions of those techniques to the subjective evaluation of echo cancellers.

Recommendation P.832 Subjective performance evaluation of hands-free terminals [235].

This ITU-T Recommendation describes methods and procedures for conducting subjective performance evaluations of hands-free terminals. The use of hands-free terminals in communication has numerous advantages for the telephone users, especially for all 'non-traditional' types of terminals such as car phones, computer/laptop-type terminals and others. Owing to the complex acoustical situation, a big variety of signal processing, which may be non-linear and/or time variant, is expected. ITU-T Recommendation P.340 describes measurement techniques for hands-free terminals, ITU-T Recommendation P.581 describes the use of the HATS for the evaluation of terminals and ITU-T Recommendations P.501 and P.502 describe measurement signals and analysis procedures. Using these methods, a minimum performance of hands-free terminals should be ensured. However, there is always the possibility that those tests do not fully address the impact of all kinds of signal processing in a hands-free terminal and their impact on speech transmission quality. Subjective testing is a commonly used method of assessing the performance of terminals, including digital speech codecs, voice-operated signal processing, echo cancellation, noise reduction and other types of signal processing. This ITU-T Recommendation defines methods for the subjective evaluation all kinds of hands-free terminals.

Recommendation P.835 Subjective test methodology for evaluating speech communication systems that include noise suppression algorithm [239].

This recommendation describes a methodology for evaluating the subjective quality of speech in noise and is particularly appropriate for the evaluation of noise suppression algorithms. The methodology uses separate rating scales to independently estimate the subjective quality of the speech signal alone, the background noise alone, and the overall quality.

Recommendation P.840 Subjective listening test method for evaluating circuit multiplication equipment [240].

This recommendation describes a subjective listening test method that can be used to evaluate the speech quality of circuit multiplication equipment (CME). It is intended for use with CME systems, such as those described in ITU-T Recs G.763, G.767, G.768 (DCME), G.765 (PCME) and G.769/Y.1242 (IP-CME), which use digital speech interpolation (DSI) techniques. In this version, the scope is expended to more recent speech coders implemented in CME. Updating the recommendation includes tandemming situations and comfort noise test configuration. The new Appendix I gives guidance for conversation tests.

Recommendation P.851 Subjective quality evaluation of telephone services based on spoken dialogue systems [241].

This recommendation describes methods and procedures for conducting subjective evaluation experiments for telephone services that are based on spoken dialogue systems. The respective systems enable a natural interaction via spoken language and possess speech recognition and interpretation, dialogue management and speech output capabilities. The set-up and running of appropriate interaction experiments is described, and questionnaires for quantifying the relevant quality dimensions perceived by the user are given.

Recommendation P.862 Perceptual evaluation of speech quality (PESQ), an objective method for end-to-end speech quality assessment of narrowband telephone networks and speech codecs [236].

This recommendation describes an objective method for predicting the subjective quality of 3.1 kHz (narrowband) handset telephony and narrowband speech codecs. This recommendation presents a high-level description of the method, advice on how to use it, and part of the results from a Study Group 12 benchmark carried out in the period 1999–2000. An ANSI-C reference implementation, described in Annex A, is provided in separate files and forms an integral part of this recommendation. A conformance testing procedure is also specified in Annex A to allow a user to validate that an alternative implementation of the model is correct. This ANSI-C reference implementation shall take precedence in case of conflicts between the high-level description as given in this recommendation and the ANSI-C reference implementation. This recommendation includes an electronic attachment containing an ANSI-C reference implementation of PESQ and conformance testing data.

Recommendation P.862.1 Mapping function for transforming P.862 raw result scores to MOS-LQO [242].

ITU-T Recommendation P.862 provides raw scores in the range −0.5 to 4.5. It is desired to provide a MOS-LQO (P.800.1) score from P.862 to allow a linear comparison with MOS. This recommendation presents the mapping function and its performance for a single mapping from raw P.862 scores to the MOS-LQO (P.800.1). This will allow MOS-LQO scores from ITU-T Recommendation P.862 to be comparable independent of the implementation of ITU-T Recommendation P.862. The function for transformation presented in this recommendation has been optimised on a large corpus of subjective data representing different applications and languages.

Recommendation P.862.2 Wideband extension to Recommendation P.862 for the assessment of wideband telephone networks and speech codecs [245].

This recommendation describes a simple extension to the perceptual evaluation of listening speech quality (PESQ) algorithm defined in ITU-T Recommendation P.862. It allows this algorithm to be applied to the evaluation of conditions, such as speech codecs, where the listener uses wideband headphones. (In contrast, Recommendation P.862 assumes a standard IRS-type narrowband telephone handset, which attenuates strongly below 300 Hz and above 3100 Hz.) This recommendation is mainly intended for use with wideband audio systems (50–7000 Hz), although it may also be applied to systems with a narrower bandwidth.

Recommendation P.862.3 Application guide for objective quality measurement based on Recommendations P.862, P.862.1 and P.862.2 [246].

This recommendation provides some important remarks that should be taken into account in the objective quality evaluation of speech conforming to Recommendations P.862, P.862.1 and P.862.2. Users of Recommendation P.862 should understand and follow the guidance given in this recommendation.

This recommendation forms a supplementary guide for users of Recommendation P.862, which recommends a means of estimating listening speech quality by using reference and degraded speech samples. It cannot be used for the assessment of talking quality or interaction quality. It assumes that an objective quality estimation algorithm strictly conforms to Recommendation P.862. This can be confirmed by the conformance test provided as an annex to Recommendation P.862.

The scope of Recommendation P.862 is clearly defined in the recommendation itself. This recommendation does not extend or narrow the scope, but provides necessary and important information for obtaining stable, reliable and meaningful objective measurement results in practice.

Applications and limitations associated with the wideband extension of Recommendation P.862 are as defined in Recommendation P.862.2.

Recommendation P.880 Continuous evaluation of time-varying speech quality [244].

This recommendation describes a methodology called *continuous evaluation of time-varying speech quality (CETVSQ)* that can be used for evaluating the impact of time fluctuations of speech quality on the instantaneous perceived quality (that is perceived at any instant of a speech sequence) and on the overall perceived quality (at the end of the speech sequence). The method uses a two-part task: first, an instantaneous judgment on a continuous scale with a slider during the speech sequence, and second, an overall judgment on a standard five-category scale at the end of the speech sequence.

Recommendation P.910 Subjective video quality assessment methods for multimedia applications [227].

This recommendation describes non-interactive subjective assessment methods for evaluating the one-way overall video quality for multimedia applications such as videoconferencing, storage and retrieval applications, tele-medical applications, and so on. These methods can be used for several purposes including, but not limited to, selection of algorithms, ranking of audio-visual system performance and evaluation of the quality level during an audio-visual connection. This recommendation also outlines the characteristics of the source sequences to be used, like duration, kind of content, number of sequences, and so on.

Recommendation P.911 Subjective audio-visual quality assessment methods for multimedia applications [232].

This recommendation describes non-interactive subjective assessment methods for evaluating the one-way overall audio-visual quality for multimedia applications such as videoconferencing, storage and retrieval applications, tele-medical applications, and so on. These methods can be used for several purposes including, but not limited to, selection of algorithms, ranking of audio-visual system performance and evaluation of the quality level during an audio-visual connection. When interactive aspects are to be assessed, conversation test methods described in Recommendation P.920 should be used. This recommendation also outlines the characteristics of the source sequences to be used, like duration, kind of content, number of sequences, and so on. Finally, it provides indications about the relation between audio, video and audio-visual quality, as they are derived from results of tests carried out independently in different laboratories.

Recommendation P.920 Interactive test methods for audio-visual communications [228].

This ITU-T Recommendation is intended to define interactive evaluation methods for quantifying the impact of terminal and communication link performance on point-to-point or multi-point audio-visual communications. This methodology is based upon conversation-opinion tests, and can be considered to be an extension of the methods defined in Annex A/P.800.

A.5.2 Radiocommunication Sector

The Radiocommunication Sector of the International Telecommunications Union provide a helpful means to navigate around their recommendations relating to listening tests and perceptual evaluation of audio in ITU-R Recommendation BS.1283 [209], in the form of a flow diagram. To assist the reader, this diagram has been reproduced in Figure A.1.

The Radiocommunication Sector focuses upon applications relating to audio for radiocommunication. As a result the sector considers the following aspects:

- Full-band audio
  - 20 Hz to 20 kHz
- Basic audio quality

Recommendation BS.775-1 Multichannel stereophonic sound system with and without accompanying picture [205].

This recommendation outlines the basic physical loudspeaker configurations to be employed in domestic 5.1 multichannel sound reproduction in the form of 3/2 (3 frontal, 2 surround channels) and 3/4 (3 frontal, 4 surround channels) systems.

Additionally, information regarding screen sizes and locations is provided.

Recommendation BS.1116-1 Methods for the subjective assessment of small impairments in audio systems including multichannel sound systems [207].

This recommendation is intended for use in the assessment of systems that introduce impairments so small as to be undetectable without rigorous control of the experimental conditions and appropriate statistical analysis. If used for systems that introduce relatively large and easily detectable impairments, it leads to excessive expenditure of time and effort and may also lead to less reliable results than a simpler test. This recommendation forms the base reference for the other recommendations, which may contain additional special conditions or relaxations of the requirements included in BS.1116-1.

Recommendation BS.1283 A guide to ITU-R Recommendations for subjective assessment of sound quality [209].

This recommendation provides an overview of the assembly's five main recommendations on this topic. A flow diagram is provided to guide experimenters through these standards as found in Figure A.1. Reference to other related standards and recommendations is also provided, nearly all of which are presented or referenced in this Appendix.

Recommendation BS.1284 General methods for the subjective assessment of sound quality [210].

This recommendation provides a short guide to general requirements for performing listening tests. A broad overview of experimental design, selection of subjects, test methods, through to statistical analysis and reporting is provided, often referring to details within ITU-R Recommendation BS.1116-1.

Recommendation BS.1285 Pre-selection methods for the subjective assessment of small impairments in audio systems [211].

This short recommendation document provides a means of pre-screening test systems providing large impairments. This is of value as ITU-R Recommendation BS.1116-1 is intended only for evaluation of small impairments, which is a complex and involved method. The recommendation advises on the use

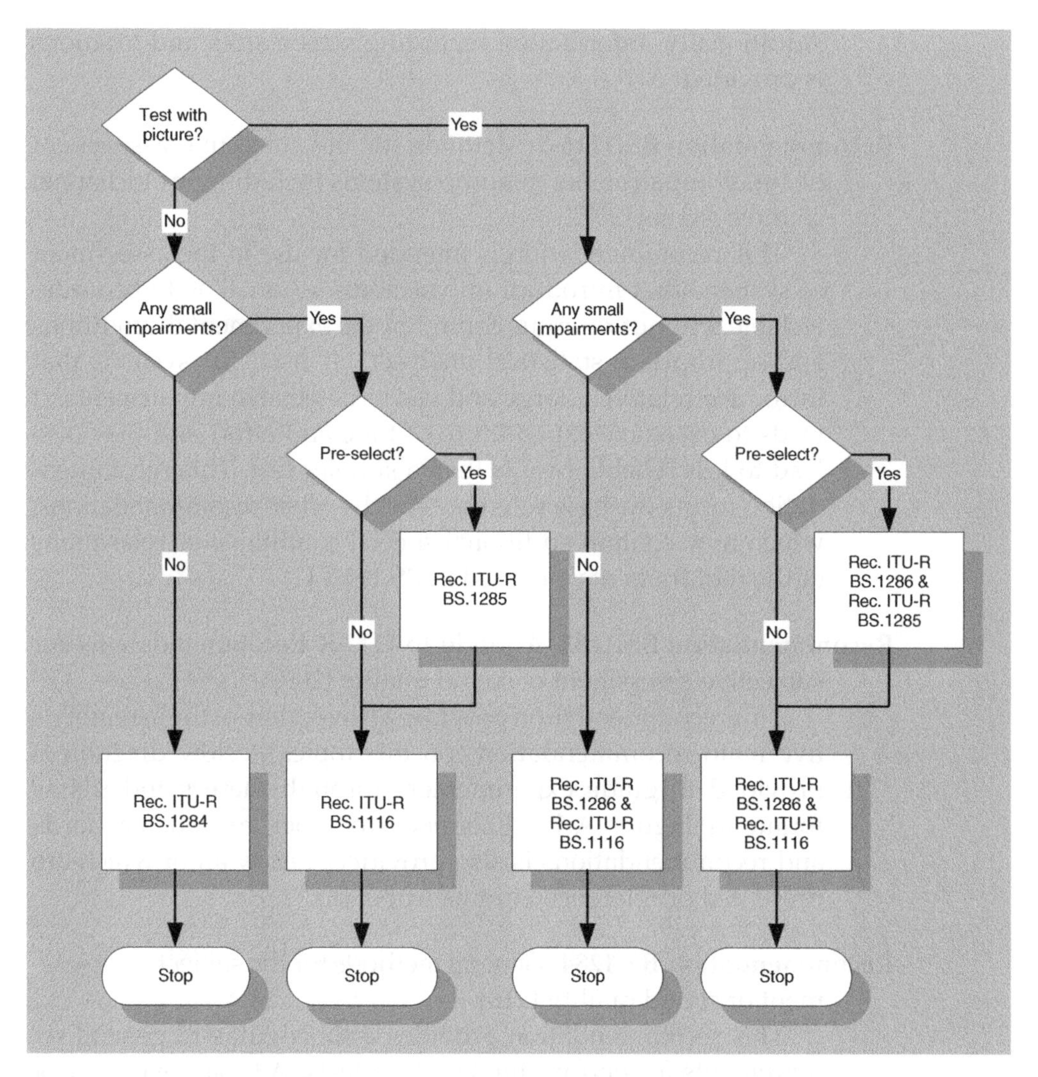

Figure A.1: A flow diagram providing an overview of ITU-R standards and how best to select the appropriate recommendation. Extract from [209]. Reproduced with the kind permission of ITU.

of headphones, if possible allowing for parallel testing of subjects without audible interference employing a fixed order test paradigm. In this case, different orders of presentation are suggested between groups of subjects. In most other respects, the recommendation is similar to ITU-R Recommendation BS.1116-1.

Recommendation BS.1286 Methods for the subjective assessment of audio systems with accompanying picture [212].

This recommendation provides valuable input on how to handle testing of audio systems in the presence of an accompanying image. This recommendation should be used in conjunction with one of the audio-only recommendations, that is, ITU-R Recommendations BS.1284, BS.1116-1 or BS.1285. A full list of viewing distances and conditions is provided in Appendix 1 of the recommendation for different image sizes, aspect ratios and image definitions.

Recommendation BS.1387 Method for objective measurements of perceived audio quality [213].

This recommendation specifies a method for objective measurement of the perceived audio quality of a device under test, for example, a low bit-rate codec. It is divided into two annexes. Annex 1 gives the user a general overview of the method and includes four appendices. Appendix 1 describes applications and test signals.

Appendix 2 lists the model output variables and discusses limitations of use and accuracy. Appendix 3 gives the outline of the model, while Appendix 4 describes the principles and characteristics of objective perceptual audio quality measurement methods in general. Annex 2 provides the implementer with a detailed description of the method using two versions of the psycho-acoustic model that were developed during the integration phase where six models were combined. In Appendix 1 of Annex 2, the validation process of the objective measurement method is described. Appendix 2 of Annex 2 gives an overview of all the databases that were used in the development and validation of the method.

Recommendation BS.1534 Method for the subjective assessment of intermediate quality levels of coding systems [217].

This recommendation describes a new method for the subjective assessment of intermediate audio quality. This method mirrors many aspects of recommendation ITU-R Recommendation BS.1116 and uses the same grading scale as used for the evaluation of picture quality (i.e. Recommendation ITU-R Recommendation BT.500).

The method, called 'MUlti-Stimulus test with Hidden Reference and Anchor (MUSHRA)' is intended to give a reliable and repeatable measure of systems having audio quality that would normally fall in the lower half of the impairment scale used by Recommendation ITU-R Recommendation BS.1116. In the MUSHRA test method, a high-quality reference signal is used and the systems under test are expected to introduce significant impairments. If the systems under test can improve the subjective quality of a signal, then other test methods should be used.

Recommendation BS.1679 Subjective assessment of the quality of audio in large-screen digital imagery applications intended for presentation in a theatrical environment [218].

This annex provides a summary of the provisions that should be implemented when performing subjective assessment tests of audio quality or audio impairment for LSDI applications designed for programme presentation in a theatrical environment. The recommendations provided are derived and refined on the basis of ITU-R Recommendation BS.775-1, ITU-R Recommendation BS.1116, ITU-R Recommendation BS.1284 and ITU-R Recommendation BS.1286.

Recommendation BT.500 Methodology for the subjective assessment of the quality of television pictures [216].

This recommendation provides extensive details on methodologies for evaluation of television picture quality. To the audio researcher, this document provides important information for audio-visual testing in terms of viewing distances, screen sizes for different resolutions and aspect ratio displays, illuminations levels, and so on.

APPENDIX

B

Attribute lists

THIS appendix provides a number of attribute lists and associated descriptions. Several of the lists have been developed through descriptive analysis techniques. The descriptions provided are those developed by each of the listening panels for each experiment. As a result, the terminology is sometimes very specific to the application and may not be simply understood. Therefore, the re-application of such attributes should be handled with great care as the meaning of the words is not self-evident and subjects will require significant training to learn the meaning and application of specific attributes. Anchor words provided are to be applied to the end points line scales. These line scales are typically 15 cm in length with arbitrary numerical values associated. Certain attribute lists also have additional anchor points, for example, mid-point anchor words, to provide further guidance to listeners.

Perceptual Audio Evaluation – Theory, Method and Application Søren Bech and Nick Zacharov

B.1 Speech quality

The following attributes are associated with speech quality in narrowband digital communication systems, according to Mattila [285–288]. The attributes are divided into speech and background noise attributes. Attribute descriptions are provided below and original Finnish and translated attributes and scale anchors are shown in Tables B.1 and B.2.

Tense/sharp The speech is tense or sharp as opposed to relaxed speech.

Dark–bright The speech is dark if its low frequency components dominate, and becomes gradually brighter by the inclusion of its high frequency components and the attenuation of its low frequency components.

Mechanic The speech has a mechanic-like characteristic and sounds unnatural such as produced by a synthesiser. In contrast to mechanic, speech sounds natural.

Metallic The speech has a metallic-like characteristic that makes it sound cold. As having a harsh resonance in the voice. As with the *mechanic* attribute, the *metallic* attribute relates to the general naturalness of speech.

Nasal/whining speech is nasal or whining. This speech is produced through the nose.

Clear–muffled The speech is clear and easily intelligible as opposed to speech with muffled or smothered phonemes, reducing its intelligibility.

Smooth–interrupted The speech is smooth as opposed to interrupted speech caused by occasional attenuations applied to it.

Rough The speech is rough as opposed to smooth speech.

Rustling The speech is continuously rustling.

Scratching The speech is occasionally scratching.

Open–distant The speech is distant or thin, and sounds like it has been occluded. In contrast, when speech is open, there is an impression of a speaker near the listener.

No.	Negative end-word	Speech attribute	Positive end-word
1	Erittäin vähän kireä *Very little sharp*	Kireä / pingottunut *Tense/sharp*	Kireä *Sharp*
2	Erittäin tumma *Very dark*	Tumma–kirkas *Dark–bright*	Kirkas *Bright*
3	Erittäin vähän konemainen *Very little mechanic*	Konemainen / mekaaninen *Mechanic*	Konemainen *Mechanic*
4	Erittäin vähän metallinen *Very little metallic*	Metallinen *Metallic*	Metallinen *Metallic*
5	Erittäin vähän nasaalinen *Very little nasal*	Nasaalinen / inisevä *Nasal/whining*	Nasaalinen *Nasal*
6	Selkeä *Clear*	Selkeä–puuroutuvä *Clear–muffled*	Puuroutuva *Muffled*
7	Ehjä *Smooth*	Ehjä–rikkonainen / katkonainen *Smooth–interrupted*	Erittäin rikkonainen *Very interrupted*
8	Erittäin vähän käheä *Very little coarse*	Käheä–karhea–rosoinen *Rough*	Rosoinen *Strongly rough*
9	Erittäin vähän särisevä *Very little rustling*	Särisevä *Rustling*	Särisevä *Rustling*
10	Erittäin harvoin särähtelevä *Very rarely scratching*	Särähtelevä / rasahteleva (frekvenssi) *Scratching (frequency)*	Särähtelevä *Scratching*
11	Erittäin harvoin särähtelevä *Very rarely scratching*	Särähtelevä / rasahteleva (voimakkuus) *Scratching (intensity)*	Särähtelevä *Scratching*
12	Avoin *Open*	Avoin–etäinen / tukahtunut *Open–distant*	Tukahtunut *Distant*

Table B.1: Speech quality attribute scales, part 1, as developed in Finnish and translated into English (*italic text*). Reproduced by permission of V-V. Mattila

No.	Negative end-word	Background noise attribute	Positive end-word
13	Erittäin vähän humiseva *Very little humming*	Humiseva *Humming*	Humiseva *Humming*
14	Vähän kitisevä *Very little creaking*	Kitisevä *Creaking*	Kitisevä *Creaking*
15	Erittäin vähän kohiseva *Very little noisy*	Kohiseva *Noisy*	Voimakkaasti kohiseva *Very noisy*
16	Matala *Low*	Matala–korkea (taajuus) *Low–high (frequency content)*	Korkea *High*
17	Pulpahtava *Very rarely bubbling*	Kupliva (frekvenssi) *Bubbling (frequency)*	Kupliva *Bubbling*
18	Erittäin vähän suhiseva *Very little hissing*	Suhiseva *Hissing*	Suhiseva *Hissing*
19	Erittäin vähän poriseva *Very little boiling*	Poriseva *Boiling*	Poriseva *Boiling*
20	Erittäin vähän ritisevä *Very little crackling*	Ritisevä *Crackling*	Ritisevä *Crackling*
21	Tasainen *Steady*	Tasainen–vaihteleva–ryöpsähtelevä *Steady–fluctuating*	Ryöpsähtelevä *Strongly fluctuating*

Table B.2: Speech quality attribute scales, part 2, as developed in Finnish and translated into English (*italic text*). Reproduced by permission of V-V. Mattila

Humming The presence of low frequency noise.

Creaking The background noise is continuously creaking.

Noisy The presence of noise.

Low–high The background noise is low, if its low frequency components dominate, and becomes higher with the inclusion of its high frequency components and the attenuation of its low frequency components.

Bubbling A bubbling sound alternating with the speech signal, but perceived as being in the background.

Hissing A Gaussian type of continuous noise.

Boiling The background noise has the sound of continuous boiling.

Crackling The background noise is continuously crackling.

Steady–fluctuation The background noise is fluctuating with the speech signal as opposed to a steady continuation of noise.

B.2 *Spatial sound quality*

The attributes in the following tables are focused upon the qualification of spatial sound systems of different kinds.

B.2.1 *Loudspeakers*

The Attributes described here were developed for spatial sound reproduction over loudspeakers, according to Koivuniemi and Zacharov [259]. The attributes are divided into spatial and timbral attributes. Attribute descriptions are provided below and original Finnish and translated attributes and scale anchors are shown in Tables B.3 and B.4.

Sense of direction Describes how easily the locations of events can be discriminated. This also measures whether several sound sources can be distinguished. A negative value of this attribute implies that the location of a sound event is ill-defined or enveloping.

No.	Negative end-word	Spatial attribute	Positive end-word
1	Huonosti välittyvä *Ill-defined*	Suunnan tuntu *Sense of direction*	Hyvin välittyvä *Well defined*
2	Huonosti välittyvä *Ill-defined*	Syvyyden tuntu *Sense of depth*	Hyvin välittyvä *Well defined*
3	Huonosti välittyvä *Ill-defined*	Tilantuntu *Sense of space*	Hyvin välittyvä *Well defined*
4	Huonosti välittyvä *Ill-defined*	Liikkuvuuden tuntu *Sense of movement*	Hyvin välittyvä *Well defined*
5	Olematon *Non-penetrating*	Pistävyys *Penetration*	Runsas *Penetrating*
6	Lähellä *Close*	Tapahtumien etäisyys *Distance to events*	Kaukana *Distant*
7	Suppea *Narrow*	Laajuus *Broadness*	Laaja *Broad*
8	Epäluonnollinen *Unnatural*	Luonnollisuus *Naturalness*	Luonnollinen *Natural*

Table B.3: Loudspeaker spatial attribute scales as developed in Finnish and translated into English (*italic text*), according to Koivuniemi and Zacharov [259].

No.	Negative end-word	Timbral attribute	Positive end-word
9	Vähäinen *Thin*	Täyteläisyys *Richness*	Täyteläinen *Rich*
10	Pehmeä *Soft*	Kovuus *Hardness*	Kova *Hard*
11	Neutraali *Neutral*	Korostuneisuus *Emphasis*	Korostunut *Emphasised*
12	Tumma *Dark*	Tummuus *Tone colour*	Kirkas *Bright*

Table B.4: Loudspeaker timbral attribute scales as developed in Finnish and translated into English (*italic text*), according to Koivuniemi and Zacharov [259].

Sense of depth Describes how strongly the sensation of distance is perceived, or how ambiguous the sensation of distance is. Once again, this assesses whether several sound events can be discriminated in terms of distance. A negative value could mean that distances for all events are ambiguous except those originating from the transducer's position.

Sense of space This attribute scales how well the space where the recording was made is perceived. A positive value could mean a strong sensation of being in a certain kind of environment, for example, in a room.

Sense of movement This describes whether a sound source is perceived to actually move in the sound space. A negative value could indicate a sound source simply disappearing from its original location and reappearing in another without moving through any intermediate position.

Penetration This describes the sensation often found in cross-talk cancelled binaural reproduction. A positive value means that spatial information in the sample seems artificial. The sounds sometimes seem to originate very close to, or even inside, ones head.

Distance to events This attribute simply describes the actual distance from where the sound events appear to originate. A positive value implies that the sound sources are sensed to be far from the listening point.

Broadness This attribute describes how wide an area the perceived sound event seems to have. A strong positive value would mean that sounds are coming from all around the listener that is, envelope the listener.

Naturalness Naturalness describes how well the perceived events conform to what the subjects consider as realism. Perception of something that is not possible in reality yields a negative value, for example, a train rising straight up.

Richness This describes the homogeneity of the timbre of a sample. If a sound lacks some of its timbral aspects, it would be graded with a negative value.

Hardness This describes how aggressive the sound is perceived to be. A soft sound is mellow and does not offer any surprises.

Emphasis An emphasised sound is somehow, partly or in whole, too loud. Some part of the sample might be playing too loud and this would result in a positive value.

Tone colour Tone colour describes the spectral content of a sample. A dark sample lacking treble and having bass boost is graded as negative. A bright sample with more treble and/or less bass is graded as positive.

B.2.2 Headphones

This section will present attributes developed by Lorho [274] associated with headphone reproduction for stereo and stereo enhanced sound reproduction. The attributes are divided into location, spatially and timbrally related attributes. Attribute descriptions are provided below and in original Finnish and translated attributes and scale anchors are shown in Tables B.5, B.6 and B.7.

Localisation related attributes

Sense of distance This attribute describes how well the distance between the sound source(s) and the listener can be defined.

Sense of direction This attribute describes how well the direction of the sound source(s) can be defined.

Sense of movement This attribute describes how well the movement of the sound source(s) can be defined.

Ratio of localisability Localisability describes how well the direction and the distance of a sound source(s) can be defined. The attribute ratio of localisability describes how many sound events can be localised from those present in the audio sample.

Spatially related attributes

Quality of echo This attribute describes how well the echoes relate to their sound source(s) in a qualitative way.

Amount of echo This attribute describes how the listener experiences the amount of echo in relation to the sound sources.

Sense of space This attribute describes how well the space represented in the audio sample can be defined.

Balance of space This attribute relates to the space represented by the audio sample in relation to the listener's inner reference.

No.	Negative end-word	LOCALISATION RELATED ATTRIBUTES Mid-point anchor word	Positive end-word
1	 Ei määriteltävissä *Not definable*	Etäisyyden tunne/*Sense of distance* Jokseenkin määriteltävissä *Somewhat definable*	 Hyvin määriteltävissä *Well definable*
2	 Ei määriteltävissä *Not definable*	Suunnattavuus/*Sense of direction* Jokseenkin määriteltävissä *Somewhat definable*	 Hyvin määriteltävissä *Well definable*
3	 Ei määriteltävissä *Not definable*	Liikkeen tunne/*Sense of movement* Jokseenkin määriteltävissä *Somewhat definable*	 Hyvin määriteltävissä *Well definable*
4	 Ei yhtään *None*	Paikallistettavien osuus/*Ratio of localisability* Jonkin verran *Some*	 Kaikki *All*

Table B.5: Headphone localisation related attribute scales for headphones as developed in Finnish and translated into English (*italic text*), according to Lorho [274]. Reproduced by permission of Gaetan Lorho.

		SPATIALLY RELATED ATTRIBUTES	
No.	**Negative end-word**	**Mid-point anchor word**	**Positive end-word**
5		Kaiun laatu/*Quality of echo*	
	Epämiellyttävä *Unpleasant*	Ei kaikua *No echo*	Miellyttävä *Pleasant*
6		Kaiun määrä/*Amount of echo*	
	Kaiuton *No echo*	Sopivasti kaikua *Adequate echo*	Liikaa kaikua *Excessive echo*
7		Tilan määriteltävyys/*Sense of space*	
	Ei määriteltävissä *Not definable*	Jokseenkin määriteltävissä *Somewhat definable*	Hyvin määriteltävissä *Well definable*
8		Tilan tasapino/*Balance of space*	
	Ei tasapainossa *Out of balance*	Jokseenkin tasapainossa *Somewhat balanced*	Hyvin tasapainossa *In balance*
9		Laajuus/*Broadness*	
	Pään sisäinen *Inside head*	Läheinen *Close by*	Avara *Broad*

Table B.6: Headphone spatially related attribute scales for headphones as developed in Finnish and translated into English (*italic text*), according to Lorho [274]. Reproduced by permission of Gaetan Lorho.

No.	Negative end-word	TIMBRALLY RELATED ATTRIBUTES Mid-point anchor word	Positive end-word
10	Ei yhtään *None*	Eroteltavuus/*Separability* Jonkin verran *Some*	Kaikki *All*
11	Matalat korostuneet *Lower sounds emphasised*	Äänen sävy/*Tone Colour* Ei osaa sanoa *Cannot say*	Korkeat korostuneet *Higher sounds emphasised*
12	Lattea *Flat*	Värikkyys/*Richness* Neutraali *Neutral*	Rikas *Rich*
13	Vääristynyt *Distorted*	Vääristyneisyys/*Distortion* Jokseenkin vääristynyt *Somewhat distorted*	Vääristymätön *Not distorted*
14	Häiriöinen *Disrupted*	Häirioisyys/*Disruption* Jokseenkin häiriöinen *Somewhat disrupted*	Häiriötön *Not disrupted*
15	Tukahtunut *Muffled*	Selkeys/*Clarity* – –	Kirkas *Clear*
16	Ei tasapainossa *Out of balance*	Äänien voimakkuuksien tasapaino/*Balance of Sounds* Jokseenkin tasapainossa *Somewhat balanced*	Hyvin tasapainossa *Well balanced*

Table B.7: Headphone timbre related attribute scales for headphones as developed in Finnish and translated into English (*italic text*), according to Lorho [274]. Reproduced by permission of Gaetan Lorho.

A negative value means that the space is weighted in some direction. If no space is perceived, the space is out of balance.

Broadness This attribute describes the perceived extent of the soundscape relative to the listener's head.

Timbre related attributes

Separability This attribute describes how well the sound events can be separated out in the audio sample.

Tone Colour This attribute describes the spectral content of the audio sample.

Richness This attribute describes how rich and nuanced the audio sample is overall, and relates to a combination of harmonics and dynamics perceived in the sample.

Distortion This attribute describes the possible metallic, machine-like, electrical-like artifacts in the audio sample.

Disruption This attribute describes how much hiss, snap/crackle/pop is perceived in the audio sample.

Clarity This attribute describes if the sound sample appears clear or muffled, for example, if the sound source is perceived as covered by something.

Balance of Sounds This attribute describes the possible difference in loudness between the sound sources present in the audio sample. The sound sample is well balanced if it contains only one sound source.

B.3 Other quality attributes

Attributes as outlines in ITU-R Recommendation BS.1116-1 [207]:

Basic audio quality This single, global attribute is used to judge any and all detected differences between the reference and the object. Applicable to monophonic, stereophonic and multichannel systems.

Stereophonic image quality This attribute is related to differences between the reference and the object in terms of sound image locations and sensations of depth and reality of the audio event. Applicable to stereophonic systems.

Front image quality This attribute is related to the localisation of the frontal sound sources. It includes stereophonic image quality and losses of definition. Applicable to multichannel systems.

Impression of surround quality This attribute is related to spatial impression, ambience, or special directional surround effects. Applicable to multichannel systems.

APPENDIX

Audio source and demonstration material

The following list provides some examples of audio demonstrations, training material for samples for preparing listening tests. The body of this list was compiled by Ronald Aarts and Armin Kohlrausch and provided via the Audio Engineering Society, Technical Council, Technical committee on Perception and Subjective Evaluation of Audio Signals. Amendments have been provided by the authors.

- *Gehörschutz–Das Gehör schützen.* Fachausschuss Persönliche Schutzausrüstung, Arbeitskreis Gehörschutz, Nürnberg Fa. Hamann Consult, Dresden, 1992 [142].
- *Lärm und Gehörschutz. Ein Lehr- und Lernprogramm auf Audio-CD.* 1995 [267].

Perceptual Audio Evaluation – Theory, Method and Application Søren Bech and Nick Zacharov

- Speech databases recorded in vehicles. http://www.speechdat.com/SP-CAR/, 1999 [308]. More information available from Moreno *et al.* [309].

- Acoustical Society of America. Auditory demonstrations. Compact Disc, 1989 [171].

- Audio Engineering Society. Perceptual audio coders: What to listen for. Compact Disc, 2001 [18].

- Bang and Olufsen. Music for Archimedes. Compact disc CD B&O 101, 1992 [21].

- Bolia, R. S., Nelson, W. T., Ericson, M. A., and Simpson, B. D. A speech corpus for multitalker communications research. *Journal of the Acoustical Society of America 107*, 2 (February 2000), 1065–1066 [59].

- Bregman, A. S., and Ahad, P. Demonstrations of auditory scene analysis. Compact disc available from http://www.psych.mcgill.ca/labs/auditory/home.html, 1995 [66].

- Brungart, D. S. Evaluation of speech intelligibility with the coordinate response measure. *Journal of the Acoustical Society of America 109*, 5 (May 2001), 2276–2279 [75].

- Butler, D. *Musician's Guide to Perception and Cognition.* Schirmer Books, 1992 [78].

- Delta. Hør engang! Compact Disc, 1998 [104].

- Deutsch, D. Musical illusions and paradoxes. Compact disc available from http://www.philomel.com, 1995 [105].

- Deutsch, D. Phantom words and other curiosities. Compact disc available from http://www.philomel.com, 2003 [106].

- European Broadcasting Union. *Sound Quality Assessment Material, (recordings for subjective tests).* No. 422 204-2. 1988 [110].

- ITU-T. *Supplement 23, ITU-T coded speech database.* International Telecommunications Union, Telecommunications Standardisation Sector, 1998 [233].

- ITU-T. *Recommendation P.501, Test signals for use in telephonometry.* International Telecommunications Union, Telecommunications Standardisation Sector, 2000 [234].

- KRAMER, G., Ed. *Auditory Display: Sonification, Audification, and Auditory Interfaces.* Addison Wesley Publishing Company, 1993 [260].

- LUCE, R. D. *Sound and Hearing Demonstration.* Lawrence Erlbaum Assoc. Inc., 1992 [276].

- MOORE, B. C. J. *Compact disc with book: Perceptual consequences of cochlear damage.* Oxford University Press, 1995 [301].

- MOULTON, D. Golden ears. `http://www.moultonlabs.com/gold.html`, 2002 [310].

- NAKAJIMA, Y., SASAKI, T., AND TEN HOOPEN, G. Demonstrations of auditory illusions and tricks. Compact disc available from `http://www.kyushu-id.ac.jp/~ynhome/ENG/Demo/illusions2nd.html`, 2000 [315].

- NASA GLENN RESEARCH CENTER ACOUSTICAL TESTING LABORATORY. Auditory demonstrations in acoustics and hearing conservation. Compact disc available from `http://www.grc.nasa.gov/WWW/AcousticalTest/HearingConservation/Resource%s/Auditory_Demonstrations1.html`, 1998 [324]. More information available from Moreno *et al.* [323].

- NASA GLENN RESEARCH CENTER ACOUSTICAL TESTING LABORATORY. Auditory demonstrations II: Challenges to speech communication and music listening. Compact disc available from `http://www.grc.nasa.gov/WWW/AcousticalTest/HearingConservation/Resources/Auditory_Demonstrations2.html`, 2004 [316]. More information available from Cooper *et al.* [97].

- PLOMP, R. Hoe wij horen. over de toon die de muziek maakt. (in dutch: How we hear. On the tone that makes music), 1998 [347].

- TERHARDT, E. *Compact disc with book: Akustische Kommunikation.* Springer, New York, 1998 [412].

- WARREM, R. M. *Compact disc with book: Auditory Perception. A new analysis and synthesis.* Cambridge University Press, 1999 [443].

APPENDIX

D

A-, B-, C- and D-weighting curves

Perceptual Audio Evaluation – Theory, Method and Application Søren Bech and Nick Zacharov

Centre frequency (Hz)	A-weighting curve [179] (dB)	B-weighting curve [179] (dB)	C-weighting curve [179] (dB)	D-weighting curve [178] (dB)
10	−70.4	−38.1	−14.3	−26.6
12.5	−63.4	−33.2	−11.2	−24.6
16	−56.7	−28.5	−8.5	−22.6
20	−50.5	−24.2	−6.2	−20.6
25	−44.7	−20.4	−4.4	−18.7
31.5	−39.4	−17.1	−3.0	−16.7
40	−34.6	−14.2	−2.0	−14.7
50	−30.2	−11.6	−1.3	−12.8
63	−26.2	−9.3	−0.8	−10.9
80	−22.5	−7.4	−0.5	−9.0
100	−19.1	−5.6	−0.3	−7.2
125	−16.1	−4.2	−0.2	−5.5
160	−13.4	−3.0	−0.1	−4.0
200	−10.9	−2.0	0.0	−2.6
250	−8.6	−1.3	0.0	−1.6
315	−6.6	−0.8	0.0	−0.8
400	−4.8	−0.5	0.0	−0.4
500	−3.2	−0.3	0.0	−0.3
630	−1.9	−0.1	0.0	−0.5
800	−0.8	0.0	0.0	−0.6
1000	0.0	0.0	0.0	0.0
1250	0.6	0.0	0.0	2.0
1600	1.0	0.0	−0.1	4.9
2000	1.2	−0.1	−0.2	7.9
2500	1.3	−0.2	−0.3	10.4
3150	1.2	−0.4	−0.5	11.6
4000	1.0	−0.7	−0.8	11.1
5000	0.5	−1.2	−1.3	9.6
6300	−0.1	−1.9	−2.0	7.6
8000	−1.1	−2.9	−3.0	5.5
10000	−2.5	−4.3	−4.4	3.4
12500	−4.3	−6.1	−6.2	1.4
16000	−6.6	−8.4	−8.5	−0.7
20000	−9.3	−11.1	−11.2	−2.7

Table D.1: One-third octave band free-field relative frequency responses for A-, B-, C- and D- weighting curves according to [178, 179]. Note that for each relative frequency response point there is an associated measurement tolerance, which can be found from the associated recommendation. Reproduced by permission of International Electrotechnical Commission (IEC).[60]

[60]The authors thank the International Electrotechnical Commission (IEC) for permission to reproduce: "weighting curve" from its International Standard IEC 61672-1 and IEC 60537. All such extracts are copyright of IEC, Geneva, Switzerland. All rights reserved. Further information on the IEC is available from `www.iec.ch`. IEC has no responsibility for the placement and context in which the extracts and contents are reproduced by the author; nor is IEC in any way responsible for the other content or accuracy therein.

APPENDIX

E

DRP-ERP compensation curves

Table E.1 provides DRP to ERP correction data for application to artificial ears of type 2, 3.3 and 3.4 in accordance to ITU-T Recommendation P.57 [237].

Where S_{DE} is the transfer function from DRP to ERP, given by

$$S_{DE} = 20 \log_{10} \frac{p_E}{p_D} \tag{E.1}$$

and p_E is the sound pressure level at the ERP (dB) and p_D the sound pressure level at the DRP (dB).

Frequency (Hz)	S_{DE} **(dB)**
100	0.0
125	0.0
160	0.0
200	0.0
250	−0.3
315	−0.2
400	−0.5
500	−0.6
630	−1.7
800	−1.1
1000	−1.7
1250	−2.6
1600	−4.2
2000	−6.5
2500	−9.4
3150	−10.3
4000	−6.6
5000	−3.2
6300	−3.3
8000	−16.0
(10000)	(−14.4)

Table E.1: Third octave DRP to ERP correction data according to ITU-T Recommendation P.57 [237] to be applied type 2, 3.3 and 3.4 artificial ears. Reproduced with the kind permission of ITU.

APPENDIX

F

Abbreviations

2AFC Two-alternative forced choice

3AFC Three-alternative forced choice

AAC Advanced audio coding

ACR Absolute category rating [224]

A/D Analogue-to-digital converter

ADC Analogue-to-digital converter

AES Audio Engineering Society

AES/EBU Audio Engineering Society / European Broadcasting Union digital audio interface

AIFF (Apple) audio interchange file format

Perceptual Audio Evaluation – Theory, Method and Application Søren Bech and Nick Zacharov

ANOVA Analysis of variance

ANSI American National Standards Institute, Inc.

APA American Psychological Society

ASA Acoustical Society of America

ASTM American Society for Testing and Materials

BAQ Basic audio quality [207]

BBC British Broadcasting Corporation

BER Bit error rate

BTL Bradley-Terry-Luce model

CB Critical bandwidth

CCITT International Telegraph and Telephone Consultative Committee

CD Compact disc

CELP Code excited linear prediction

CETVSQ Continuous evaluation of time varying speech quality [244]

CIPIC Center for Image Processing and Integrated Computing

CMOS Comparison mean opinion score [224]

CCR Comparison category rating [224]

CQS Continuous quality scale

CRT Cathode ray tubes

D/A Digital-to-analogue

DA Descriptive analysis

DAC Digital-to-analogue converter

DCME Digital circuit multiplication equipment

DCR Degradation category rating [224]

DF Diffuse field

DF Degrees of freedom

DL Differenz Limen

DMOS Degradation mean opinion score [224]

DRP Eardrum reference point [237]

DTS Digital Theatre Systems

DVD Digital versatile disc

EC Echo canceller

EBU European Broadcast Union

EEP Ear canal entrance point [237]

EMS Expected mean square

ERB Equivalent rectangular bandwidth

ERP Ear reference point [237]

ETSI European Telecommunications Standards Institute

FA Factor analysis

FEC *F*ree air *e*quivalent *c*oupling [298]

FF Free field

FCP Free-choice profiling

FP Flash profile

GAL Graphical assessment language

GAS Gasto-Acoustical Society

GD Suffix for Zwicker's loudness model, referring to *G*roup (Frequenzgruppen or critical band) and *D*iffuse field measurement

GF Suffix for Zwicker's loudness model, referring to *G*roup (Frequenzgruppen or critical band) and *F*ree field measurement

GLM Generalised linear model

GLS Generalised listeners selection [290]

GP2 Guinea Pig subjective test system [175, 176]

GPL General Public License

GPA Generalised procrustes analysis

GSM Global system for mobile communication

GSM-EFR Global system for mobile communication enhanced full rate codec

GSM-FR Global system for mobile communication full rate codec

GSM-HR Global system for mobile communication half rate codec

GUI Graphical user interface

HATS Head and torso simulator

HFT Hands-free terminal

HRTF Head related transfer function

HSD (Tukey's) honestly significant difference

HTML Hypertext markup language

HVAC Heating, ventilation and air conditioning

IEC International Electrotechnical Commission

IBU International Broadcast Union

ILD Interaural level difference

INDSCAL Individual differences scaling

IP Internet protocol

IRS Intermediate reference system [219]

ISO International Organisation for Standards

ITD Interaural time difference

ITU-R International Telecommunications Union, Radiocommunication Sector

ITU-T International Telecommunications Union, Telecommunications Standardisation Sector

JND Just noticeable difference

LCD Liquid crystal display

LD-CELP Low-delay code excited linear prediction

LFE Low frequency effects

LSDI Large screen digital imagery

MAF Minimum audible field

MAP Minimum audible pressure

MAPD Minimum audible pressure at the eardrum

MANOVA Multivariate analysis of variance

MCL Most comfortable level

MDS Multidimensional scaling

MDU Multidimensional unfolding

MNRU Modulated noise reference unit [225]

MOA Method of adjustment

MOS Mean opinion score [229]

MOS-LQS MOS subjective listening quality [238]

MOS-LQO MOS objective listening quality [238]

MOS-LQE MOS estimated listening quality [238]

MOS$_c$ Arithmetic mean of any collection of conversation opinion scores [224]

MOS-CQS MOS subjective conversational quality [238]

MOS-CQO MOS objective conversational quality [238]

MOS-CQE MOS estimated conversational quality [238]

MOS_{le} Listening effort scale [224]

MOS_{lp} Loudness preference scale [224]

MPANL Maximum permissible ambient noise levels [14]

MPEG Motion picture expert group

MUSHRA Multistimulus test with hidden reference and anchor [217]

NDA Non-disclosure agreement

OD Suffix for Stevens' loudness model, referring to *O*ctave band and *D*iffuse field measurement

ODG Objective difference grade [213]

PC Paired comparison

PCA Principle components analysis

PCM Pulse-code modulation

PEAQ Perceptual evaluation of audio quality [213]

PESQ Perceptual evaluation of audio quality [236]

PR Paired rating

PSA Perceptual structure analysis

PSTN Public switched telephone network

QDA Quantitative Descriptive Analysis®

RGT Repertory grid technique

RMS Root mean square

RSM Response surface method

RTA Real-time analyser

SDG Subjective difference grade [213]

SMPTE The Society of Motion Picture and Television Engineers

S/N Signal-to-noise

SNR Signal-to-noise ratio

S/PDIF Sony/Philips digital interface

SPL Sound pressure level

SS Single stimulus [216]

SS Sum of squares

SSCQE Single stimulus continuous quality evaluation [216]

SSR Single stimulus rating

TAFC Two-alternative forced choice

TET Timbral ear training

UI User interface

VoIP Voice-over-IP

VPA Verbal protocol analysis

Bibliography

[1] 3GPP. Technical specification group services and system aspects; AMR-WB speech codec performance characterisation. Technical Report 3GPP TR 26.976, 3rd Generation Partnership Project, 2001.

[2] Aarts, R. M. Calculation of the loudness of loudspeakers during listening tests. *Journal of the Audio Engineering Society 39* (1991), 27–28.

[3] Aarts, R. M. Comparison of some loudness measures for loudspeaker listening tests. *Journal of the Audio Engineering Society 40* (1992), 142–146.

[4] Aarts, R. M. *On the Design and Psychophysical Assessment of Loudspeaker Systems*. PhD thesis, Delft Technical University, 1995.

[5] AES. *20, AES Recommended Practice for Professional Audio–Subjective Evaluation of Loudspeakers*. Audio Engineering Society, 1996.

[6] AES. *17, AES Standard Method for Digital Audio Engineering–Measurement of Digital Audio Equipment*. Audio Engineering Society, 1998.

[7] AES. *6id, AES Information Document for Digital Audio–Personal Computer Audio Quality Measurement*. Audio Engineering Society, 2000.

Perceptual Audio Evaluation – Theory, Method and Application Søren Bech and Nick Zacharov

[8] AES. *47, AES Standard on Interconnections–Grounding and EMC Practices–Shields of Connectors in Audio Equipment Containing Active Circuitry*. Audio Engineering Society, 2005.

[9] Agresti, A. *Categorical Data Analysis*. Wiley, 2002.

[10] Airo, E., Pekkarinen, J., and Olkinuora, P. Listening to music over earphones: an assessment of noise exposure. *Acta Acustica united with Acustica 82* (1996), 885–894.

[11] Algazi, V. R., Duda, R. O., Thompson, D. M., and Avendano, C. The CIPIC HRTF database. Available from: `http://interface.cipic.ucdavis.edu/CIL_html/CIL_HRTF_database.htm`, 2002.

[12] Algazi, V. R., Duda, R. O., Thompson, D. M., and Avendano, C. The CIPIC HRTF database. In *Proceedings of the IEEE Workshop of Applications of Signal Processing to Audio and Acoustics* (New Plaza, New York, USA, 2002), IEEE.

[13] Altman, D. G., Gore, S. M., Gardner, M. J., and Pocock, S. J. Statistical guidelines for contributors to medical journals. In *Statistics with Confidence–Confidence Intervals and Statitical Guidelines* M. J., Gardner and D. G., Altman, Eds. Universities Press, Great Britain, United Kingdom, 1998, ch. 9, pp. 83–100.

[14] ANSI. *S3.1-1991, Maximum Permissible Ambient Noise Levels for Audiometric Test Rooms*. American National Standards Institute, 1991.

[15] APA. *Publication Manual of the American Psychological Association*, 5th ed. American Psychological Association, Washington, DC, USA, 2002.

[16] ASTM. *STP 758. Guidelines for the Selection and Training of Sensory Panel Members*. American Society for Testing and Materials, 1981.

[17] Atkinson, C. and Giddings, P. Grounding systems and their implementation. *Journal of the Audio Engineering Society 43*, 6 (1995), 465–471.

[18] Audio Engineering Society. Perceptual audio coders: What to listen for. Compact Disc, 2001.

[19] Audio Research Labs. STEP–Subjective training and evaluation program. `http://www.audioresearchlabs.com/step_info.html`, 2004.

[20] Ballas, J. A., Brock, D., Stroup, J., and Fouad, H. The effect of auditory rendering on perceived movement: loudspeaker density and HRTF. In *Proceedings of the 2001 International Conference on Auditory Display* (Espoo, Finland, July 2001), pp. 235–238.

[21] Bang & Olufsen. Music for Archimedes. Compact Disc CD B&O 101, 1992.

[22] Barr, D. R. Using confidence intervals to test hypotheses. *Journal of Quality Technology 1*, 4 (October 1969), 256–258.

[23] Barrett, P. A., Campos-Neto, S. F., and Sharpley, A. D. Fixed-point subjective selection test plan for the ITU-T 4-kbit/s speech coding algorithm. Technical report TD 13 (WP2/12), Dakar, Senegal, 22–26 October, 2001 SG12 Meeting of the 2001–2004 Study Period, International Telecommunications Union, Telecommunications Standardization Sector, 2001.

[24] Barton, R. R. Design-plots for factorial and fractional-factorial designs. *Journal of Quality Technology 30*, 1 (1998), 40–54.

[25] Beattie, R. C., Zentil, A., and Svihovec, D. A. Effects of white noise on the most comfortable level for speech with normal listeners. *Journal of Auditory Research 22*, 1 (January 1982), 71–76.

[26] Bech, S. Listening tests on loudspeakers: A discussion of experimental procedures and evaluation of the response data. In *Proceedings of the Audio Engineering Society 8th International Conference* (1990), Audio Engineering Society, pp. 101–115.

[27] Bech, S. Selection and training of subjects for listening tests on sound-reproducing equipment. *Journal of the Audio Engineering Society 40*, 7/8 (1992), 590–610.

[28] Bech, S. Training of subjects for auditory experiments. *Acta Acustica united with Acustica 1* (1993), 89–99.

[29] Bech, S. Perception of timbre of reproduced sound in small rooms: Influence of room and loudspeaker position. *Journal of the Audio Engineering Society 42*, 12 (1994), 999–1007.

[30] Bech, S. Timbral aspects of reproduced sound in small rooms. I. *Journal of the Acoustical Society of America 97*, 3 (March 1995), 1717–1726.

[31] Bech, S. Timbral aspects of reproduced sound in small rooms. II. *Journal of the Acoustical Society of America 99*, 6 (June 1996), 3539–3549.

[32] Bech, S. Calibration of relative level differences of a domestic multichannel sound reproduction system. *Journal of the Audio Engineering Society 46* (1998), 304–313.

[33] Bech, S. The influence of stereophonic width on the perceived quality of an audio-visual presentation using a multichannel sound system. *Journal of the Audio Engineering Society 46* (1998), 314–322.

[34] Bech, S. Spatial aspects of reproduced sound in small rooms. *Journal of the Acoustical Society of America 103*, 1 (January 1998), 434–445.

[35] Bech, S. Methods for subjective evaluation of spatial characteristics of sound. In *Proceedings of the Audio Engineering Society 16th International Conference on Spatial Sound Reproduction* (1999), Audio Engineering Society, pp. 487–504.

[36] Bech, S. Requirements for low-frequency sound reproduction, part I: The audibility of changes in passband amplitude and lower system cutoff frequency and slope. *Journal of the Audio Engineering Society 50*, 7/8 (2002), 564–580.

[37] Bech, S. and Cutmore, N. Multichannel level alignment, part V: The effects of reproduction level, reproduction room, step size, and symmetry. In *Proceedings of the Audio Engineering Society 108th International Convention* (Paris, France, Feb. 2000).

[38] Bech, S., Gulbol, M.-A., Martin, G., Ghani, J., and Ellermeier, W. A listening test system for automotive audio part 2: Initial verification. In *Proceedings of the 118th Convention of the Audio Engineering Society*, (Barcelona, Spain, May 2005).

[39] Bech, S., Hamberg, R., Nijenhuis, M., Teunissen, C., de Jong, H. L., Houben, P., and Pramanik, S. K. The rapid perceptual

image description method (RaPID). In *SPIE Proceedings* (1996), Vol. 2657, SPIE, pp. 317–328.

[40] Bech, S. and Zacharov, N. Multichannel level alignment, part III: The effects of loudspeaker directivity and reproduction bandwidth. In *Proceedings of the Audio Engineering Society 106th International Convention* (Munich, Germany, May 1999).

[41] Beerends, J. G., Hekstra, A. P., Rix, A. W., and Hollier, M. P. Perceptual evaluation of speech quality (PESQ). The new ITU standard for end-to-end speech quality assessment part II–psychoacoustic model. *Journal of the Audio Engineering Society 50*, 10 (October 2002), 765–778.

[42] Begault, D. R. *3-D Sound for Virtual Reality and Multimedia*. Academic Press, Cambridge, MA, USA, 1994.

[43] Benjamin, E. Characteristics of musical signals. In *Proceedings of the 97th Convention of the Audio Engineering Society* (San Francisco, CA, USA, November 1994).

[44] Benjamin, E. Signal characteristics of matrix and discrete multichannel recordings. In *Proceedings of the 105th Convention of the Audio Engineering Society* (San Francisco, CA, USA, September 1998).

[45] Benjamin, E. Preferred listening levels and acceptance windows for dialog reproduction in the domestic environment. In *Proceedings of the 117th Convention of the Audio Engineering Society* (San Francisco, CA, USA, October 2004).

[46] Benjamin, E. and Gannon, B. The effect of room acoustics on subwoofer performance and level setting. In *Proceedings of the 109th Convention of the Audio Engineering Society* (San Francisco, CA, USA, September 2000).

[47] Beranek, L. L. *Acoustical Measurements*, revised ed. Acoustical Society of America, 1988.

[48] Berg, J. *Systematic Evaluation of Perceived Spatial Quality in Surround Sound Systems*. PhD thesis, School of Music, Luleå University of Technology, May 2002.

[49] Berg, J. OPAQUE – a tool for the elicitation and grading of audio quality attributes. In *Proceedings of the 118th Convention of the Audio Engineering Society*, (Barcelona, Spain, May 2005).

[50] Berg, J. and Rumsey, F. Identification of perceived spatial attributes of recordings by repertory grid technique and other methods. In *Proceedings of the Audio Engineering Society 106th International Convention* (Munich, Germany, May 1999).

[51] Berg, J. and Rumsey, F. Spatial attribute identification and scaling by repertory grid technique and other methods. In *Proceeding of the Audio Engineering Society 16th International Conference* (Munich, Germany, May 1999), Audio Engineering Society.

[52] Berg, J. and Rumsey, F. Correlation between emotive, descriptive and naturalness attributes in subjective data relating to spatial sound reproduction. In *Proceedings of the Audio Engineering Society 109th International Convention* (Los Angeles, CA, USA, Sept. 2000).

[53] Berg, J. and Rumsey, F. In search of the spatial dimensions of reproduced sound: Verbal protocol analysis and cluster analysis of scaled verbal descriptors. In *Proceedings of the Audio Engineering Society 108th International Convention* (Paris, France, Feb. 2000).

[54] Bernstein, I. H. *Applied Multivariate Analysis*. Springer, New York, 1988.

[55] Bies, D. A. and Hansen, C. H. *Engineering Noise Control*, 2nd ed. E & FN Spon, 1996.

[56] Blauert, J. *Spatial Hearing. The Psychophysics of Human Sound Localisation*. MIT Press, Cambridge, MA, USA, 1997.

[57] Blauert, J., Brueggen, M., Bronkhorst, A. W., Drullman, R., Reynaud, G., Pellieux, L., Krebber, W., and Sottek, R. The AUDIS catalog of human HRTFs. Information from: `http://www.eaa-fenestra.org/Products/Documenta/Publications/09-de2`, 1998.

[58] Block, M. G., Killion, M. C., and Tillman, T. W. The "missing 6 dB" of Tillman, Johnson and Olsen was found–30 years ago. *Seminars in Hearing 25*, 1 (2004), 7–16.

[59] Bolia, R. S., Nelson, W. T., Ericson, M. A., and Simpson, B. D. A speech corpus for multitalker communications research. *Journal of the Acoustical Society of America 107*, 2 (February 2000), 1065–1066.

[60] Borenius, J. Perceptibility of direction and time delay errors in subwoofer reproduction. In *Proceedings of the 79th Convention of the Audio Engineering Society* (New York, USA, October 1985).

[61] Borg, I. and Groenen, P. *Modern Multidimensional Scaling: Theory and Applications*. Springer-Verlag, New York, 1997.

[62] Bosi, M., Brandenburg, K., Quackenbush, S., Fielder, L., Akagiri, K., Fuchs, H., and Dietz, M. ISO/IEC MPEG-2 advanced audio coding. *Journal of the Audio Engineering Society 45*, 10 (October 1997), 789–814.

[63] Box, G. E. P., Hunter, W. G., and Hunter, J. S. *Statistics for Experimenters*. Wiley, 1978.

[64] Brandenburg, K. and Bosi, M. Overview of MPEG audio: Current and future standards for low bit-rate audio coding. *Journal of the Audio Engineering Society 45*, 1/2 (January/February 1997), 4–21.

[65] Brandt, M. A., Skinner, E. Z., and Coleman, J. A. Texture profile method. *Journal of Food Science 28* (1963), 404–409.

[66] Bregman, A. S. and Ahad, P. Demonstrations of auditory scene analysis. Compact Disc available from `http://www.psych.mcgill.ca/labs/auditory/home.html`, 1995.

[67] Brixen, E. B. Spectral ear training. In *Proceedings of the Audio Engineering Society 94th International Convention* (Berlin, Germany, March 1993).

[68] Brixen, E. B. and Søndergaard, M. Rapport vedrørende: niveauopfattelse i hovedtelefoner (in Danish). Technical Report KKDK 068-1-ebb-1, Delta Akustik and Vibration, Denmark, 2001.

[69] Brockhoff, P. *Statistical Analysis of Sensory Data*. PhD thesis, The Royal Veterinary and Agricultural University Copenhagen, 1994.

[70] Brockhoff, P. M. Assessor modeling. *Food Quality and Preference 9*, 3 (1998), 87–89.

[71] Brockhoff, P. M. Statistical testing of individual differences in sensory profiling. *Food Quality and Preference 14*, 5–6 (2003), 425–434.

[72] Brockhoff, P. M. and Skovgaard, I. M. Modelling individual differences between assessors in sensory evaluations. *Food Quality and Preference 5* (1994), 215–224.

[73] Brookes, T., Kassier, R., and Rumsey, F. A simplified scene-based paradigm for use in spatial audio listener training applications. In *Proceedings of the Audio Engineering Society 117th International Convention* (San Francisco, CA, USA, Oct. 2004).

[74] Brown, H. and Prescott, R. *Applied Mixed Models in Medicine*. Wiley, London, 1999.

[75] Brungart, D. S. Evaluation of speech intelligibility with the coordinate response measure. *Journal of the Acoustical Society of America 109*, 5 (May 2001), 2276–2279.

[76] BSI. *BS 5969: Specification for Sound Level Meters*. British Standards Institute, 1981.

[77] Bunning, H. and Wilkens, H. Mehrdimensionale Verknüpfung der Höreindrücke von Lautsprechern mit deren physikalischen Daten. Abschlussbericht Bo 421/7, Heinrich-Hertz-Institut für Nachrichtentechnik Berlin GmbH, 1979.

[78] Butler, D. *Musician's Guide to Perception and Cognition*. Schirmer Books, 1992.

[79] Cabot, R. C. Fundamentals of modern audio measurement. *Journal of the Audio Engineering Society 47*, 9 (September 1999), 738–744, 746–762.

[80] Canévet, G. An auditory test battery for professional screening. *Acta Acustica united with Acustica 85*, 5 (1999), 667–673.

[81] Cardozo, B. L. Adjusting the method of adjustment: SD vs DL. *Journal of the Acoustical Society of America 37*, 5 (1965), 786–792.

[82] Carroll, J. D. Individual differences and multidimensional scaling. *Multidimensional Scaling: Theory*, Vol. 1. Seminar Press, New York, 1972, pp. 105–155.

[83] Carroll, J. D. and Chang, J. J. Analysis of individual differences in multidimensional scaling via an N-way generalization of "Eckart-Young" decomposition. *Psychometrika 35* (1970), 283–319.

[84] CCITT. *Handbook of Telephonometry*. International Telecommunications Union, 1992.

[85] Civille, G. V. and Liska, I. H. Modifications and applications to foods of the general foods sensory texture profile technique. *Journal of Texture Studies 6* (1975), 19–31.

[86] Chalupper, H. and Fastl, H. Dynamic loudness model (DLM) for normal and hearing-impaired listeners. *Acta Acustica united with Acustica 88*, 3 (May/June 2002), 378–386.

[87] Chapman, P. J. Programme material analysis. In *Proceedings of the Audio Engineering Society 100th International Convention* (Copenhagen, Denmark, May 1996).

[88] Choisel, S. Pointing technique with visual feedback for sound source localization experiments. In *Proceedings of the 115th Convention of the Audio Engineering Society* (New York, NY, USA, October 2003).

[89] Choisel, C. and Wickelmaier, F. Extraction of auditory features and elicitation of attributes for the assessment of multichannel reproduced sound. In *Proceedings of the 118th Convention of the Audio Engineering Society*, (Barcelona, Spain, May 2005).

[90] Choisel, S. and Wickelmaier, F. Scaling auditory attributes underlying listener preference of multichannel reproduced sound. *Journal of the Acoustical Society of America* (2006), submitted.

[91] Chouard, N. *Perception of Overall Loudness and Unpleasantness*. PhD thesis, Université du Maine, 1998.

[92] Chouard, N. Perception of overall loudness and unpleasantness. In *Proceedings of Internoise Conference* (Christchurch, New Zealand, November 1998).

[93] Cochran, W. G. Some consequences when the assumptions for the analysis of variance are not satisfied. *Biometrics 3*, 1 (1947), 22–38.

[94] Cochran, W. G. and Cox, G. M. *Experimental Design*, 2nd ed. Wiley, 1992.

[95] Communications Research Center of Canada. CRC-SEAQ: System for the evaluation of audio quality. http://www.crc.ca/en/html/aas/home/products/products#CRC_SEAQ, 2005.

[96] Condamines, R. Relationship between the passband and the preferred listening level for music. *EBU Review Part A 139* (June 1973), 124–127.

[97] Cooper, B. A., Nelson, D. A., and Danielson, R. W. Auditory demonstrations II: Challenges to speech communication and music listening. In *Noise-Con* (Baltimore, MD, USA, 2004), Noise-Con.

[98] Coren, S. Most comfortable listening level as a function of age. *Ergonomics 37*, 7 (1994), 1269–1274.

[99] Corey, J. An ear training system for identifying parameters of artificial reverberation in multichannel audio. In *Proceedings of the Audio Engineering Society 117st International Convention* (San Francisco, CA, USA, Oct. 2004).

[100] Cowles, M. and Davis, C. On the origins of the .05 level of statistical significance. *American Psychologist 37* (1982), 553–558.

[101] Database of HRTFs. Available from: http://www.ais.riec.tohoku.ac.jp/lab/db-hrtf/, 2001.

[102] de Bruijn, W. P. J. and Boone, M. M. Subjective experiments on the effects of combining spatialized audio and 2D video projection in audio-visual systems. In *Proceedings of the 112th Convention of the Audio Engineering Society* (Munich, Germany, May 2002).

[103] Delarue, J. and Sieffermann, J. M. Sensory mapping using flash profile. Comparison with a conventional descriptive method for

the evaluation of the flavour of fruit dairy products. *Food Quality and Preference 15*, 4 (June 2004), 383–392.

[104] Delta. Hør engang! Compact Disc, 1998.

[105] Deutsch, D. Musical illusions and paradoxes. Compact Disc available from http://www.philomel.com, 1995.

[106] Deutsch, D. Phantom words and other curiosities. Compact Disc available from http://www.philomel.com, 2003.

[107] Dijksterhuis, G. Procrustes analysis in sensory research. *Multivariate Analysis of Data in Sensory Science, Vol. 16 of Data Handling in Science and Technology* [314]. Elsevier, Amsterdam, The Netherlands, 1996, ch. 7, pp. 185–217.

[108] Dunn, J. Digital-to-analog converter measurements. Available from: http://www.audioprecision.com/index.php?page=resources\&id=1000000161, 2001.

[109] Dunn, J. *Measurement Techniques for Digital Audio*. Audio Precision Inc., 2004.

[110] EBU. *Sound Quality Assessment Material, (recordings for subjective tests)*. No. 422 204-2. 1988.

[111] EBU. *Technical Document Tech 3286–Assessment Methods for the Subjective Evaluation of the Quality of Sound Programme Material–Music*. European Broadcast Union, August 1997.

[112] EBU. *Technical Document Tech 3276–Listening Conditions for the Assessment of Sound Programme Material: Monophonic and Two-Channel Stereophonic*, 2nd ed. European Broadcast Union, May 1998.

[113] EBU. *Technical Document Tech 3276: Supplement 1–Listening Conditions for the Assessment of Sound Programme Material: Multichannel Sound*. European Broadcast Union, February 1999.

[114] EBU. *Technical Document Tech 3286: Supplement 1–Assessment Methods for the Subjective Evaluation of the Quality of Sound Programme Material–Multichannel*. European Broadcast Union, July 2000.

[115] Ekman, G. and Sjöberg, L. Scaling. *Annual Review of Psychology 16* (1965), 451–474.

[116] Ellermeier, W. and Zimmer, K. Using psychological choice models to investigate overall sound quality. In *Proceedings of the First ISCA Tutorial and Research Workshop on Auditory Quality of Systems* (Germany, Bochum, April 23–25 2003), Acadamie Mont-Cenis.

[117] Engeldrum, P. G. *Psychometric Scaling: A Toolkit for Imaging Systems Development*. Imcotek Press, Winchester, MA, USA, 2000.

[118] European Union. Directive 2003/10/EC of the European parliament and of the council of 6 February 2003 on the minimum health and safety requirements regarding the exposure of workers to the risks arising from physical agents (noise). *Official Journal of the European Union* (2003), Vol. L42, pp. 38–44.

[119] Exner, S. Zur Lehre von den Gehörsempfindungen. *Pflügers Archiv 13* (1876), 228–253.

[120] Fastl, H. Loudness evaluation by subjects and by a loudness meter. In *Sensory Research, Multimodal Perspectives* (Hillsdale NJ, USA, 1993), R. T. Verrillo, Ed. Lawrence Erlbaum Associates, pp. 199–210.

[121] Fause, K. R. Fundamentals of grounding, shielding, and interconnection. *Journal of the Audio Engineering Society 43*, 6 (1995), 498–516.

[122] Fielder, L. D. and Benjamin, E. M. Subwoofer performance for accurate reproduction of music. *Journal of the Audio Engineering Society 36*, 8 (June 1988), 443–457.

[123] Fjelland, R. and Gjengedal, E. Videnskab på egne praemisser. *Videnskabsteori og etik for sundhedspersonale*, 1st ed. Munksgaard, 1996.

[124] Ford, N. *Developing a Graphical Language to Represent Listeners' Experiences of Spatial Attributes in Reproduced Sound*. PhD thesis, University of Surrey, Department of Music & Sound Recording, School of Performing Arts, Surrey, GB, 2005.

[125] Ford, N., Rumsey, F., and de Bruyn, B. Graphical elicitation techniques for subjective assessment of the spatial attributes of loudspeaker reproduction–a pilot investigation. In *Proceedings of the Audio Engineering Society 110th International Convention* (Amsterdam, The Netherlands, May 2001).

[126] Ford, N., Rumsey, F., and Nind, T. Evaluating influences of a central automotive loudspeaker on perceived spatial attributes using a graphical assessment language. In *Proceedings of the Audio Engineering Society 113th International Convention* (Los Angeles, CA, USA, Oct. 2002).

[127] Ford, N., Rumsey, F., and Nind, T. Subjective evaluation of perceived spatial differences in car audio systems using a graphical assessment language. In *Proceedings of the Audio Engineering Society 112th International Convention* (Munich, Germany, May 2002).

[128] Ford, N., Rumsey, F., and Nind, T. Evaluating spatial attributes of reproduced audio events using a graphical assessment language–understanding differences in listener depictions. In *Proceedings of the Audio Engineering Society 24th International Conference* (2003), Audio Engineering Society, pp. 171–183.

[129] Ford, N., Rumsey, F., and Nind, T. Communicating listeners' auditory spatial experiences: A method for developing a descriptive language. In *Proceedings of the 118th Convention of the Audio Engineering Society*, (Barcelona, Spain, May 2005).

[130] Foxlin, E. *Motion Tracking Requirements and Technologies.* Lawrence Erlbaum Associates, Mahwah, NJ, USA, 2002, pp. 163–210.

[131] Frank, T., Durrant, J. D., and Lovrinic, J. H. Maximum permissible ambient noise levels for audiometric test rooms. *American Journal of Audiology* 2, 1 (1993), 414–422.

[132] Gabrielsson, A. Dimension analyses of perceived quality of sound reproduction systems. *Scandinavian Journal of Psychology 20* (1979), 159–169.

[133] Gabrielsson, A. Statistical treatment of data for listening tests on sound reproduction systems. Technical Report TA 92,

Department of Technical Audiology, Karolinska Institutet, Stockholm, Sweden, 1979.

[134] Gabrielsson, A., Hagerman, B., Bech-Kristensen, T., and Lundberg, G. Perceived sound quality of reproduction with different frequency responses and sound levels. *Journal of the Acoustical Society of America 88* (1990), 1359–1366.

[135] Gabrielsson, A. and Lindström, B. Perceived sound quality of high-fidelity loudspeakers. *Journal of the Audio Engineering Society 33*, 1/2 (January/February 1985), 33–53.

[136] Gabrielsson, A., Rosenberg, V., and Sjögren, H. Judgement and dimension analyses of perceived sound quality of sound reproducing systems. *Journal of the Acoustical Society of America 55*, April (1974), 854–861.

[137] Gabrielsson, A. and Sjögren, H. Perceived sound quality of sound-reproduction systems. *Journal of the Acoustical Society of America 65*, 4 (1979), 1019–1033.

[138] Gaito, J. and Yokubynas, R. An empirical basis for the statement that measurement scale properties (and meaning) are irrelevant in statistical analysis. *Bulletin of the Psychonomic Society 24*, 6 (1986), 449–450.

[139] Gardner, W. G. *3-D Audio Using Loudspeakers*. Kluwer Academic Publishers, Boston, MA, 1998.

[140] Gardner, M. J. and Altman, D. G., Eds. *Statistics with Confidence–Confidence Intervals and Statistical Guidelines*. Universities Press, Great Britain, United Kingdom, 1998.

[141] Gardner, B. and Martin, K. HRTF measurements of a KEMAR dummy-head microphone. `http://sound.media.mit.edu/KEMAR.html`, 2000.

[142] *Gehörschutz–Das Gehör schützen*. Fachausschuss Persönliche Schutzausrüstung, Arbeitskreis Gehörschutz, Nürnberg Fa. Hamann Consult, Dresden, 1992.

[143] Gescheider, G. A. *Psychophysics. Method, Theory, and Application*. Lawrence Erlbaum Associates, 1985.

[144] Gescheider, G. A. *Psychophysics: The Fundamentals*, 3rd ed. Lawrence Erlbaum Associated, 1997.

[145] Ghani, J., Ellermeier, W., and Zimmer, K. A test battery measuring auditory capabilities of listening panels. In *Proceedings of the Forum Acusticum 2005 Congress* (Budapest, Hungary, 2005).

[146] Glasberg, B. R. and Moore, B. C. J. A model of loudness applicable to time-varying sounds. *Journal of the Audio Engineering Society 50*, 5 (May 2002), 331–342.

[147] Gleiss, N. Preferred listening levels in telephony. *TELE, English Edition*, 2 (1974), 24–27.

[148] Green, D. M. and Swets, J. A. *Signal Detection Theory and Psychophysics*. Peninsula Publishing, 1988.

[149] Greenhouse, S. W. and Geisser, S. On methods in the analysis of profile data. *Psychometrica 24* (1959), 95–112.

[150] Grey, J. M. Multidimensional perceptual scaling of musical timbres. *Journal of the Acoustical Society of America 61*, 5 (1977), 1270–1277.

[151] Griesinger, D. Spaciousness and envelopment in musical acoustics. In *Proceedings of the 101st Convention of the Audio Engineering Society* (Los Angeles, CA, USA, November 1996).

[152] Griesinger, D. Spatial impression and envelopment in small rooms. In *Proceedings of the 103rd Convention of the Audio Engineering Society* (New York, USA, September 1997).

[153] Griesinger, D. Objective measures of spaciousness and envelopment. In *Proceedings of the Audio Engineering Society 16th International Conference on Spatial Sound Reproduction* (1999), Audio Engineering Society, pp. 27–41.

[154] Guastavino, C. and Katz, B. F. G. Perceptual evaluation of multidimensional spatial audio reproduction. *Journal of the Acoustical Society of America 116*, 2 (August 2004), 1105–1115.

[155] Hair, J. F., Anderson, R. E., Tatham, R. L., and Black, W. C. *Multivariate Data Analysis*, 5th ed. Prentice-Hall International Inc., London, 1998.

[156] Hammershøi, D. and Møller, H. *Binaural Technique–Basic Methods for Recording, Synthesis, and Reproduction*. Springer, 2005, ch. 9, pp. 223–254.

[157] Hansen, V. Establishing a panel of listeners at Bang & Olufsen: A report. In *Proceedings of the Symposium on Perception of Reproduced Sound* (Gammel Avernæs, Denmark, 1987), 89–90.

[158] Hansen, V. and Munch, G. Making recordings for simulation tests in the Archimedes project. *Journal of the Audio Engineering Society 39* (1991), 768–774.

[159] Härmä, A. HUT ear matlab toolbox version 2.0. `http://www.acoustics.hut.fi/software/HUTear/HUTear.html`, 2000.

[160] Hassall, J. R. and Zavari, K. *Acoustic Noise Measurement*. Brüel and Kjær, 1979.

[161] Hellman, R. and Zwicker, E. Why can a decrease in dB(A) produce an increase in loudness. *Journal of the Acoustical Society of America 82*, 5 (November 1987), 1700–1705.

[162] Hicks, C. R. *Fundamental Concepts in the Design of Experiments*. Holt, Rhinehart and Winston, New York, USA, 1973.

[163] Hoaglin, D. C., Mosteller, F., and Tukey, J. W. *Fundamentals of Exploratory Analysis of Variance*. Wiley, 1991.

[164] Hofer, B. Measuring switch-mode power amplifiers. Technical Report, Audio Precision Inc., Beaverton, OR, USA, 2003.

[165] Hollander, M. and Wolfe, D. A. *Nonparametrical Statistical Methods*, 2nd ed. Wiley, 1999.

[166] Holman, T. *5.1 Surround Sound up and Running*. Focal press, Boston, MA, 2000.

[167] Holman, T. LFE level setting. *Surround Professional 3*, 7 (August 2000), 43–47.

[168] Holman, T. The number of loudspeaker channels. In *Proceedings of the 19th International Conference of the Audio Engineering Society* (Schloss Elmau, Germany, May 2001).

[169] Holman, T. Setting up multichannel systems. You've got questions, we've got answers.... Surround Professional. Available from http://surroundpro.com/articles/publish/inside_surround/article_30.shtm%1, 2003.

[170] Hoth, D. F. Room noise spectra at subscribers' telephone locations. *Journal of the Acoustical Society of America 12* (1941), 499–504.

[171] Houtsma, A. J. M., Rossing, T. D., Wagenaars, W. M. *Auditory Demonstrations*. Audio CD, Philips/Acoustical Society of America, 1987.

[172] Hugonnet, C. and Walder, P. *Stereophonic Sound Recording–Theory and Practice*. Wiley, Chichester, England and New York, 1998.

[173] Hunter, E. A. Experimental design. *Multivariate Analysis of Data in Sensory Science, Vol. 16 of Data handling in science and technology*. Elsevier, Amsterdam, The Netherlands, 1996, ch. 2, pp. 37–69.

[174] Huynh, H. and Feldt, L. S. Estimation of the box correction for degrees of freedom in the randomised block and split plot designs. *Journal of Educational Statistics 1* (1976), 69–82.

[175] Hynninen, J. *A Software-Based System for Listening Tests*. Master's thesis, Helsinki University of Technology, Helsinki, Finland, May 2001. http://www.acoustics.hut.fi/publications/files/theses/hynninen_mst.pdf.

[176] Hynninen, J. and Zacharov, N. Guinea Pig–A generic subjective test system for multichannel audio. In *Proceedings of the Audio Engineering Society 106th International Convention* (Munich, Germany, May 1999).

[177] IEC. *60318, An IEC Artificial Ear, of Wideband Type, for the Calibration of Earphones Used in Audiometry*. International Electrotechnical Commission, 1970.

[178] IEC. *60537, Frequency Weighting for Measurement of Aircraft Noise (D-Weighting)*. International Electrotechnical Commission, 1976, withdrawn.

[179] IEC. *61672-1, Electroacoustics*. Sound level meters. Specifications. International Electrotechnical Commission, 2003

[180] IEC. *60711, Occluded-Ear Simulator for the Measurement of Earphones Coupled to the Ear and Ear Inserts*. International Electrotechnical Commission, 1981.

[181] IEC. *60581-10, High Fidelity Audio Equipment and Systems; Minimum Performance Requirements. Part 10: Headphones*. International Electrotechnical Commission, 1986.

[182] IEC. *60959, Provisional Head and Torso Simulator for Acoustic Measurements and air Conduction Hearing Aids*. International Electrotechnical Commission, 1990.

[183] IEC. *60268-7, Sound System Equipment–Part 7: Headphones and Earphones*. International Electrotechnical Commission, 1996.

[184] IEC. *60268-13, Sound System Equipment–Part 13: Listening Tests on Loudspeakers*. International Electrotechnical Commission, 1998.

[185] IEC. *60268-3, Sound System Equipment–Part 3: Amplifiers*. International Electrotechnical Commission, 2001.

[186] Illényi, A. and Korpássy, P. Correlation between loudness and quality. *Acustica 49* (1981), 334–336.

[187] Isherwood, D., Lorho, G., Mattila, V-V., and Zacharov, N. Augmentation, application and verification of the generalized listener selection procedure. In *Proceedings of the 115th Convention of the Audio Engineering Society* (New York, NY, USA, Oct. 2003).

[188] ISO. *532, Acoustics–Method for Calculating Loudness Levels*. International Organization for Standards, 1975.

[189] ISO. *6189. Acoustics–Pure Tone Conduction Threshold Audiometry for Hearing Conservation Purposes*. International Organization for Standards, 1983.

[190] ISO. *7029. Acoustics–Threshold of Hearing by Air Conduction as a Function of Age and Sex for Otologically Normal Persons*. International Organization for Standards, 1984.

[191] ISO. *1999. Acoustics–Determination of Occupational Noise Exposure and Estimation of Noise-Induced Hearing Impairment*. International Organization for Standards, 1990.

[192] ISO. *8586-1. Sensory Analysis–General Guidance for the Selection, Training and Monitoring of Assessors–Part 1: Selected Assessors.* International Organization for Standards, 1993.

[193] ISO. *11035. Sensory Analysis–Identification and Selection of Descriptors for Establishing a Sensory Profile by a Multidimensional Approach.* International Organization for Standards, 1994.

[194] ISO. *8586-2. Sensory Analysis–General Guidance for the Selection, Training and Monitoring of Assessors–Part 2: Experts.* International Organization for Standards, 1994.

[195] ISO. *717-1. Acoustics–Rating of Sound Insulation in Buildings and of Building Elements–Part 1: Airborne Sound Insulation.* International Organization for Standards, 1996.

[196] ISO. *3382. Acoustics–Measurement of Reverberation Time of Rooms with Reference to Other Acoustics Parameters.* International Organization for Standards, 1997.

[197] ISO. *3745. Acoustics–Determination of Sound Power Levels of Noise Sources. Precision Methods for Anechoic and Semi-Anechoic Rooms.* International Organization for Standards, 1997.

[198] ISO. *140-4. Acoustics–Measurement of Sound Insulation in Buildings and of Building Elements–Part 4: Field Measurements of Airborne Sound Insulation Between Rooms.* International Organization for Standards, 1998.

[199] ISO. *389-1. Acoustics–Reference Zero for the Calibration of Audiometric Equipment–Part 1: Reference Equivalent Threshold Sound Pressure Levels for Pure Tones and Supra-Aural Earphones.* International Organization for Standards, 1998.

[200] ISO. *3741. Acoustics–Determination of Sound Power Levels of Noise Sources. Precision Methods for Reverberation Rooms.* International Organization for Standards, 1999.

[201] ISO. *13300-1. Sensory Analysis–General Guidance for the Staff of a Sensory Evaluation Laboratory–Part 1: Staff Responsibilities.* International Organization for Standards, 2002.

[202] ISO. *13300-2. Sensory Analysis–General Guidance for the Staff of a Sensory Evaluation Laboratory–Part 2: Recruitment and Training of Panel Leaders*. International Organization for Standards, 2002.

[203] ITU-R. *Recommendation BS.562-3, Subjective Assessment of Sound Quality*. International Telecommunications Union Radiocommunication Assembly, 1990.

[204] ITU-R. *Recommendation BS.708, Determination of the Electro-Acoustical Properties of Studio Monitor Headphones*. International Telecommunications Union Radiocommunication Assembly, 1990.

[205] ITU-R. *Recommendation BS.775-1, Multichannel Stereophonic Sound Systems with and without Accompanying Picture*. International Telecommunications Union Radiocommunication Assembly, 1994.

[206] ITU-R. *Recommendation BT.811, The Subjective Assessment of Enhanced PAL and SECAM Systems*. International Telecommunications Union Radiocommunication Assembly, 1994.

[207] ITU-R. *Recommendation BS.1116-1, Methods for the Subjective Assessment of Small Impairments in Audio Systems Including Multichannel Sound Systems*. International Telecommunications Union Radiocommunication Assembly, 1997.

[208] ITU-R. *Recommendation BT.1128, Subjective Assessment of Conventional Television Systems*. International Telecommunications Union Radiocommunication Assembly, 1997.

[209] ITU-R. *Recommendation BS.1283, Subjective Assessment of Sound Quality–A Guide to Existing Recommendations*. International Telecommunications Union Radiocommunication Assembly, 1998.

[210] ITU-R. *Recommendation BS.1284, Methods for the Subjective Assessment of Sound Quality–General Requirements*. International Telecommunications Union Radiocommunication Assembly, 1998.

[211] ITU-R. *Recommendation BS.1285, Pre-selection methods for the subjective assessment of small impairments in audio systems*.

International Telecommunications Union Radiocommunication Assembly, 1998.

[212] ITU-R. *Recommendation BS.1286, Methods for the subjective assessment of audio systems with accompanying picture*. International Telecommunications Union Radiocommunication Assembly, 1998.

[213] ITU-R. *Recommendation BS.1387, Method for Objective Measurement of Perceived Audio Quality*. International Telecommunications Union Radiocommunication Assembly, 1998.

[214] ITU-R. *Recommendation BT.1129-2, Subjective Assessment of Standard Definition Digital Television (SDTV) Systems*. International Telecommunications Union Radiocommunication Assembly, 1998.

[215] ITU-R. *Recommendation BT.710-4, Subjective Assessment Methods for Image Quality in High-Definition Television*. International Telecommunications Union Radiocommunication Assembly, 1998.

[216] ITU-R. *Recommendation BT.500-11, Methodology for the Subjective Assessment of the Quality of Television Pictures*. International Telecommunications Union Radiocommunication Assembly, 2002.

[217] ITU-R. *Recommendation BS.1534-1, Method for the Subjective Assessment of Intermediate Quality Level of Coding Systems*. International Telecommunications Union Radiocommunication Assembly, 2003.

[218] ITU-R. *Recommendation BS.1679, Subjective assessment of the quality of audio in large screen digital imagery applications intended for presentation in a theatrical environment*. International Telecommunications Union Radiocommunication Assembly, 2004.

[219] ITU-T. *Recommendation P.48, Specification for an Intermediate Reference System*. International Telecommunications Union, Telecommunications Standardization Sector, 1993.

[220] ITU-T. *Recommendation P.56, Telephone Transmission Quality. Objective Measuring Apparatus. Objective Measurement of Active*

Speech Level. International Telecommunications Union, Telecommunications Standardization Sector, 1993.

[221] ITU-T. *Recommendation P.84, Subjective Listening Test Method for Evaluating Digital Circuit Multiplication and Packetized Voice Systems–Telephone Transmission Quality Subjective Opinion Tests*. International Telecommunications Union, Telecommunications Standardization Sector, 1993.

[222] ITU-T. *Recommendation P.85, A method for subjective performance assessment of the quality of speech voice output devices*. International Telecommunications Union, Telecommunications Standardization Sector, 1994.

[223] ITU-T. *Recommendation P.58, Head and Torso Simulator for Telephonometry*. International Telecommunications Union, Telecommunications Standardization Sector, 1996.

[224] ITU-T. *Recommendation P.800, Methods for Subjective Determination of Transmission Quality*. International Telecommunications Union, Telecommunications Standardization Sector, 1996.

[225] ITU-T. *Recommendation P.810, Telephone Transmission Quality. Methods for Objective and Subjective Assessment of Quality; Modulated Noise Reference Unit (MNRU)*. International Telecommunications Union, Telecommunications Standardization Sector, 1996.

[226] ITU-T. *Recommendation P.830, Subjective Performance Assessment of Telephone-Band and Wideband Digital Codecs*. International Telecommunications Union, Telecommunications Standardization Sector, 1996.

[227] ITU-T. *Recommendation P.910, Subjective Video Quality Assessment Methods for Multimedia Applications*. International Telecommunications Union, Telecommunications Standardization Sector, 1996.

[228] ITU-T. *Recommendation P.920, Interactive Test Methods for Audiovisual Communications*. International Telecommunications Union, Telecommunications Standardization Sector, 1996.

[229] ITU-T. *Recommendation P.10, Vocabulary of Terms on Telephone Transmission Quality and Telephone Sets*. International Telecommunications Union, Telecommunications Standardization Sector, 1998.

[230] ITU-T. *Recommendation P.341, Transmission Characteristics For Wideband (150–7000 Hz) Digital Hands Free Telephony Terminals*. International Telecommunications Union, Telecommunications Standardization Sector, 1998.

[231] ITU-T. *Recommendation P.831, Subjective Performance Evaluation of Network Echo Cancellers*. International Telecommunications Union, Telecommunications Standardization Sector, 1998.

[232] ITU-T. *Recommendation P.911, Subjective Audiovisual Quality Assessment Methods for Multimedia Applications*. International Telecommunications Union, Telecommunications Standardization Sector, 1998.

[233] ITU-T. *Supplement 23, ITU-T Coded Speech Database*. International Telecommunications Union, Telecommunications Standardization Sector, 1998.

[234] ITU-T. *Recommendation P.501, Test Signals for Use in Telephonometry*. International Telecommunications Union, Telecommunications Standardization Sector, 2000.

[235] ITU-T. *Recommendation P.832, Subjective Performance Evaluation of Handsfree Terminals*. International Telecommunications Union, Telecommunications Standardization Sector, 2000.

[236] ITU-T. *Recommendation P.862, Perceptual Evaluation of Speech Quality (PESQ), an Objective Method for End-to-End Speech Quality Assessment of Narrow-Band Telephone Networks and Speech Codecs*. International Telecommunications Union, Telecommunications Standardization Sector, 2001.

[237] ITU-T. *Recommendation P.57, Telephone Transmission Quality. Objective Measuring Apparatus. Artificial Ears*. International Telecommunications Union, Telecommunications Standardization Sector, 2002.

[238] ITU-T. *Recommendation P.800.1, Mean Opinion Score (MOS) Terminology*. International Telecommunications Union, Telecommunications Standardization Sector, 2003.

[239] ITU-T. *Recommendation P.835, Subjective Test Methodology for Evaluating Speech Communication Systems that Include Noise Suppression Algorithm*. International Telecommunications Union, Telecommunications Standardization Sector, 2003.

[240] ITU-T. *Recommendation P.840, Subjective Listening Test Method for Evaluating Circuit Multiplication Equipment*. International Telecommunications Union, Telecommunications Standardization Sector, 2003.

[241] ITU-T. *Recommendation P.851, Subjective Quality Evaluation of Telephone Services Based on Spoken Dialogue Systems*. International Telecommunications Union, Telecommunications Standardization Sector, 2003.

[242] ITU-T. *Recommendation P.862.1, Mapping Function for Transforming P.862 raw Result Scores to MOS-LQO*. International Telecommunications Union, Telecommunications Standardization Sector, 2003.

[243] ITU-T. *Recommendation P.563, Single Ended Method for Objective Speech Quality Assessment in Narrow-Band Telephony Applications*. International Telecommunications Union, Telecommunications Standardization Sector, 2004.

[244] ITU-T. *Recommendation P.880, Continuous Evaluation of Time Varying Speech Quality*. International Telecommunications Union, Telecommunications Standardization Sector, 2004.

[245] ITU-T. *Recommendation P.862.2, Wideband Extension to Recommendation P.862 for the Assessment of Wideband Telephone Networks and Speech Codecs*. International Telecommunications Union, Telecommunications Standardization Sector, 2005.

[246] ITU-T. *Recommendation P.862.3, Application Guide for Objective Quality Measurement Based on Recommendations P.862, P.862.1 and P.862.2*. International Telecommunications Union, Telecommunications Standardization Sector, 2005.

[247] Järvinen, A. Design of a reference listening room–a case study. In *Proceedings of the Audio Engineering Society 103rd International Convention* (New York, NY, USA, Sept. 1997).

[248] Johansson, I. and Lynoe, N. *Medicin og Filosofi. En Introduktion*, 2nd ed. FADL, 1999.

[249] Kabel, P. An examination and interpretation of ITU-R BS.1387: perceptual evaluation of audio quality. Technical Report, McGill University, Montreal, Canada, 2003.

[250] Kaplan, H., Gladstone, V. S., and Lloyd, L. L. *Audiometric Interpretation: A Manual of Basic Audiometry*, 2nd ed. Allyn & Bacon, 1993.

[251] Kelly, G. *The Psychology of Personal Constructs*. Norton, New York, 1955.

[252] Kelly, M. C. and Tew, A. I. A novel method for the efficient comparison of spatialization conditions. In *Proceedings of the Audio Engineering Society 114th International Convention* (Amsterdam, The Netherlands, March 2003).

[253] Keppel, G. and Wickens, T. D. *Design and Analysis. A Researcher's Handbook*, 4th ed. Pearson Prentice Hall, 2004.

[254] Killion, M. C. Revised estimate of minimum audible pressure: Where is the 'missing 6 dB'. *Journal of the Acoustical Society of America 63* (1978), 1501–1510.

[255] Kirby, D. G. ISO/MPEG subjective tests on multichannel audio systems. In *Proceedings of the Audio Engineering Society 99th International Convention* (New York, NY, USA, Oct. 1995).

[256] Kirk, R. E. Learning, a major factor influencing preferences for high-fidelity reproducing systems. *Journal of the Acoustical Society of America 28* (1956), 1113–1116.

[257] Kish, L., Ed. *Statistical Design for Research*. Wiley Interscience, 2004.

[258] Kohlrausch, A. G. and van de Par, S. L. J. Auditory-visual interaction: From fundamental research in cognitive psychology to (possible) applications. In *Proceedings of SPIE: Human Vision and Electronic Imaging IV* (San Jose, CA, USA, January 1999).

[259] Koivuniemi, K. and Zacharov, N. Unravelling the perception of spatial sound reproduction: Language development, verbal protocol analysis and listener training. In *Proceedings of the Audio Engineering Society 111th International Convention* (New York, NY, USA, Nov. 2001).

[260] Kramer, G., Ed. *Auditory Display: Sonification, Audification, and Auditory Interfaces.* Addison Wesley Publishing Company, 1993.

[261] Kreiman, J., Garratt, B. R., Precoda, K., and Berke, G. S. Individual differences in voice quality perception. *Journal of Speech and Hearing Research 35* (June 1992), 512–520.

[262] Krueger, A. B. PCABX. `http://www.pcabx.com`, 2003.

[263] Kügler, C. and Theile, G. Loudspeaker reproduction: study on the subwoofer concept. In *Proceedings of the 92nd Convention of the Audio Engineering Society* (Vienna, Austria, March 1992).

[264] Kuhn, T. S. *The Structure of Scientific Revolutions*, 3rd ed. The University of Chicago Press, 1996.

[265] Kylliäinen, M., Helimki, H., Zacharov, N., and Cozens, J. Compact high performance listening spaces. In *Proceedings of Euronoise* (2003), Euronoise.

[266] Langron, S. P. The application of Procrustes statistics to sensory profiling. *Sensory Quality in Foods and Beverages: Definition, Measurement and Control.* Horwood, Chichester, UK, 1983, pp. 89–95.

[267] *Lärm und Gehörschutz. Ein Lehr- und Lernprogramm auf Audio-CD.* Fachausschuss Personliche Schutzausrustung, Arbeitskreis gehorschutz, Nurnberg Fa. Hamann Consult, Dresden, 1995.

[268] Lawless, H. T. and Heyman, H. *Sensory Evaluation of Food.* Chapman & Hall, 1998.

[269] Letowski, T. Development of technical skills: Timbre solfeggio. *Journal of the Audio Engineering Society 33* (1985), 240–243.

[270] Levene, H. Robust tests for equality of variance. *Contributions to Probability and Statistics: Essays in Honor of Harold Hotelling.* Stanford University Press, 1960.

[271] Leventhal, L. Type 1 and Type 2 error in the statistical analysis of listening tests. *Journal of the Audio Engineering Society 34*, 6 (1986), 437–453.

[272] Levitt, H. Transformed up-down methods in psychoacoustics. *Journal of the Acoustical Society of America 49*, 2, part 2 (1971), 467–477.

[273] Listen HRTF database. `http://recherche.ircam.fr/equipes/salles/listen/`, 2002–2003.

[274] Lorho, G. Evaluation of spatial enhancement systems for stereo headphone reproduction by preference and attribute rating. In *Proceedings of the 118th Convention of the Audio Engineering Society*, (Barcelona, Spain, May 2005).

[275] Lorho, G. Individual vocabulary profiling of spatial enhancement systems for stereo headphone reproduction. In *Proceedings of the 119th Convention of the Audio Engineering Society*, (New York, USA, October 2005).

[276] Luce, R. D. *Sound and Hearing Demonstration*. Lawrence Erlbaum Associates Inc., 1992.

[277] Macatee, S. R. Considerations in grounding and shielding audio devices. *Journal of the Audio Engineering Society 43*, 6 (1995), 472–483.

[278] Marchall, R. J. and Kirby, S. P. J. Sensory measurement of food texture by free-choice profiling. *Journal of Sensory Studies 3* (1988), 63–80.

[279] Martens, W. L. The impact of decorrelated low-frequency reproduction on auditory spatial imagery: Are two subwoofers better than one. In *Proceedings of the Audio Engineering Society 16th International Conference on Spatial Sound Reproduction* (1999), Audio Engineering Society, pp. 67–77.

[280] Martens, W. L., Braasch, J., and Ryan, T. J. Spatial auditory display using multiple subwoofers in two different reverberant reproduction environments. In *Proceedings of the International Conference on Auditory Display* (Limerick, Ireland, 2005), ICAD.

[281] Martens, W. L. and Zacharov, N. Multidimensional perceptual unfolding of spatially processed speech I: Deriving stimulus space using INDSCAL. In *Proceedings of the 109th Convention of the Audio Engineering Society* (Los Angeles, CA, USA, September 22–25 2000).

[282] Martin, G. and Bech, S. Attribute identification and quantification in automotive audio–part 1: Introduction to the descriptive analysis technique. In *Proceedings of the 118th Convention of the Audio Engineering Society*, (Barcelona, Spain, May 2005).

[283] Mason, R., Ford, N., Rumsey, F., and Bryun, B. Verbal and non-verbal elicitation techniques in the subjective assessment of spatial sound reproduction. *Journal of the Audio Engineering Society 49*, 5 (2001), 366–384.

[284] Mathers, C. D. and Lansdowne, K. F. L. Hearing risk to wearers of circumaural headphones: an investigation. Technical Report BBC RD 1979/3, The British Broadcasting Corporation, Research Department, 1979.

[285] Mattila, V-V. Descriptive analysis of speech quality in mobile communications: Descriptive language development and external preference mapping. In *Proceedings of the 111th Convention of the Audio Engineering Society* (New York, NY, USA, Nov. 2001).

[286] Mattila, V-V. *Perceptual Analysis of Speech Quality in Mobile Communications*. PhD thesis, Tampere University of Technology, Tampere, Finland, October 2001.

[287] Mattila, V-V. Descriptive analysis and ideal point modelling of speech quality in mobile communications. In *Proceedings of the 113th Convention of the Audio Engineering Society* (Los Angeles, CA, USA, Oct. 2002).

[288] Mattila, V-V. Semantic analysis of speech quality in mobile communications: descriptive language development and mapping to acceptability. *Food Quality and Preference 14*, 441–453 (2003).

[289] Mattila, V-V. and Kurittu, A. Practical issues in objective speech quality assessment with ITU-T P.862. In *Proceedings of the 117th Convention of the Audio Engineering Society* (San Francisco, CA, USA, Oct. 2004).

[290] Mattila, V-V. and Zacharov, N. Generalized listener selection (GLS) procedure. In *Proceedings of the 110th Convention of the Audio Engineering Society* (Amsterdam, The Netherlands, 2001).

[291] Meilgaard, M., Civille, G. V., and Carr, B. T. *Sensory Evaluation Techniques*. CRC Press, 1991.

[292] Merimaa, J. and Hess, W. Training of listeners for evaluation of spatial attributes of sound. In *Proceedings of the Audio Engineering Society 117th International Convention* (San Francisco, CA, USA, Oct. 2004).

[293] Metzler, B. *Audio Measurement Handbook*. Audio Precision Inc., 1993.

[294] Michaelson, H. How an author can avoid the pitfalls of practical ethics. *IEEE Transactions on Professional Communication 33*, 2 (June 1990), 58–61.

[295] Miller, G. A. The magical number seven, plus or minus two: some limits on our capacity for processing information. *The Psychological Review 63*, 2 (1956), 81–97.

[296] Miller, R. G. Texts in statistical science. *Beyond ANOVA. Basics of Applied Statistics*, 1st ed. Chapman & Hall, 1997.

[297] Miskiewicz, A. Timbre solfege: A course in technical listening for sound engineers. *Journal of the Audio Engineering Society 40* (1992), 621–625.

[298] Møller, H., Hammershøi, D., Jensen, C. B., and Sørensen, M. F. Transfer characteristics of headphones measured on human ears. *Journal of the Audio Engineering Society 43*, 4 (April 1995), 203–217.

[299] Møller, H., Jensen, C. B., Hammershøi, D., and Sørensen, M. F. Design criteria for headphones. *Journal of the Audio Engineering Society 43*, 4 (1995), 218–232.

[300] Montgomery, D. *Design and Analysis of Experiments*. Wiley, 2000.

[301] Moore, B. C. J. *Compact Disc with Book: Perceptual Consequences of Cochlear Damage*. Oxford University Press, 1995.

[302] Moore, B. C. J. *An Introduction to the Psychology of Hearing*, 4th ed. Academic Press, 1997.

[303] Moore, B. C. J. and Glasberg, B. R. A model of loudness perception applied to cochlear hearing loss. *Auditory Neuroscience 3* (1997), 289–311.

[304] Moore, B. C. J. and Glasberg, B. R. A model of loudness perception applied to cochlear hearing loss. `http://hearing.psychol.cam.ac.uk/Demos/ansloud.zip`, 1997.

[305] Moore, B. C. J., Glasberg, B. R., and Baer, T. A model for the prediction of thresholds, loudness, and partial loudness. *Journal of the Audio Engineering Society 45* (1997), 224–239.

[306] Moore, B. C. J., Glasberg, B. R., and Baer, T. A model for the prediction of thresholds, loudness and partial loudness. Available from: `http://hearing.psychol.cam.ac.uk/Demos/aesloud.zip`, 1997.

[307] Moore, B. C. J., Peters, R. W., and Glasberg, B. R. A revision of Zwicker's loudness model. *Acta Acustica united with Acustica 82* (1996), 335–345.

[308] Moreno, A., Lindberg, B., Draxler, C., Richard, G., Choukri, K., Euler, S., and Allen, J. Speech databases recorded in vehicles. `http://www.speechdat.com/SP-CAR/`, 1999.

[309] Moreno, A., Lindberg, B., Draxler, C., Richard, G., Choukri, K., Euler, S., and Allen, J. SpeechDat-Car A large speech database for automotive environments. In *2nd International Conference on Language Resources and Evaluation* (Athens, 2000).

[310] Moulton, D. Golden ears. `http://www.moultonlabs.com/gold.htm`, 2002.

[311] Muncy, N. Noise susceptibility in analog and digital signal processing systems. *Journal of the Audio Engineering Society 43*, 6 (1995), 435–453.

[312] Munson, W. A. and Weiner, F. M. In search of the missing 6 dB. *Journal of the Acoustical Society of America 24* (1952), 498–501.

[313] Myers, R. H. and Montgomery, D. C. *Response Surface Methodology: Process and Product Optimization Using Designed Experiments*, Series in Probability and Statistics, 2nd ed. Wiley-Interscience, 2002.

[314] Næs, T. and Risvik, E. *Multivariate Analysis of Data in Sensory Science, Vol. 16 of Data Handling in Science and Technology*. Elsevier, Amsterdam, The Netherlands, 1996.

[315] Nakajima, Y., Sasaki, T., and Ten Hoopen, G. Demonstrations of auditory illusions and tricks. Compact Disc available from http://www.kyushu-id.ac.jp/~ynhome/ENG/Demo/illusions2nd.html, 2000.

[316] NASA. Auditory demonstrations II: Challenges to speech communication and music listening. Compact Disc available from http://www.grc.nasa.gov/WWW/AcousticalTest/HearingConservation/Resource%s/Auditory_Demonstrations2.html, 2004.

[317] National Advisory Board on Research Ethics, Finland. Good scientific practise and procedures for handling misconduct and fraud in science. http://pro.tsv.fi/tenk/htkoengl.pdf, 2002.

[318] Neher, T. *Towards a Spatial Ear Trainer*. PhD thesis, Institute of Sound Recording, University of Surrey, Great Britain, United Kingdom, 2004.

[319] Neher, T., Brookes, T., and Rumsey, F. J. Unidimensional simulation of the spatial attribute 'ensemble width' for training purposes. In *Proceedings of the Audio Engineering Society 24th International Conference* (2003), Audio Engineering Society.

[320] Neher, T., Rumsey, F. J., and Brookes, T. Training of listeners for the evaluation of spatial sound reproduction. In *Proceedings of the Audio Engineering Society 112th International Convention* (Munich, Germany, May 2002).

[321] Neher, T., Rumsey, F. J., Brookes, T., and Craven, P. Unidimensional simulation of the spatial attribute 'ensemble width' for training purposes. In *Proceedings of the Audio Engineering Society 114th International Convention* (Amsterdam, The Netherlands, March 2003).

[322] Nelson, L. S. Evaluating overlapping confidence intervals. *Journal of Quality Technology 21*, 2 (April 1989), 140–141.

[323] Nelson, D. A. and Cooper, B. A. Auditory demonstrations in acoustics and hearing conservation. *Noise-Con*. Noise-Control Foundation, 1998.

[324] Nelson, D., and Taylor, A. Auditory demonstrations in acoustics and hearing conservation. Compact Disc available from `http://www.grc.nasa.gov/WWW/AcousticalTest/HearingConservation/Resource%s/Auditory_Demonstrations1.html`, 1998.

[325] Nijenhuis, M. *Sampling and Interpolation of Static Images: a Perceptual View*. PhD thesis, Institute of Perception Research (IPO), Eindhoven University of Technology, Eindhoven, The Netherlands, 1993.

[326] Noble, A. C., Arnold, R. A., Buechsenstein, J., Leach, E. J., Schmidt, J. O., and Stern, P. M. Modification of a standardized system of wine aroma technology. *American Journal of Enology and Viticulture 38*, 2 (1987), 143–146.

[327] Nousaine, T. Multiple subwoofers for home theater. In *Proceedings of the 103rd Convention of the Audio Engineering Society* (New York, NY, USA, Sept. 1997).

[328] Nunnally, J. C. and Bernstein, I. H. *Psychometric Theory*, 3rd ed. McGraw-Hill, 1994.

[329] Olive, S. E. A method of training of listeners and selecting program material for listening tests. In *Proceedings of the Audio Engineering Society 97th International Convention* (San Francisco, CA, USA, Oct. 1994).

[330] Olive, S. E. A new listener training software application. In *Proceedings of the Audio Engineering Society 110th International Convention* (Amsterdam, The Netherlands, May 2001).

[331] Olive, S. Differences in performance and preference of trained versus untrained listeners in loudspeaker tests: A case study. *Journal of the Audio Engineering Society 51*, 9 (2003), 806–825.

[332] Olive, S. E., Castro, B., and Toole, F. E. A new laboratory for evaluating multichannel audio components and systems. In *Proceedings of the Audio Engineering Society 105th International Convention* (San Francisco, CA, USA, Sept. 1998).

[333] Olive, S. E., Schuck, P. L., Sally, S. L., and Bonneville, M. E. The effects of loudspeaker placement on listener preference ratings. *Journal of the Audio Engineering Society 42*, 12 (1994), 651–668.

[334] Olive, S. E., Schuck, P. L., Sally, S. L., and Bonneville, M. The variability of loudspeaker sound quality among four domestic-sized rooms. In *Proceedings of the Audio Engineering Society 99th International Convention* (New York, NY, USA, Oct. 1995).

[335] Olive, S. E. and Toole, F. E. The detection of reflections in typical rooms. *Journal of the Audio Engineering Society 37*, 7/8 (1989), 539–553.

[336] Patterson, R. D., Allerhand, M. H., and Giguere, C. Time-domain modelling of peripheral auditory processing: A modular architecture and a software platform. *Journal of the Acoustical Society of America 98* (1995), 1890–1894.

[337] Pearson, K. S., Bennett, R. L., and Fidell, S. Speech levels in various noise environments. Technical Report EPA-600/1-77-025, Environmental Protection Agency, Washington, DC, USA, May 1977.

[338] Pedersen, T. H. and Fog, C. L. Optimisation of perceived product quality. In *Euronoise II* (1998), pp. 633–638.

[339] Pedersen, J. A. and Mäkivirta, A. Requirements for low-frequency sound reproduction, part II: Generation of stimuli and listening system equalization. *Journal of the Audio Engineering Society 50*, 7/8 (July/August 2002), 581–593.

[340] Perkins, C. Automated test and measurement of common impedance coupling in audio system shield terminations. *Journal of the Audio Engineering Society 43*, 6 (1995), 488–497.

[341] Pernaux, J.-M., Emerit, M., and Nicol, R. Perceptual evaluation of binaural sound synthesis: The problem of reporting localization judgments. In *Proceedings of the Audio Engineering Society 114th*

International Convention (Amsterdam, The Netherlands, March 2003).

[342] Peryam, D. R. and Girardot, N. F. Advanced taste test method. *Food Engineering 24*, 194 (1952), 58–61.

[343] Petrinovich, L. F. and Hardyck, C. D. Error rates for multiple comparison methods: Some evidence concerning the frequency of erroneous conclusion. *Psychological Bulletin 71* (1969), 43–54.

[344] Piggott, J. R. Selection of terms for descriptive analysis. *Sensory Science Theory and Applications in Foods*. Dekker, New York, 1991.

[345] Plack, C. J. and Carlyon, R. P. Loudness perception and intensity coding. *Hearing. Handbook of Perception and Cognition*. Academic Press, 1995, ch. 4, pp. 123–205.

[346] Plomp, R. *Aspects of Tone Sensation: a Psychophysical Study*. Academic Press, London, 1976.

[347] Plomp, R. Hoe wij horen. over de toon die de muziek maakt. (in Dutch: How we hear. on the tone that makes music), 1998.

[348] Poldy, C. A. Philosophy of modern headphone design. *Australian Sound and Recording 3* (1981), 31–36.

[349] Poldy, C. A. *Headphones*, 4th ed. Butterwoths, 1988, ch. 12, pp. 466–548.

[350] Popper, K. R. *The Logic of Scientific Discovery*. Routledge, 1992.

[351] Poulton, E. C. *Bias in Quantifying Judgements*. Lawrence Erlbaum Associates, 1989.

[352] Powers, D. A. and Xie, Y. *Statistical Methods for Categorical Data Analysis*. Academic Press, 2000.

[353] Precoda, K. and Meng, T. H. Listener differences in audio compression evaluations. *Journal of the Audio Engineering Society 45*, 9 (1997), 708–715.

[354] Quesnel, R. Timbral ear trainer: Adaptive, interactive training of listening skills for evaluation of timbre difference. In *Proceedings of the Audio Engineering Society 98th International Convention* (Paris, France, Feb. 1996).

[355] Quesnel, R. *A Computer-Assisted Method for Training and Researching Timbre Memory and Evaluation Skills*. PhD thesis, McGill University, Montreal, Canada, 2002.

[356] Quesnel, R. and Woszczyk, W. R. A computer-aided system for timbral ear training. In *Proceedings of the Audio Engineering Society 96th International Convention* (Amsterdam, The Netherlands, Feb. 1994).

[357] Reiner, R. This is Spinal Tap. DVD, 1984.

[358] Rhenius, V. Kopfhörer: Messverfahren im Vergleich: Wer hat das richtige Maß. *Funkschau 17* (1983), pp. 44–46.

[359] Riederer, K. A. J. Head-related transfer function measurements. Master's thesis, Helsinki University of Technology, Laboratory of Acoustics and Audio Signal Processing, 1998.

[360] Riederer, K. A. J. Repeatability analysis of head-related transfer function measurement. In *Proceedings of the Audio Engineering Society 105th International Convention* (San Francisco, CA, USA, Sept. 1998).

[361] Riederer, K. A. J. *HRTF Analysis: Objective and Subjective Evaluation of Measured Head-Related Transfer Functions*. PhD thesis, Helsinki University of Technology, Espoo, Finland, 2005.

[362] Rioux, V. LISE–Listening interface for sound experiments. `http://www.ta.chalmers.se/homepages/vincent/doc/lise/lise.html`, 2000.

[363] Rix, A. W., Hollier, M. P., Hekstra, A. P., and Beerends, J. G. Perceptual evaluation of speech quality (PESQ). The new ITU standard for end-to-end speech quality assessment part I–time-delay compensation. *Journal of the Audio Engineering Society 50*, 10 (October 2002), 755–764.

[364] Robinson, D. W. and Whittle, L. S. The loudness of directional sound fields. *Acustica 10* (1960), 74–80.

[365] Roessler, E. B., Pangborn, R. M., Sidel, J. L., and Stone, H. Expanded statistical tables for estimating statistical significance in paired-preference, paired-difference, duo-trio and triangle tests. *Journal of Food Science 43* (1978), 940–947.

[366] Rumsey, F. *Spatial Audio*. Focal Press, 2001.

[367] Rumsey, F. Spatial quality evaluation for reproduced sound: terminology, meaning and a scene-based paradigm. *Journal of the Audio Engineering Society 50* (2002), 651–666.

[368] Rumsey, F. J., Neher, T., and Brookes, T. Unidimensional simulation of the spatial attribute 'ensemble depth' for training purposes–part 2: Creation and validation of reference stimuli. In *Proceedings of the Audio Engineering Society 116th International Convention* (Berlin, Germany, May 2004).

[369] Ryan, T. A. Multiple comparisons in psychological research. *Psychological Bulletin 56* (1959), 26–47.

[370] Salava, T. Low-frequency performance of listening rooms for steady-state and transient signals. *Journal of the Audio Engineering Society 39*, 11 (November 1991), 853–863.

[371] Salava, T. Subwoofers in small listening rooms. In *Proceedings of the 106th Convention of the Audio Engineering Society* (Munich, Germany, May 1999).

[372] Samoylenko, E., McAdams, S., and Nosulenko, V. Systematic analysis of verbalisations produced in comparing musical timbres. *International Journal of Psychology 31*, 6 (1996), 255–278.

[373] Schlich, P. GRAPES: A method and SAS program for graphical representation of assessor performance. *Journal of Sensory Science 9* (1994), 157–169.

[374] Schonemann, P. H. A generalized solution of the orthogonal procrustes problem. *Psychometrika 31* (1966), 1–10.

[375] Shanefield, D., Clark, D., and Nousaine, T. Comments on "Type 1 and Type 2 error in the statistical analysis of listening tests". *Journal of the Audio Engineering Society 35*, 7/8 (1987), 567–572.

[376] Shiffman, S. S., Reynolds, M. L., and Young, F. W. *Introduction of Multidimensional Scaling*. Academic Press, 1981.

[377] Shively, R. E. and House, W. N. Listener training and repeatability for automobiles. In *Proceedings of the Audio Engineering Society 104th International Convention* (Amsterdam, The Netherlands, May 1998).

[378] Shlien, S. Auditory models for gifted listeners. *Journal of the Audio Engineering Society 48*, 11 (November 2000), 1032–1044.

[379] Shlien, S. and Soulodre, G. Measuring the characteristics of "expert" listeners. In *Proceedings of the 101st Convention of the Audio Engineering Society* (Los Angeles, CA, USA, November 1996).

[380] Sieffermann, J. M. Flash profiling. a new method of sensory descriptive analysis. In *AIFST Proceedings, 35th Convention* (Sidney, Australia, July 21–24 2002).

[381] Sivian, L. J. and White, S. D. On minimum audible sound fields. *Journal of the Acoustical Society of America 4* (1933), 288–321.

[382] Sivonen, V. P. and Ellermeier, W. Directional loudness measured in a real sound field: Implications for sound quality evaluation. In *the 33rd International Congress and Exposition on Noise Control Engineering* (Prague, Czech Republic, 2004).

[383] Sivonen, V. P., Minnaar, P., and Ellermeier, W. Effect of direction on loudness in individual binaural synthesis. In *Proceedings of the 118th Convention of the Audio Engineering Society* (Barcelona, Spain, May 2005).

[384] Skovenborg, E. and Nielsen, S. H. Evaluation of different loudness models with music and speech material. In *Proceedings of the 117th Convention of the Audio Engineering Society* (San Francisco, CA, USA, Oct. 2004).

[385] Skovenborg, E., Quesnel, R., and Nielsen, S. H. Loudness assessment of music and speech. In *Proceedings of the 116th Convention of the Audio Engineering Society* (Berlin, Germany, May 2004).

[386] Slaney, M. Auditory toolbox. `http://rvl4.ecn.purdue.edu/~malcolm/interval/1998-010/`, 1998.

[387] SMPTE. *For Motion-Pictures–Dubbing Theatres, Review Rooms and Indoor Theatres–B-Chain Electroacoustic Response*. No. RP 202M-1998, The Society of Motion Picture and Television Engineers, 1998.

[388] SMPTE. *Relative and Absolute Sound Pressure Levels for Motion Picture Sound Systems–Application for Analog Photographic Film*

Audio, Digital Photographic Film Audio and D-Cinema. No. RP 200-2002, The Society of Motion Picture and Television Engineers, 2002.

[389] Sørensen, M. F., Lydolf, M., Frandsen, P. C., and Møller, H. Directional dependence of loudness cues and binaural summation. *Proceedings of the 15th International Congress on Acoustics 3* (1995), 293–296.

[390] Soulodre, G. A. Evaluation of objective loudness meters. In *Proceedings of the 116th Convention of the Audio Engineering Society* (Berlin, Germany, May 2004).

[391] Soulodre, G. A., and Norcross, S. G. Objective measures of loudness. In *Proceedings of the 115th Convention of the Audio Engineering Society* (New York, NY, USA, Oct. 2003).

[392] Soulodre, G. A. and Lavoie, M. C. Subjective evaluation of MPEG layer II with spectral band replication. In *Presented at the 117th Convention of the Audio Engineering Society*, (San Francisco, CA, USA, October 2004).

[393] Spikofski, G. The diffuse field probe transfer function of studio-quality headphones. EBU Technical Review 229, EBU, 1988, pp. 111–126.

[394] Spreen, O. and Strauss, E. *A Compendium of Neuropsychological Tests*. Oxford University Press, New York, 1998.

[395] Sprent, P. *Applied Nonparametric Statistical Methods*, 2nd ed. Chapman & Hall, 1993.

[396] SPSS. *SPSS Base 10.0 Applications Guide*. SPSS, 1999.

[397] SPSS. *SPSS 13 Base Users Guide*. SPSS, 2004.

[398] Staffeldt, H. Correlation between subjective and objective data for quality loudspeakers. *Journal of the Audio Engineering Society 22*, 6 (July-August 1974), 402–415.

[399] Staffeldt, H. Difference in the perceived quality of loudspeaker sound reproduction caused by loudspeaker-room-listener interactions. In *Proceedings of the 90th Convention of the Audio Engineering Society*, (Paris, France, February 1991).

[400] Steinke, G. Surround sound: Relations of listening and viewing configurations–the useful assignment of loudspeaker basis width to video picture dimension. In *Proceedings of the Audio Engineering Society 116th International Convention* (Berlin, Germany, May 2004).

[401] Stevens, S. S. A scale for the measurement of a psychological magnitude: loudness. *Psychological Review 43* (1936), 405–416.

[402] Stevens, S. S. Calculation of the loudness of complex noise. *Journal of the Acoustical Society of America 28* (1956), 807–832.

[403] Stevens, S. S. On the psychophysical law. *Psychological Review 64* (1957), 153–181.

[404] Stevens, S. S. Ratio scales, partition scales and confusion scales. *Psychological Scaling: Theory and Applications*. Wiley, New York, 1960.

[405] Stevens, S. S. Procedure for calculating loudness: Mark VII. *Journal of the Acoustical Society of America 33* (1961), 1577–1585.

[406] Stevens, S. S. The surprising simplicity of sensory metrics. *American Psychologist 17* (1962), 29–39.

[407] Stone, M. A., Moore, B. C. J., and Glasberg, B. R. A real-time DSP based loudness meter. In *Proceedings of the Seventh Oldenburg Symposium on Psychological Acoustics* (Germany, 1997), pp. 587–601.

[408] Stone, H. and Sidel, J. L. *Sensory Evaluation Practices*, 2nd ed. Academic Press, 1993.

[409] Stonehouse, J. M. and Forrester, G. J. Robustness of the t and u tests under combined assumption violations. *Journal of Applied Statistics 25*, 1 (1998), 63–74.

[410] Suokuisma, P., Zacharov, N., and Bech, S. Multichannel level alignment, part I: signals and methods. In *Proceedings of the Audio Engineering Society 105th International Convention* (San Francisco, CA, USA, Sept. 1998).

[411] *Technical Recommendation N 12-A–Sound Control Rooms and Listening Room*, 2nd ed. The Nordic Public Broadcasting Corporations, August 1992.

[412] Terhardt, E. *Compact Disc with Book: Akustische Kommunikation.* Springer, New York, 1998.

[413] Teunissen, K. The validity of quality indicators along a graphical scale. Report 1077, Institute of Perception Research, Technical University of Eindhoven, Eindhoven, The Netherlands, 1995.

[414] The database of head related transfer functions. Available from: `http://www.itakura.nuee.nagoya-u.ac.jp/HRTF/`, 1999.

[415] Theile, G. On the standardisation of the frequency response of high-quality studio headphones. *Journal of the Audio Engineering Society 34* (1986), 956–969.

[416] Thiede, T., Treurniet, W. C., Bitto, R., Schmidmer, C., Sporer, T., Beerends, J. G., and Colomes, C. PEAQ–the ITU standard for objective measurement of perceived audio quality. *Journal of the Audio Engineering Society 48*, 1/2 (Jan-Feb 2000), 3–29.

[417] Thomas, L. and Krebs, C. J. A review of statistical power analysis software. *Ecological Society of America 78*, 2 (April 1997), 126–139.

[418] Thurstone, L. L. A law of comparative judgment. *Psychological Review 34* (1927), 273–286.

[419] Thurstone, L. L. Psychophysical analysis. *American Journal of Psychology 38* (1927), 368–369.

[420] Tillman, T. W., Johnson, R. M., and Olsen, W. O. Earphone versus sound-field threshold sound-pressure levels for spondee words. *Journal of the Acoustical Society of America 39* (1966), 125–133.

[421] Toole, F. E. Listening tests-turning opinion into fact. *Journal of the Audio Engineering Society 30*, 6 (1982), pp. 431–445.

[422] Toole, F. E. Subjective measurements of loudspeaker sound quality and listener performance. *Journal of the Audio Engineering Society 33*, 1/2 (1985), pp. 2–32.

[423] Toole, F. E. Loudspeaker measurements and their relationship to listener preferences: Part 1. *Journal of the Audio Engineering Society 34*, 4 (1986), pp. 227–235.

[424] Toole, F. E. Subjective evaluation: Identifying and controlling the variables. In *Proceedings of the Audio Engineering Society 8th International Conference* (1990), Audio Engineering Society, pp. 95–100.

[425] Toole, F. Hearing is believing vs. believing is hearing: blind vs. sighted listening tests, and other interesting things. In *Proceedings of the 97th Convention of the Audio Engineering Society* (San Francisco, CA, USA, Oct. 1994).

[426] Toole, F. The acoustics and psychoacoustics of loudspeakers and rooms–the stereo past and the multichannel future. In *Proceedings of the 109th Convention of the Audio Engineering Society* (Los Angeles, CA, USA, Sept. 2000).

[427] Treurnietand, W. C. and Soulodre, G. A. Evaluation of the ITU-R objective audio quality measurement method. *Journal of the Audio Engineering Society 48*, 3 (March 2000), 164–173.

[428] Tuffy, M. A. Establishing a reference playback level for video games. In *Proceedings of the 117th Convention of the Audio Engineering Society* (San Francisco, CA, USA, October 2004).

[429] Tuomi, O. and Zacharov, N. A real-time binaural loudness model. In *Presented at the 139th Meeting of the Acoustical Society of America* (Atlanta, USA, May/June 2000).

[430] University of Chicago Press Staff. *The Chicago Manual of Style*, 15th ed. University Of Chicago Press, 2003.

[431] Usher, J. and Woszczyk, W. Design and testing of a graphical mapping tool for analyzing spatial audio scenes. In *Proceedings of the Audio Engineering Society 24th International Conference* (2003), Audio Engineering Society, pp. 157–170.

[432] Usher, J. and Woszczyk, W. Visualizing auditory spatial imagery of multi-channel audio. In *Proceedings of the 116th Convention of the Audio Engineering Society*, (Berlin, Germany, May 2004).

[433] Vainio, M., Suni, A., Järvelinen, H., Järvikivi, and Mattila, V-V. Developing a speech intelligibility test based on measuring speech reception thresholds in noise for English and Finnish. *Journal of the Acoustical Society of America 118*, 3 (September 2005), 1742–1750.

[434] van Belle, G. *Statistical Rules of Thumb*. Wiley, New York, USA, 2002.

[435] Ventry, I. M. and Woods, R. W. Most comfortable loudness for pure tones, noise and speech. *Journal of the Acoustical Society of America 49*, 6, part 2 (1971), 1805–1813.

[436] Vincent, E. MUSHRAM–A Matlab interface for MUSHRA listening tests. `http://www.elec.qmul.ac.uk/digitalmusic/downloads/#mushram`, 2005.

[437] Walker, R. Low frequency room response: part 1–background and qualitative consideration. Research Department Report BBC RD 1992/8, The British Broadcasting Corporation, 1992.

[438] Walker, R. Low frequency room response: part 1–calculation methods and experimental results. Research Department Report BBC RD 1992/9, The British Broadcasting Corporation, 1992.

[439] Walker, R. Optimum dimension ratios for studios, control rooms and listening rooms. Research Department Report BBC RD 1993/8, The British Broadcasting Corporation, 1993.

[440] Walker, R. Optimum dimension ratios for small rooms. In *Proceedings of the 100th Convention of the Audio Engineering Society* (Copenhagen, Denmark, May 1996).

[441] Walker, R. A controlled reflection listening room for multichannel audio. In *Proceedings of the 104th Convention of the Audio Engineering Society*, (Amsterdam, The Netherlands, May 1998).

[442] Walker, R. Equalisation of room acoustics and adaptive systems in the equalisation of small room acoustics. In *Proceedings of the Audio Engineering Society 15th International Conference* (1998), Audio Engineering Society, pp. 32–47.

[443] Warrem, R. M. *Compact Disc with Book: Auditory Perception. A New Analysis and Synthesis*. Cambridge University Press, 1999.

[444] Welti, T. How many subwoofers are enough? In *Proceedings of the 112th Convention of the Audio Engineering Society* (Munich, Germany, May 2002).

[445] Whitlock, B. Balanced lines in audio systems: Fact, fiction, and transformers. *Journal of the Audio Engineering Society 43*, 6 (1995), 454–464.

[446] Wickelmaier, F. and Choisel, S. Selecting participants for listening tests of multichannel reproduced sound. In *Proceedings of the 118th Convention of the Audio Engineering Society* (Barcelona, Spain, May 2005).

[447] Wickelmaier, F. and Ellermeier, W. Deriving auditory features from triadic comparisons. *Perception and Psychophysics* (2006). Is accepted for publication.

[448] Wickelmaier, F. and Schmid, C. A Matlab function to estimate choice-model parameters from paired-comparison data. *Behaviour Research Methods, Instruments, and Computers* 36(1), (2004), 29–40.

[449] Wilkens, H. *Mehrdimensionale Beschreibung subjektiver Beurteilungen der Akustik von Konzertsälen*. PhD thesis, TU Berlin, 1975.

[450] Wilksinson, A. M. *The Scientist's Handbook for Writing Papers and Dissertations*. Prentice Hall, 1991.

[451] Williams, A. A. and Arnold, G. M. A comparison of the aromas of six coffees characterised by conventional profiling, free-choice profiling and similarity scaling methods. *Journal of the Science of Food and Agriculture 36* (1985), 204–214.

[452] Williams, A. A. and Langron, S. P. The use of free-choice profiling for the evaluation of commercial ports. *Journal of Food an Agriculture 35* (1984), 558–568.

[453] Wilson, W. A note on the inconsistency inherent in the necessity to perform multiple comparisons. *Psychological Bulletin 59* (1962), 296–300.

[454] Windt, J. An easily implemented procedure for identifying potential electromagnetic compatibility problems in new equipment and existing systems: The hummer test. *Journal of the Audio Engineering Society 43*, 6 (1995), 484–487.

[455] Wischgolf, K. J. Der halboffene stereokopfhörer. *Funkschau 4* (1981), 73–74.

[456] Yendrikhovskij, S. N. *Color Reproduction and the Naturalness Constraint*. PhD thesis, Institute of Perception Research (IPO), Eindhoven University of Technology, Eindhoven, The Netherlands, 1998.

[457] Zacharov, N. *Subjective Testing of Loudspeaker Directivity for Multichannel Audio*. Master's thesis, Helsinki University of Technology, Laboratory of Acoustics and Audio Signal Processing, 1997.

[458] Zacharov, N. An overview of multichannel level alignment. In *Proceedings of the Audio Engineering Society 15th International Conference* (1998), Audio Engineering Society, pp. 174–186.

[459] Zacharov, N. Subjective appraisal of loudspeaker directivity for multichannel reproduction. *Journal of the Audio Engineering Society 46*, 4 (1998), 288–303.

[460] Zacharov, N. *Perceptual Studies on Spatial Sound Reproduction Systems*. PhD thesis, Helsinki University of Technology, Helsinki, Finland, December 2000.

[461] Zacharov, N. and Bech, S. Multichannel level alignment, part IV: The correlation between physical measures and subjective level calibration. In *Proceedings of the Audio Engineering Society 109th International Convention* (Los Angeles, CA, USA Sept. 2000).

[462] Zacharov, N., Bech, S., and Meares, D. The use of subwoofers in the context of multichannel surround sound reproduction. *Journal of the Audio Engineering Society 46*, 4 (1998), 276–287.

[463] Zacharov, N., Bech, S., and Suokuisma, P. Multichannel level alignment, part II: The influence of signals and loudspeaker placement. In *Proceedings of the Audio Engineering Society 105th International Convention* (San Francisco, CA, USA, Sept. 1998).

[464] Zacharov, N., Huopaniemi, J., and Hämäläinen, M. Round robin subjective evaluation of virtual home theatre sound systems at the AES 16th international conference. In *Proceedings of the Audio Engineering Society 16th International Conference on Spatial Sound Reproduction* (1999), Audio Engineering Society, pp. 544–556.

[465] Zacharov, N. and Koivuniemi, K. Perceptual audio profiling and mapping of spatial sound displays. In *Proceedings of the International Conference on Auditory Display* (2001), ICAD, pp. 95–104.

[466] Zacharov, N. and Koivuniemi, K. Unravelling the perception of spatial sound reproduction: Analysis & external preference mapping. In *Proceedings of the Audio Engineering Society 111th International Convention* (New York, NY, USA, Nov. 2001).

[467] Zacharov, N. and Koivuniemi, K. Unravelling the perception of spatial sound reproduction: Techniques and experimental design. In *Proceedings of the Audio Engineering Society 19th International Conference on Surround Sound* (2001), Audio Engineering Society.

[468] Zacharov, N. and Lorho, G. Sensory analysis of sound (in telecommunications). In *European Sensory Network Conference* (Madrid, Spain, 2005), ESN.

[469] Zacharov, N., Tuomi, O., and Lorho, G. Auditory periphery, HRTFs and directional loudness perception. In *Proceedings of the Audio Engineering Society 110th International Convention* (Amsterdam, The Netherlands, May 2001).

[470] Zielinski, S. K., Rumsey, F., and Bech, S. Subjective audio quality trade-offs in consumer multichannel audio-visual delivery systems. part I: Effects of high frequency limitation. In *Proceedings of the Audio Engineering Society 112th International Convention* (Munich, Germany, May 2002).

[471] Zielinski, S. K., Rumsey, F., and Bech, S. Comparison of quality degradation effects caused by limitation of bandwidth and by down-mix algorithms in consumer multichannel audio delivery systems. In *Proceedings of the 114th Convention of the Audio Engineering Society* (Amsterdam, The Netherlands, March 2003).

[472] Zwicker, E. Über psychologische und methodische Grundlagen der Lautheit. *Acustica 8* (1958), 237–258.

[473] Zwicker, E. Ein Verfahren zur Berechnung der Lautstärke. *Acustica 10* (1960), 74–80.

[474] Zwicker, E. Temporal effects in simultaneous masking by white-noise burst. *Journal of the Acoustical Society of America 37* (1965), 653–663.

[475] Zwicker, E. Procedure for calculating loudness of temporally variable sounds. *Journal of the Acoustical Society of America 62* (1977), 675–682.

[476] Zwicker, E. and Fastl, H. *Psychoacoustics: Facts and Models*. Springer-Verlag, Heidelberg, Germany, 1990.

[477] Zwicker, E., Fastl, H., and Dallmayr, C. BASIC-program for calculating the loudness of sound from their 1/3-oct band spectra according to ISO 532B. *Acustica 55* (1984), 63–67.

[478] Zwicker, E. and Scharf, B. A model of loudness summation. *Psychological Review 72* (1965), 3–26.

Index

Printed and bound by CPI Group (UK) Ltd, Croydon, CR0 4YY

16/04/2025

14658475-0002